Molecular Entomology

Molecular Entomology

Krishna Chaitanya Tirunagaru
Department of Agricultural Entomology, School of Agricultural Sciences & Technology (SAST), Narsee Monjee Institute of Management Studies (NMIMS), Shirpur, Dhule District, Maharashtra

Arshad Ali
Department of Zoology, Gandhi Faiz-e-Aam College, Shahjahanpur

Kennedy Ningthoujam
College of Postgraduate Studies in Agricultural Sciences, Central Agricultural University (Imphal) Umiam, Meghalaya

Anup Chandra
Division of Crop Protection,
ICAR-Indian Institute of Pulses Research, Kanpur, Uttar Pradesh

Mahesh Pathak
School of Crop Protection, College of Post Graduate Studies in Agricultural Sciences, Central Agricultural University (Imphal), Umiam, Meghalaya

Rojeet Thangjam
Department of Entomology, College of Horticulture,
Central Agricultural University (Imphal), Thenzawl, Mizoram

CRC Press
Taylor & Francis Group
Boca Raton London New York

CRC Press is an imprint of the
Taylor & Francis Group, an **informa** business

-EPH-

Elite Publishing House

First published 2026
by CRC Press
4 Park Square, Milton Park, Abingdon, Oxon, OX14 4RN

and by CRC Press
2385 NW Executive Center Drive, Suite 320, Boca Raton FL 33431

CRC Press is an imprint of Informa UK Limited

British Library Cataloguing-in-Publication Data
A catalogue record for this book is available from the British Library

Print edition not for sale in India

ISBN: 9781041247333 (hbk)
ISBN: 9781041280286 (pbk)
ISBN: 9781003757405 (ebk)

DOI: 10.1201/9781003757405

Typeset in A Caslon Pro
by Elite Publishing House

-EPH-

Contents

	Preface	*vii*
	About the Authors	*ix*
1	Molecular Biology Techniques	1
2	Analysis of DNA and RNA in Insects	103
3	Genetic Improvement of Natural Enemies	160
4	Genes of Interest in Entomology	181
5	Transgenic Plants for Pest Resistance	239
6	Insect Gene Transformation	301
7	DNA Fingerprinting for Taxonomy and Phylogeny	344
8	DNA Diagnostics and Sf Transgenic Technology	402
	References	*458*

Preface

The realm of molecular entomology is a dynamic and rapidly advancing field that merges molecular biology with entomological research to address pressing challenges in agriculture, health, and environmental conservation. Insects, as one of the most diverse and ecologically significant groups of organisms, play pivotal roles as pollinators, decomposers, pests, and disease vectors. Understanding their biology at a molecular level offers innovative pathways to manage pest populations, enhance beneficial insect traits, and mitigate the impacts of vector-borne diseases. This book, designed for researchers, students, and professionals, provides an in-depth exploration of the molecular mechanisms underpinning insect biology and biotechnology applications.

Molecular entomology is built upon the foundations of molecular biology, which provides the tools and techniques to analyze and manipulate the genetic material of organisms. Unit I, titled Introduction to Molecular Biology, sets the stage by introducing the core concepts of molecular biology and the fundamental techniques used in this field. Readers will gain insights into the tools that form the backbone of molecular research, such as DNA extraction, PCR, gel electrophoresis, and sequencing. This unit establishes the technical framework necessary for understanding the more advanced topics discussed later in the book.

Moving into the specific molecular mechanisms in insects, Unit II, titled DNA and RNA Analysis in Insects, delves into the processes of transcription and translation, the cornerstone of gene expression. The unit explains how these molecular processes operate in insects, highlighting their unique aspects compared to other organisms. It also introduces DNA recombinant technology, a revolutionary tool for identifying and manipulating genes or nucleotide sequences of interest. By focusing on genetic improvement, this unit addresses the potential of molecular techniques to enhance natural enemies of pests, such as predators and parasitoids, making them more effective in biological control. Additional topics include the development of insect cell lines and the genetic engineering of entomopathogens like baculoviruses, *Bacillus thuringiensis (Bt)*, and entomopathogenic fungi, offering sustainable alternatives to chemical pesticides.

Entomological research has benefited immensely from the identification and utilization of specific genes. Unit III, titled Genes of Interest in Entomological Research, explores the molecular basis of key traits in insects. Marker genes for sex identification have streamlined the study of insect populations and pest control programs. Neuropeptides, juvenile hormone esterase, and venom-related toxins such as St toxins have opened new avenues for understanding insect physiology and behavior. Additionally, genes such as chitinase, CPTI, lectins, and proteases hold potential for developing innovative pest management strategies. The unit also discusses the application of transgenic plants, engineered for pest and disease resistance, as a cornerstone of sustainable agriculture.

As the book progresses, Unit IV, titled Insect Gene Transformation and Biotechnology, dives deeper into genetic modification techniques and their implications. Gene transformation in insects has enabled groundbreaking innovations, including the enhancement of traits in silkworms for superior silk production and honey bees for improved disease resistance. The introduction of lectin genes for pest suppression showcases the application of molecular tools in controlling insect populations while minimizing environmental impacts. This unit also highlights the use of DNA fingerprinting in taxonomy and phylogeny, offering precise methods for identifying species and understanding evolutionary relationships. Additionally, the genetic improvement of natural enemies for inebriate tolerance underlines the potential of molecular biotechnology in pest management systems.

The final section, Unit V, titled Advanced Applications in Molecular Entomology, focuses on cutting-edge topics that highlight the integration of molecular biology with practical applications. DNA-based diagnostics have become indispensable for detecting pathogens, identifying pest species, and monitoring resistance traits. This unit also provides a comparative analysis of insect immune systems and vertebrates, shedding light on the evolutionary divergence of immune mechanisms. The molecular basis of metamorphosis, driven by hormones and genetic pathways, is explored for its relevance in pest control and developmental biology. Sf transgenic technology, molecular biology of baculoviruses, and the genetic basis of insecticide resistance are discussed in detail, emphasizing their roles in sustainable agriculture and integrated pest management strategies. The unit concludes with a focus on resistance management strategies in transgenic crops, a critical aspect of ensuring the longevity and efficacy of genetically engineered solutions.

This book not only compiles the latest knowledge in molecular entomology but also contextualizes its applications within broader challenges in pest management, agriculture, and conservation. The integration of molecular techniques with entomology offers transformative solutions for challenges that traditional methods alone cannot address. From the development of transgenic plants for pest resistance to the use of genetically engineered natural enemies, molecular entomology provides the tools to enhance agricultural productivity while minimizing environmental harm.

Moreover, this book acknowledges the challenges and ethical considerations associated with molecular entomology. The deployment of genetically modified organisms, whether plants or insects, raises ecological, regulatory, and societal concerns. Addressing these requires transparent communication, rigorous safety assessments, and a commitment to sustainability. The discussions in this book aim to equip readers with the knowledge and critical perspectives necessary to navigate these complex issues.

Each unit is structured to provide a comprehensive understanding of its topics, with an emphasis on practical applications and future prospects. The chapters are written to cater to diverse audiences, from students new to molecular entomology to seasoned researchers looking for the latest advancements. Case studies and real-world examples are included to bridge the gap between theory and practice, ensuring that readers can relate the molecular concepts to tangible outcomes.

In closing, this book serves as a guide to the fascinating world of molecular entomology, where the smallest genetic details unlock vast possibilities for innovation. It is a testament to the power of interdisciplinary science, combining molecular biology, entomology, and biotechnology to address some of the most pressing challenges of our time. It is hoped that the knowledge and insights presented here will inspire future researchers to explore the untapped potential of molecular entomology, contributing to a more sustainable and resilient world.

Authors

About the Authors

Dr. Krishna Chaitanya Tirunagaru is an esteemed expert in Agricultural Entomology, with a Master's degree from Tamil Nadu Agricultural University and a Ph.D. from Annamalai University, India. He specializes in insect taxonomy and has made significant contributions to the field, having discovered, and described 26 new species of insects within the Encyrtidae family of Hymenoptera. With six years of teaching experience as an Assistant Professor at NMIMS – Deemed to be University and two years of research experience at Dr. YSRHU, Dr. Tirunagaru has built a distinguished career. His academic work includes three patents, 16 research articles published in peer-reviewed journals, and six book chapters. Dr. Tirunagaru has received several prestigious awards, including being named a Best Faculty Award (2022-2023) from SVKM's NMIMS, Best Young Scientist, Best Article, and Research Excellence awards from various prestigious foundations. His accolades include a Fellow of the Royal Entomological Society, UK, and a life member of the Institute of Scholars. He is also the faculty in charge of the World Wide Fund (WWF) Project and the Faculty representative for the Institution's Innovation Council (IIC) at NMIMS. His passion for entomology and commitment to education continue to inspire future generations.

Dr. Arshad Ali is an Assistant Professor at the Department of Zoology, Gandhi Faiz-e-Aam College, Shahjahanpur, India. He has received his M.Sc. (Entomology) and Ph.D. (Agricultural Entomology) from Aligarh Muslim University, Aligarh. After that, he received Dr. D.S. Kothari's Postdoctoral Fellowship from UGC, New Delhi. Dr. Ali has been teaching Zoology at the undergraduate level and Entomology at the postgraduate level for the last 12 years and has research experience of 16 years. In addition to teaching, he is engaged in guiding students for their doctoral degree on various aspects of insect pest problems and their management, especially in Entomology. He has published over 50 research papers including original research articles, book chapters, and research notes in various national and international publications, and also presented his research findings in relevant conferences, seminars, symposia, and workshops. The area of his research activities focuses on population dynamics, economic entomology, and insect pest management mainly through biological pest control and bio-pesticides.

Dr. Kennedy Ningthoujam (M.Sc., Ph.D.) is currently serving as Assistant Professor (Sr. Scale) Entomology in School of Crop Protection, CPGS-AS, CAU (I) Umiam, Meghalaya. He completed his M.Sc. in Agricultural Entomology from TNAU, Coimbatore in 2008 from where he was awarded TNAU Gold Medal in Best Thesis Award; Ph.D. in Entomology from IARI, New Delhi in 2015. He was awarded ICAR-JRF (with Fellowship) during his M.Sc.; ICAR-SRF (with Fellowship) and IARI (Merit Fellowship) during Ph.D. degree period and he is a Life time member of Entomological Society of India, IARI, Pusa, New Delhi-110012.

He is specialized in Molecular Entomology and has wide knowledge on Bt resistance monitoring and detection; molecular markers associated with Bt resistance in Lepidopteran pests; novel insecticidal

Bt genes detection. He is also currently serving as Principle Investigator (PI) of 3 externally funded projects (1 DBT and 2 DST-SERB). He has isolated and submitted 32 Bt strains from Northeast region (of which 4 are novel strains) in NAIMCC, ICAR-NBAIM, U.P. He has reported 12 species of fruit flies for the first time in India of which 6 fruit flies are new species belonging to the tribe Dacini and Acanthonevrini.

He is currently mentoring 3 M.Sc. and 2 Ph.D. students, and 15 M.Sc. students have completed degree under his chairmanship. Dr. Kennedy has authored over 50 research articles in peer-reviewed national and international journals; along with several book chapters and popular articles.

Dr. Anup Chandra graduated with BSc (Agriculture) from the University of Agricultural Sciences, Dharwad, and later pursued his Masters and Doctoral degree in 'Entomology' from the Indian Agricultural Research Institute, New Delhi. He joined the Agricultural Research Service (ARS-ICAR) in the discipline of 'Agricultural Entomology' and serving in the field since 2011. He has expertise in the field of Pulse Entomology, especially on sucking insect pests, and had been a member of several expert committees constituted for the diagnosis of insect pests infestation. He is a member of several scientific societies as well as a Life Fellow of the Entomological Society of India, New Delhi. He is recipient of awards viz., IAHF-Young Scientist Award-2017, Best Scientist Award-2014 (ICAR RC for NEH Region, Arunachal Pradesh Centre), 4th e North East Award 2013 (IMD, Govt. of India), Excellence for Outstanding Dissemination of Agromet Advisories for Farming Community Award-2014 (IMD, Govt. of India), Best Team Award-2023 (ICAR-IIPR Kanpur). Apart from this, he has also edited books in the field of agriculture in addition to his contribution of several Research articles/Book chapters/Technical articles, etc. Presently, he is working as a Senior Scientist in the Division of Crop Protection, ICAR-Indian Institute of Pulses Research, Kanpur, Uttar Pradesh.

Dr. Mahesh Pathak obtained his Ph.D. in Plant Protection from Aligarh Muslim University, Aligarh, India in the year 2008. At present, he is serving as Professor of Entomology and Incharge School of Crop Protection, College of Post Graduate Studies in Agricultural Sciences, Central Agricultural University (Imphal), Umiam, Meghalaya. He is a Fellow of the Entomological Society of India, Society of Plant Protection Sciences and Applied Zoologist Research Association, Life member of the Indian Association of Hill Farming, Society of Extension Education, and Society for Advancement of Agricultural Innovations. He also has experience in establishing the Integrated Beekeeping Development Centre (IBDC), and Centre of Excellence in Beekeeping in Arunachal Pradesh. He is also actively involved in the research works on biocontrol-based Integrated Pest Management. He has published more than 101 research papers in National and International journals of scientific repute, 8 books, and resource manuals, 40 book chapters, 28 technical bulletins, and 140 popular articles. He is the recipient of several awards including Pandit Deendayal Upadhyay Rashtriya Krishi Vigyan Protshahan Puraskar, Young Scientist Award, Eminent Scientist Award, Best Extension Professional Award, and Aryabhatt Samman Pattar.

Dr. Rojeet Thangjam completed his B.Sc. (Agriculture) from AAU, Jorhat, Assam, and continued his M.Sc. in Agricultural Entomology (2012) at the UAS, Dharwad, Karnataka, and Ph.D. in Entomology (2018) from the Department of Entomology, AAU, Jorhat, Assam. Dr. Thangjam qualified for ASRB ICAR-NET during 2013 and 2014. His significant contribution to entomological fields is that he worked on insect vectors, IPM, and apiculture. During his Master's and PhD Thesis, he worked on the detection of phytoplasma and viruses from their vectors respectively, through molecular techniques. He has identified 6 new species of stingless bee under the genus Tetragonula and Lepidotrigona, recorded many insects from NE India for the first time, and also documented many edible insects from the region. He has published 35 peer-reviewed research articles in international and national journals, 17 extended abstracts, 5 book chapters, 7 extension leaflets, and one short communication. He has also received many awards in Best Poster Presentation, Best Oral Presentation, and Young Achiever Award 2020 at National and International Conferences. He has organized many Capacity Building Training programs sponsored by ICAR-ATARI, Zone –VII, ICAR-NBAIR, Karnataka, etc, and also acts as a resource person in many training programs.

Chapter 1

Molecular Biology Techniques

Introduction

Molecular biology is a branch of science focused on understanding the structure, function, and interactions of biological molecules that govern cellular processes and life's fundamental mechanisms. At its core, molecular biology explores the intricate relationships between nucleic acids (DNA and RNA), proteins, and other macromolecules within cells, emphasizing how these interactions drive biological functions and enable cellular development, differentiation, and replication.

The origins of molecular biology can be traced to foundational discoveries in genetics and biochemistry, where scientists first began to understand heredity and the biochemical nature of cellular functions. Molecular biology emerged as a distinct discipline in the mid-20th century, propelled by landmark discoveries like the DNA double helix structure by James Watson and Francis Crick in 1953. This discovery provided the first glimpse into the molecular composition of genetic material, establishing the framework for subsequent studies on genetic coding, gene expression, and inheritance.

Central to molecular biology is the Central Dogma, a principle proposed by Francis Crick, which outlines the directional flow of genetic information from DNA to RNA to proteins. According to the Central Dogma, DNA holds the genetic blueprint of organisms, which is transcribed into RNA. The RNA then serves as a template for synthesizing proteins through a process called translation. Proteins, in turn, carry out cellular activities, serve as enzymes, and build cellular structures, thereby implementing the genetic instructions encoded in DNA.

Molecular biology seeks to uncover how this flow of information is regulated, maintained, and, in some cases, altered under specific conditions, such as during development, environmental stress, or disease. The study of DNA includes understanding its structure, replication, and repair mechanisms, as well as investigating mutations that may lead to genetic disorders. RNA research involves analyzing its types—such as messenger RNA (mRNA), ribosomal RNA (rRNA), and transfer RNA (tRNA)—and functions in various cellular contexts, including the regulation of gene expression. Protein studies in molecular biology encompass examining protein synthesis, folding, and function, as well as how proteins interact in complex networks to facilitate cellular processes.

DOI: 10.1201/9781003757405-1

Molecular biology overlaps with several related disciplines, including genetics, biochemistry, cell biology, and biotechnology. While genetics focuses primarily on heredity and the transmission of traits, molecular biology delves into the molecular structures and interactions that make genetic inheritance possible. Biochemistry examines the chemical reactions within cells, which are crucial for understanding how molecular biology's macromolecules, such as enzymes and nucleic acids, operate at a functional level. Cell biology explores cellular structure and function as a whole, while molecular biology targets the molecular components within cells to explain cellular behavior and differentiation. Biotechnology applies molecular biology knowledge to develop practical applications, from genetic engineering to pharmaceutical innovations.

Technological advancements have been instrumental in shaping modern molecular biology. Techniques such as polymerase chain reaction (PCR) allow scientists to amplify specific DNA segments, facilitating genetic analysis. DNA sequencing technologies, from traditional Sanger sequencing to high-throughput next-generation sequencing (NGS), have enabled the detailed mapping of entire genomes, providing a comprehensive view of genetic information in various organisms. Techniques like CRISPR-Cas9 gene editing allow precise modifications of genetic material, offering transformative potential in medical research, agriculture, and beyond. Additionally, the field of proteomics, which studies the entire set of proteins in a cell, leverages molecular biology to examine protein expression patterns and interactions, leading to insights into cellular processes and disease mechanisms.

The applications of molecular biology are vast and impactful. In medicine, molecular biology plays a pivotal role in understanding genetic diseases, developing targeted therapies, and advancing personalized medicine. In agriculture, molecular biology has contributed to the development of genetically modified organisms (GMOs) with traits like pest resistance or improved nutritional content. Environmental science also benefits from molecular biology, particularly in studying biodiversity, understanding evolutionary relationships, and developing conservation strategies.

In essence, molecular biology provides a molecular-level view of life, revealing how the fundamental building blocks of cells- DNA, RNA, and proteins—work in concert to sustain life and facilitate adaptation. By exploring these molecules' interactions and the regulatory mechanisms controlling them, molecular biology continues to expand our knowledge of biological systems, enabling advances in various scientific fields and enhancing our ability to address challenges in health, agriculture, and environmental sustainability. Through its precise and targeted approach, molecular biology remains a foundational discipline, driving both scientific discovery and applied innovation in our understanding of life's complexities.

The history of molecular biology traces a remarkable journey of discovery that has revolutionized our understanding of life at the molecular level. From the early 20th century to the present day, molecular biology has evolved from a few groundbreaking discoveries into a sophisticated scientific discipline that intertwines with fields as diverse as genetics, biochemistry, biotechnology, and bioinformatics. Rooted in an initial quest to understand heredity and the biochemical foundations of life, the progression of molecular biology has been marked by pioneering discoveries, theoretical advancements, and technological innovations that have deeply impacted biological sciences and redefined numerous aspects of human knowledge and applied science.

The foundation of molecular biology rests upon several pioneering discoveries. These early breakthroughs helped clarify the structure and function of genetic material, igniting curiosity about how genes influence cellular processes and hereditary traits. The iconic discovery of the DNA double helix by James Watson and Francis Crick in 1953, which built upon crucial experimental insights from Rosalind Franklin and Maurice Wilkins, was a defining moment. This structural revelation unveiled the chemical composition of genetic material and provided a framework for understanding genetic replication and inheritance, serving as a cornerstone for the emerging field of molecular biology.

Key theories and models soon followed, contributing to an expanding framework for explaining molecular processes in cells. The Central Dogma of Molecular Biology, proposed by Francis Crick, was instrumental in outlining the flow of genetic information from DNA to RNA to proteins. This concept clarified the roles of DNA and RNA in gene expression, enabling scientists to understand how genetic instructions are translated into cellular functions. Additionally, theories related to DNA replication, repair mechanisms, and gene regulation helped further contextualize cellular processes at a molecular level, transforming our approach to biological research.

The advent of molecular biology profoundly impacted biological sciences, catalyzing advancements across various branches, from evolutionary biology to medical science. By examining cellular processes at a molecular scale, researchers gained the ability to probe the intricate machinery of life. This deeper understanding of genetic and biochemical pathways paved the way for more precise studies in genetics, cell biology, and microbiology, offering insights into disease mechanisms, gene expression patterns, and metabolic processes. Consequently, molecular biology became central to modern biological research, with its applications permeating multiple scientific domains and contributing to a holistic view of life's complexity.

As molecular biology matured, it emerged as a distinct scientific discipline in the mid-20th century. Early research initiatives, such as the establishment of the Laboratory of Molecular Biology in Cambridge, and academic programs worldwide dedicated to molecular biology, facilitated its growth as an independent field. Researchers from diverse scientific backgrounds- biochemistry, physics, genetics, and more-collaborated, enabling a multidisciplinary approach to molecular studies. The establishment of specialized journals, societies, and conferences further solidified molecular biology's standing within the scientific community, fostering collaboration and knowledge-sharing among researchers globally.

The journey of molecular biology is also punctuated by key technological milestones that propelled it forward. The invention of molecular techniques such as DNA sequencing, polymerase chain reaction (PCR), and gene cloning transformed scientific research, enabling scientists to investigate and manipulate genetic material with unprecedented precision. Each new technique opened avenues for exploration, from mapping entire genomes to understanding the regulation of gene expression. Notably, the Human Genome Project, completed in 2003, marked a landmark achievement, laying the groundwork for personalized medicine, genomics, and advanced bioinformatics. More recently, innovations in gene editing technologies, such as CRISPR-Cas9, have ushered in a new era of molecular biology, facilitating precise alterations in genetic material with applications in medicine, agriculture, and environmental science.

Today, molecular biology continues to expand in scope and depth, underpinning revolutionary applications in genetics, genomics, biotechnology, and agriculture. Techniques developed in molecular biology laboratories are used to diagnose genetic disorders, engineer crops with enhanced traits, and explore evolutionary relationships between species. The field's integration with computational tools has fostered bioinformatics and computational biology, enabling large-scale data analysis and predictive modeling. Molecular biology's applications in bioinformatics and computational biology, for instance, have allowed for vast genomic data analysis, fostering breakthroughs in personalized medicine and evolutionary biology.

1. Historical Background and Evolution of Molecular Biology

1.1 Pioneering Discoveries in Molecular Biology

The field of molecular biology emerged through a series of transformative discoveries that shaped our understanding of how genetic information is stored, transmitted, and expressed within living organisms. From identifying DNA as the genetic material to elucidating its double-helical structure, each discovery has contributed foundational knowledge and driven the progress of molecular biology as a discipline.

Discovery of DNA as the Genetic Material

At the beginning of the 20th century, scientists understood that some form of molecular substance carried genetic information, but it was widely assumed that proteins, with their structural complexity and diversity, were the most likely candidates. However, a sequence of pivotal experiments in the mid-1900s gradually established DNA as the genetic material, overturning earlier hypotheses and laying the groundwork for genetics and molecular biology.

1. **Frederick Griffith and the Transforming Principle (1928)**

Frederick Griffith, a British bacteriologist, conducted groundbreaking work that led to the first hint that DNA could be the genetic material. Griffith's experiments involved two strains of *Streptococcus pneumoniae*, the bacterium responsible for pneumonia: a virulent, smooth (S) strain that caused disease and a non-virulent, rough (R) strain that did not cause disease. He discovered that when he killed the S strain with heat and mixed it with the R strain, some of the live R strain bacteria were "transformed" into the virulent form. This transformation suggested that a "transforming principle" from the heat-killed S strain enabled the R strain to acquire the virulent phenotype.

Griffith's experiment was crucial as it suggested that traits could be transferred between organisms by an unknown molecular substance, hinting that a heritable factor existed outside of proteins. However, Griffith's work did not identify the molecular basis of this transformation, leaving an unanswered question that would drive further research.

2. **The Avery-MacLeod-McCarty Experiment (1944)**

Building on Griffith's observations, Oswald Avery and his colleagues Colin MacLeod and Maclyn McCarty conducted experiments that would identify DNA as the "transforming principle." Avery and his team aimed to isolate the specific substance responsible for transformation by systematically treating bacterial extracts with enzymes that selectively degraded proteins, lipids, polysaccharides, or nucleic acids.

- By treating the bacterial extracts with protein-digesting enzymes, they found that transformation still occurred, suggesting that proteins were not the transforming principle.

- However, when they used DNase, an enzyme that breaks down DNA, the transformation was halted. This provided compelling evidence that DNA, rather than protein or any other macromolecule, was responsible for carrying genetic information.

The Avery-MacLeod-McCarty experiment was a landmark achievement, as it provided the first strong evidence that DNA, and not protein, carried genetic information. Despite initial skepticism from the scientific community, their findings laid the groundwork for understanding DNA as the genetic material.

3. **The Hershey-Chase Experiment (1952)**

The Hershey-Chase experiment, led by Alfred Hershey and Martha Chase, offered definitive proof that DNA was the genetic material. Using *bacteriophages*-viruses that infect bacteria-they designed an experiment that differentiated between the roles of DNA and protein in viral replication. Bacteriophages are composed of only protein and DNA, making them ideal for determining which of these molecules carries genetic information.

» Hershey and Chase labeled the DNA of the phage with radioactive phosphorus (^{32}P) and the protein coat with radioactive sulfur (^{35}S).

» After allowing the labeled phages to infect bacteria, they used a blender to separate the viral protein coats from the bacterial cells.

» Analysis showed that only the radioactive phosphorus (DNA) entered the bacterial cells, while the radioactive sulfur (protein) remained outside.

This experiment conclusively demonstrated that DNA was the molecule responsible for carrying genetic information, as only the DNA entered the bacterial cells to direct viral replication. Hershey and Chase's findings were pivotal in cementing DNA as the genetic material, shifting the scientific consensus and setting the stage for future research into DNA structure and function.

Discovery of the Structure of DNA: The Double Helix

With DNA established as the genetic material, the next critical question was understanding its structure. The double-helix model of DNA, proposed by James Watson and Francis Crick in 1953, was one of the most influential discoveries in biology, revealing the molecular basis of heredity and the mechanism by which genetic information could be faithfully replicated.

1. **Chargaff's Rules (1950)**

Erwin Chargaff's research into the composition of DNA provided essential clues about its structure. Through detailed chemical analysis, Chargaff discovered two key patterns in the arrangement of nucleotide bases in DNA:

- The amount of adenine (A) always equaled the amount of thymine (T), and the amount of guanine (G) always equaled the amount of cytosine (C), leading to the conclusion that A=T and G=C. This phenomenon, now known as "Chargaff's rules," suggested a complementary base-pairing mechanism.
- Chargaff's findings also revealed that the proportions of A, T, G, and C varied between species, indicating that DNA contained species-specific information.

Chargaff's observations provided Watson and Crick with critical insights, particularly the concept of complementary base pairing, which they incorporated into their structural model of DNA.

2. **X-Ray Diffraction and Rosalind Franklin's Contribution**

In the early 1950s, Rosalind Franklin, working with Maurice Wilkins at King's College London, employed X-ray diffraction techniques to study the structure of DNA. X-ray diffraction allowed scientists to infer the spatial arrangement of atoms within a molecule based on the scattering pattern of X-rays passing through it.

- Franklin's high-resolution X-ray images, particularly the famous "Photo 51," provided crucial evidence that DNA was helical in shape.
- Her images revealed a characteristic X-shaped pattern, indicating a helical structure with specific dimensions, including the spacing between base pairs and the periodicity of the helix.

Although her contributions were not widely acknowledged during her lifetime, Franklin's work was instrumental in guiding Watson and Crick's model of the double helix.

3. **Watson and Crick's Double Helix Model (1953)**

James Watson and Francis Crick, building on Chargaff's rules and Franklin's X-ray data, developed the double-helix model of DNA. Their model proposed that DNA consists of two antiparallel strands of nucleotides, twisted into a helix and held together by hydrogen bonds between complementary base pairs (A with T, and G with C).

Key aspects of the Watson-Crick model include:

- **Complementary Base Pairing:** The specific pairing between A and T and between G and C ensures that genetic information can be accurately copied during DNA replication.
- **Antiparallel Orientation:** The two strands of DNA run in opposite directions, which is essential for DNA replication and other molecular interactions.
- **Double-Helical Structure:** The twisted ladder-like shape of the DNA molecule provides stability and compactness, allowing it to store vast amounts of information in a small cellular space.

The Watson-Crick model had profound implications, providing a molecular basis for heredity and explaining how genetic information could be replicated and transmitted. Their discovery earned them the Nobel Prize in Physiology or Medicine in 1962, shared with Maurice Wilkins.

Discovery of the Mechanism of DNA Replication

The double-helix structure of DNA suggested a possible mechanism for DNA replication, but the exact process was not confirmed until further experiments were conducted in the late 1950s.

1. **The Meselson-Stahl Experiment (1958)**

Matthew Meselson and Franklin Stahl conducted an elegant experiment to elucidate the mechanism of DNA replication, demonstrating that DNA replication is "semiconservative." In semiconservative replication, each original DNA strand serves as a template for synthesizing a new complementary strand, resulting in two DNA molecules, each containing one old and one new strand.

» Meselson and Stahl grew bacteria in a medium containing heavy nitrogen (^{15}N), which labeled the DNA with the heavy isotope.

» They then transferred the bacteria to a medium containing lighter nitrogen (^{14}N) and allowed the cells to replicate.

» By analyzing the density of the DNA using cesium chloride gradient centrifugation, they observed that the DNA formed bands corresponding to intermediate and light densities, consistent with semiconservative replication.

The Meselson-Stahl experiment was significant because it confirmed that DNA replication follows a template-driven process, ensuring fidelity in the transmission of genetic information across generations.

Discovery of mRNA and the Role of RNA in Protein Synthesis

In the years following the discovery of DNA structure and replication, researchers turned their attention to the process by which DNA is used to direct protein synthesis. This investigation led to the identification of messenger RNA (mRNA) as the intermediary between DNA and protein synthesis.

1. **The Discovery of mRNA (1961)**

François Jacob and Jacques Monod, along with Sydney Brenner, demonstrated that mRNA serves as a temporary copy of the genetic information in DNA, carrying instructions from the DNA in the nucleus to the ribosomes in the cytoplasm, where proteins are synthesized.

- Their experiments revealed that mRNA was synthesized from a DNA template and that its sequence dictated the sequence of amino acids in proteins.
- The discovery of mRNA clarified the mechanism of gene expression and introduced the concept of transcription, where DNA is "transcribed" into mRNA, which then undergoes translation into protein.

2. **The Genetic Code and Protein Synthesis**

The elucidation of the genetic code, achieved by Marshall Nirenberg and Har Gobind Khorana in the 1960s, was another critical advancement. They demonstrated that the sequence of nucleotides in mRNA corresponds to specific amino acids in proteins, establishing a "triplet code" of three nucleotides (codons) per amino acid.

- Nirenberg's experiments with synthetic RNA sequences revealed that specific sequences of three nucleotides (e.g., UUU, which codes for phenylalanine) correspond to individual amino acids.
- This discovery of the triplet code unlocked a universal language of biology, providing the basis for understanding how genes dictate protein structure.

The identification of the genetic code was essential for molecular biology, as it explained how the information encoded in DNA could be translated into the diverse proteins necessary for life. This work laid the foundation for genetic engineering, synthetic biology, and biotechnology.

1.2 Key Theories and Models in Molecular Biology

The development of molecular biology has been driven by fundamental theories and models that explain the mechanisms of genetic information storage, expression, and inheritance. These theories and models provide the framework through which scientists understand the processes that govern life at the molecular level.

The Central Dogma of Molecular Biology

The **central dogma** of molecular biology, proposed by Francis Crick in 1958, describes the directional flow of genetic information within a biological system: from DNA to RNA to protein. This model is foundational in understanding how genes encode proteins, the functional molecules within cells.

1. **DNA → RNA → Protein Pathway**
    - The central dogma asserts that genetic information is stored in DNA, transcribed into RNA, and then translated into proteins.
    - This process begins with **transcription**, where DNA is used as a template to synthesize messenger RNA (mRNA).
    - The mRNA then serves as a template for **translation**, the process by which ribosomes read the mRNA sequence to assemble amino acids into a protein.
    - By dictating the sequence of amino acids, the central dogma underscores that DNA indirectly controls cellular function and phenotype.
2. **Exceptions and Extensions of the Central Dogma**
    - **Reverse Transcription:** In certain organisms, such as retroviruses, RNA can be reverse-transcribed into DNA. This process, catalyzed by the enzyme reverse transcriptase, allows RNA viruses to integrate their genetic material into the host genome.
    - **RNA Interference and Regulation:** Small RNAs, such as microRNAs (miRNAs) and small interfering RNAs (siRNAs), play crucial roles in gene regulation by binding to mRNA and influencing its translation or stability.
    - The central dogma is a guiding model that has expanded with discoveries in epigenetics, RNA processing, and non-coding RNAs, highlighting that genetic information flow is not always linear or unidirectional.

The Operon Model of Gene Regulation

Proposed by François Jacob and Jacques Monod in 1961, the **operon model** describes how genes are regulated in prokaryotic organisms. The model was formulated to explain the regulation of the *lac* operon in *Escherichia coli*, an operon responsible for lactose metabolism. This model introduced the concepts of gene clusters, regulatory proteins, and feedback mechanisms, providing insights into genetic control mechanisms.

1. **Components of the Operon**
    - An operon is a cluster of genes under the control of a single promoter, allowing multiple genes to be regulated simultaneously.
    - It typically consists of **structural genes** (coding for enzymes or proteins), a **promoter** (binding site for RNA polymerase), an **operator** (binding site for regulatory proteins), and a **regulatory gene** (coding for repressor or activator proteins).

2. **The *lac* Operon as a Model System**
   - The *lac* operon in *E. coli* is activated in the presence of lactose and repressed in its absence. In the absence of lactose, a repressor protein binds to the operator region, blocking RNA polymerase and preventing transcription of the genes involved in lactose metabolism.
   - When lactose is present, it binds to the repressor, causing it to release from the operator, thereby allowing transcription to occur.
   - This model introduced the concept of **inducible** and **repressible** systems in gene regulation and provided the first detailed understanding of transcriptional control in prokaryotes.

3. **Applications of the Operon Model**
   - The operon model has provided insights into genetic regulatory mechanisms and has been foundational in biotechnology for designing inducible gene expression systems.
   - Variations on the operon model have been identified in other organisms, demonstrating its general applicability in understanding genetic regulation.

The Theory of the Gene and Molecular Basis of Inheritance

The theory of the gene, developed throughout the early 20th century, hypothesized that genes are the fundamental units of inheritance. With the discovery of DNA as the genetic material, this theory gained a molecular foundation, explaining how traits are transmitted across generations.

1. **Classical Gene Theory**
   - Classical genetics posited that genes reside on chromosomes, which segregate and assort independently during meiosis.
   - Based on Gregor Mendel's laws of inheritance, the classical theory explained genetic traits using concepts such as dominance, recessiveness, and independent assortment.
2. **Molecular Gene Theory**
   - Following the discovery of DNA's double-helix structure, the theory expanded to include the understanding that genes are segments of DNA coding for proteins or functional RNA molecules.
   - The one gene-one enzyme hypothesis, proposed by Beadle and Tatum, evolved into the one gene-one polypeptide concept, linking genes directly to proteins and phenotypes.
3. **Impact on Modern Genetics**
   - The molecular theory of genes provided a framework for genetic engineering, allowing scientists to identify, manipulate, and transfer genes across organisms.
   - The theory also underpins many modern biotechnological applications, including gene therapy, transgenics, and functional genomics.

The Genetic Code and Its Universality

The **genetic code** is the set of rules by which the sequence of nucleotides in DNA or RNA is translated into amino acids in proteins. The cracking of the genetic code in the 1960s by scientists such as Marshall Nirenberg and Har Gobind Khorana was a pivotal moment in molecular biology, revealing how sequences of nucleotides dictate protein structure.

		Second Position								
		U		C		A		G		
First Position		code	amino acid	code	amino acid	code	amino acid	code	amino acid	Third Position
U		UUU	phe	UCU	ser	UAU	tyr	UGU	cys	U
		UUC		UCC		UAC		UGC		C
		UUA	leu	UCA		UAA	STOP	UGA	STOP	A
		UUG		UCG		UAG	STOP	UGG	trp	G
C		CUU	leu	CCU	pro	CAU	his	CGU	arg	U
		CUC		CCC		CAC		CGC		C
		CUA		CCA		CAA	gln	CGA		A
		CUG		CCG		CAG		CGG		G
A		AUU	ile	ACU	thr	AAU	asn	AGU	ser	U
		AUC		ACC		AAC		AGC		C
		AUA		ACA		AAA	lys	AGA	arg	A
		AUG	met START	ACG		AAG		AGG		G
G		GUU	val	GCU	ala	GAU	asp	GGU	gly	U
		GUC		GCC		GAC		GGC		C
		GUA		GCA		GAA	glu	GGA		A
		GUG		GCG		GAG		GGG		G

Genetic Code and Codons

1. **Triplet Code and Codons**
   - The genetic code consists of triplet sequences of nucleotides, known as codons, each of which specifies a particular amino acid.
   - The code is **degenerate**, meaning multiple codons can specify the same amino acid, which allows for some tolerance to mutations in DNA.
2. **Universality of the Code**
   - Remarkably, the genetic code is nearly universal across all known organisms, from bacteria to humans, highlighting a shared evolutionary ancestry.
   - The universality of the genetic code allows for the transfer of genes between species, a cornerstone of genetic engineering and biotechnology.

3. **Wobble Hypothesis**

   - The **wobble hypothesis**, proposed by Francis Crick, explains the flexibility in codon-anticodon pairing, particularly at the third base position. This hypothesis accounts for the degeneracy of the genetic code and its efficiency.

Theories of Evolutionary Molecular Biology

Evolutionary molecular biology provides insights into how genetic changes at the molecular level drive evolutionary processes. Key models, such as the **neutral theory of molecular evolution** and **molecular clocks**, have contributed to our understanding of genetic variation and speciation.

1. **Neutral Theory of Molecular Evolution**

   - Proposed by Motoo Kimura in 1968, the neutral theory suggests that most genetic mutations are neutral and do not affect an organism's fitness.
   - According to this theory, genetic drift, rather than natural selection, plays a primary role in the accumulation of molecular changes over time.
   - The neutral theory challenges the notion that all genetic variations are adaptive, instead proposing that many mutations are selectively neutral and accumulate at a constant rate, providing a molecular clock for dating evolutionary events.

2. **Molecular Clock Hypothesis**

   - The molecular clock hypothesis proposes that genetic mutations accumulate at a roughly constant rate, allowing scientists to estimate the time of divergence between species.
   - Molecular clocks have been invaluable for reconstructing evolutionary histories and phylogenetic relationships, particularly when fossil records are incomplete or absent.
   - By comparing DNA or protein sequences across species, molecular clocks provide estimates of evolutionary timelines, contributing to fields such as comparative genomics and evolutionary biology.

The RNA World Hypothesis

The **RNA world hypothesis** is a theoretical model that proposes that RNA was the original molecule for storing genetic information and catalyzing chemical reactions in the earliest forms of life. This theory suggests that an "RNA world" preceded the current DNA-protein world, with RNA serving both as the genetic material and as a catalyst for biochemical reactions.

1. **Ribozymes and Catalytic RNA**

   - The discovery of ribozymes, RNA molecules with catalytic activity, provided support for the RNA world hypothesis, suggesting that RNA could function as both a repository of genetic information and an enzyme.
   - Catalytic RNAs demonstrate that RNA alone could perform essential functions necessary for life, including self-replication, without the need for proteins.

2. **Implications of the RNA World Hypothesis**
   - The RNA world hypothesis provides a plausible explanation for the origin of life, suggesting a gradual evolution from simple RNA molecules to more complex systems involving DNA and proteins.
   - This hypothesis has inspired research into the origins of life and has implications for synthetic biology, where RNA-based systems are explored for their potential in creating self-sustaining biochemical networks.

The Central Dogma Variations and Epigenetics

While the central dogma remains a guiding model, modern discoveries have revealed exceptions and added layers of complexity through **epigenetics** and **non-coding RNAs**.

1. **Epigenetic Modifications**
   - Epigenetics encompasses chemical modifications to DNA and histones that do not alter the DNA sequence but can affect gene expression. These modifications, such as DNA methylation and histone acetylation, play critical roles in development, cellular differentiation, and disease.
   - Epigenetic changes can be heritable and reversible, providing a mechanism for gene regulation that operates outside of traditional genetic mutations.
2. **Non-Coding RNAs and Regulatory Networks**
   - Beyond mRNA, many types of non-coding RNAs, including microRNAs (miRNAs) and long non-coding RNAs (lncRNAs), play roles in regulating gene expression.
   - These molecules add complexity to gene regulatory networks, influencing processes from development to immune response and adding to our understanding of how genes are controlled.

Impact of Molecular Biology on Biological Sciences

The advent of molecular biology has fundamentally transformed the biological sciences, catalyzing advancements across genetics, medicine, biotechnology, ecology, and evolutionary biology. By unraveling the molecular mechanisms underlying cellular processes, molecular biology has provided researchers with powerful tools to explore, understand, and manipulate life at the most fundamental level.

Advancements in Genetics and Genomics

Molecular biology has revolutionized genetics by uncovering the structure and function of DNA, elucidating gene regulation mechanisms, and enabling gene manipulation through genetic engineering. The discovery of the genetic code and the development of sequencing technologies have unlocked a deeper understanding of genomes, leading to the following major advancements:

1. **Genome Sequencing and Genomics**
   - The Human Genome Project (HGP), a landmark international collaboration completed in 2003, successfully mapped the entire human genome, revealing over three billion

base pairs and tens of thousands of genes. This achievement marked a turning point in genomics, allowing researchers to study the genetic basis of diseases, inheritance, and human evolution.

- The introduction of next-generation sequencing (NGS) technologies has democratized genomic research, reducing costs and increasing speed. Today, sequencing entire genomes, transcriptomes, and epigenomes is feasible, enabling comprehensive studies on genetic diversity, population genetics, and comparative genomics.

2. **Functional Genomics and Gene Expression Studies**
   - Functional genomics, made possible by molecular biology techniques like gene knockout, RNA interference (RNAi), and CRISPR-Cas9, allows scientists to investigate the roles of specific genes in cellular and physiological processes.
   - Microarray and RNA sequencing (RNA-seq) technologies enable researchers to analyze global gene expression patterns, identifying gene networks involved in development, disease, and environmental response.

3. **Personalized and Precision Medicine**
   - Molecular biology has laid the foundation for personalized medicine, where treatments are tailored to an individual's genetic profile. Genetic information can be used to predict disease susceptibility, drug responses, and the potential for adverse reactions, leading to more effective and safer therapies.
   - Advances in genomics and pharmacogenomics are paving the way for precision medicine, which aims to customize healthcare based on individual molecular and genetic characteristics.

Molecular Biology's Role in Medicine and Disease Research

Molecular biology has had a transformative impact on medicine, particularly in understanding the molecular basis of diseases, identifying biomarkers, developing diagnostic tools, and creating innovative therapies.

1. **Molecular Diagnostics and Biomarkers**
   - Techniques like PCR, qPCR, and ELISA have enabled precise detection of pathogens, genetic mutations, and biomarkers associated with diseases. These tools are widely used in clinical diagnostics, from detecting infectious diseases to identifying genetic predispositions to cancer.
   - Molecular diagnostics allow for early detection of diseases, enabling timely intervention and improved patient outcomes. Biomarkers derived from molecular studies aid in disease prognosis and monitoring treatment efficacy.

2. **Understanding Cancer at the Molecular Level**
   - Molecular biology has been instrumental in understanding cancer, revealing that mutations in oncogenes and tumor suppressor genes drive tumorigenesis. The identification of

molecular pathways involved in cell cycle regulation, apoptosis, and DNA repair has provided insights into cancer's development and progression.

- Targeted therapies, such as tyrosine kinase inhibitors and immunotherapies, have been developed based on molecular mechanisms specific to cancer cells, offering more effective treatments with fewer side effects than traditional chemotherapy.

3. **Gene Therapy and Genetic Medicine**
   - Gene therapy, the introduction or correction of genes to treat diseases, is a direct application of molecular biology. Techniques like viral vector delivery, CRISPR-Cas9, and RNA-based therapies have been explored to treat genetic disorders, such as cystic fibrosis, muscular dystrophy, and certain cancers.
   - The development of RNA-based vaccines, including those used against COVID-19, highlights the potential of molecular biology to address public health challenges through novel therapeutic strategies.

4. **Pharmacogenomics and Drug Development**
   - Pharmacogenomics studies the interaction between genes and drugs, aiming to optimize drug therapies based on genetic information. Molecular biology enables researchers to identify gene variants that affect drug metabolism, efficacy, and toxicity.
   - Molecular insights into disease pathways and targets have accelerated drug discovery, with high-throughput screening and computer-aided drug design becoming integral in developing new drugs with greater specificity and reduced side effects.

Molecular Biology in Biotechnology and Agricultural Sciences

Biotechnology and agriculture have greatly benefited from molecular biology, as the field provides tools for genetic modification, bioengineering, and sustainable agricultural practices.

1. **Genetically Modified Organisms (GMOs)**
   - Molecular biology has enabled the creation of genetically modified organisms (GMOs), where specific genes are introduced, modified, or removed to enhance desirable traits. In agriculture, GMOs are designed for improved resistance to pests, diseases, herbicides, and environmental stresses, which contribute to higher yields and reduced chemical inputs.
   - Examples include *Bt* crops, which produce insecticidal proteins derived from *Bacillus thuringiensis*, and herbicide-resistant crops, such as glyphosate-resistant soybeans, which facilitate weed control without harming the crop.

2. **Bioengineering and Synthetic Biology**
   - Bioengineering uses molecular biology to develop organisms capable of producing valuable compounds, such as biofuels, pharmaceuticals, and industrial enzymes. For example, yeast and bacteria have been engineered to produce insulin, artemisinin (an antimalarial drug), and biodegradable plastics.

- Synthetic biology goes beyond traditional genetic engineering by designing entirely new biological systems and pathways, offering innovative solutions for environmental and industrial challenges, such as waste management and renewable energy.

3. **Plant and Animal Breeding**
   - Molecular biology has accelerated traditional breeding through marker-assisted selection, where genetic markers linked to desirable traits are used to identify the best breeding candidates. This approach speeds up the breeding process and increases accuracy.
   - Molecular techniques like embryo transfer, cloning, and transgenic animals have advanced livestock production, enabling the creation of animals with improved disease resistance, productivity, and nutritional quality.

Evolutionary Biology and Ecology: Molecular Insights

Molecular biology has transformed evolutionary biology and ecology, offering tools to study genetic diversity, evolutionary processes, and ecological relationships at the molecular level.

1. **Phylogenetics and Molecular Evolution**
   - Molecular biology has enabled the reconstruction of evolutionary relationships through phylogenetic analysis, where DNA and protein sequences are compared to establish evolutionary linkages and divergence times.
   - Molecular clocks, based on the accumulation of genetic mutations over time, allow scientists to estimate when species diverged, providing insights into evolutionary history, speciation events, and the origins of biodiversity.
2. **Population Genetics and Conservation Biology**
   - Molecular techniques are used to analyze population genetics, including genetic diversity, gene flow, and population structure. These studies help in understanding population dynamics, inbreeding, and the effects of genetic drift and natural selection.
   - In conservation biology, molecular tools are employed to monitor endangered species, assess genetic diversity in populations, and design breeding programs for species recovery.
3. **Ecological Genomics and Adaptation**
   - Ecological genomics studies the genetic basis of how organisms respond to environmental changes. Molecular biology enables researchers to identify genes and pathways associated with stress tolerance, adaptation to habitats, and resistance to environmental toxins.
   - By investigating molecular adaptations, ecological genomics provides insights into how species respond to climate change, habitat degradation, and other ecological pressures.

Bioinformatics and Computational Biology

The explosion of molecular data has necessitated the development of bioinformatics, a field that integrates computational tools and biological research. Bioinformatics and computational biology have become essential in managing, analyzing, and interpreting vast amounts of molecular data.

1. **Data Management and Genomics**
   - Bioinformatics databases, such as GenBank, the Protein Data Bank (PDB), and the Ensembl Genome Browser, store DNA, RNA, and protein sequences, making them accessible for researchers worldwide.
   - Computational tools are used to assemble, annotate, and compare genomes, enabling researchers to discover new genes, study gene families, and analyze genetic variations across populations.
2. **Protein Structure Prediction and Molecular Modeling**
   - Molecular biology has inspired computational models for predicting protein structures and interactions. Software like AlphaFold and Rosetta uses machine learning algorithms to predict protein folding, contributing to drug design, functional genomics, and synthetic biology.
   - Molecular dynamics simulations allow scientists to model the behavior of biomolecules, providing insights into protein function, ligand binding, and molecular interactions.
3. **Systems Biology and Network Analysis**
   - Systems biology integrates molecular data to build models of complex biological systems, enabling the study of interactions within cells and tissues. Techniques such as gene regulatory network analysis and metabolic pathway mapping provide insights into cellular function, disease mechanisms, and developmental processes.
   - Network analysis in systems biology helps in understanding emergent properties of biological systems, offering a holistic approach to studying molecular biology.

Ethical, Environmental, and Societal Implications

The rapid advancements in molecular biology have raised ethical, environmental, and societal questions, particularly regarding genetic modification, privacy of genetic information, and biodiversity conservation.

1. **Ethics of Genetic Engineering and CRISPR**
   - Gene editing technologies, especially CRISPR-Cas9, have made it possible to alter genomes with unprecedented precision. While CRISPR holds potential for curing genetic diseases and improving crops, it also raises ethical concerns about off-target effects, gene editing in embryos, and potential ecological consequences.
   - Ethical debates surrounding CRISPR and gene therapy include concerns about "designer babies," genetic privacy, and equitable access to genetic technologies.
2. **Environmental Impact of GMOs and Synthetic Biology**
   - Genetically modified organisms and synthetic biology have raised concerns about potential risks to ecosystems, such as gene flow from GMOs to wild relatives, unintended effects on non-target species, and antibiotic resistance.
   - Environmental impacts of GMOs are subject to ongoing research, regulatory oversight, and public debate to ensure that biotechnological advances do not compromise biodiversity or ecosystem stability.

3. **Social Implications of Personalized Medicine and Genomics**
   - Molecular biology's contributions to personalized medicine and genomics raise questions about genetic privacy, data security, and the potential for genetic discrimination. Policies such as the Genetic Information Nondiscrimination Act (GINA) have been established to protect individuals' genetic information.
   - The accessibility of genetic testing has sparked discussions on the ethical implications of knowing one's genetic risks, raising questions about the psychological impact and how genetic information influences personal and societal decision-making.

Emergence of Molecular Biology as a Discipline

The field of molecular biology emerged in the 20th century as a synthesis of ideas from genetics, biochemistry, microbiology, and physics, establishing itself as a unique and influential discipline that fundamentally altered our understanding of life's molecular underpinnings. The journey of molecular biology from a convergence of disciplines to a standalone scientific field reflects the intellectual curiosity, technological advancements, and groundbreaking discoveries that have characterized its development.

The Intellectual Foundations of Molecular Biology

The conceptual foundation of molecular biology was laid by developments in genetics and biochemistry, two fields that initially focused on understanding heredity and metabolic reactions, respectively. By the early 20th century, scientists began questioning how hereditary material was stored and transmitted, and how biochemical processes within cells were regulated and maintained.

1. **Genetics and the Search for the Genetic Material**
   - **Mendelian Genetics:** Gregor Mendel's work on inheritance in pea plants in the mid-19th century established the principles of dominant and recessive traits, laying the groundwork for understanding heredity. His work, however, was largely forgotten until it was rediscovered at the turn of the century.
   - **Chromosomal Theory of Inheritance:** In the early 1900s, Walter Sutton and Theodor Boveri independently proposed that genes reside on chromosomes, solidifying the idea that genetic material was organized in specific structures within cells.
   - **Discovery of DNA:** Although DNA was first identified by Friedrich Miescher in 1869, it was not recognized as the genetic material until the mid-20th century. Early work on nucleic acids by scientists like Phoebus Levene contributed to understanding DNA's chemical composition, yet DNA's role as the carrier of genetic information was still unknown.
2. **Biochemistry and the Understanding of Cellular Processes**
   - **Enzyme Function and Metabolism:** The discovery of enzymes and their catalytic role in biochemical reactions provided insights into how cells carry out metabolic processes. By studying metabolic pathways, biochemists like Hans Krebs and Archibald Hill revealed the complexity of cellular functions and introduced the idea that specific molecules drive cellular processes.

- **Protein Structure and Function:** Advances in protein chemistry, including the study of hemoglobin and other proteins, demonstrated the importance of macromolecules in cellular functions, suggesting that life's processes might be explained by interactions among specific molecules.
- Together, the fields of genetics and biochemistry laid the groundwork for molecular biology by proposing that life's mechanisms could be reduced to fundamental biological molecules and that heredity was governed by chemical principles.

Defining Molecular Biology as a Discipline:

One of the defining moments in the emergence of molecular biology was the influx of physicists into biological research during the mid-20th century. This shift brought new perspectives, methodologies, and conceptual approaches, particularly in understanding the physical structure of biological molecules.

1. **The Influence of Physics on Molecular Biology**
   - **Biophysics and Structural Biology:** Physicists like Linus Pauling, Erwin Schrödinger, and Max Delbrück applied principles of physics to biological questions, introducing a quantitative and structural approach to understanding life at the molecular level.
   - **Schrödinger's "What is Life?" (1944):** Erwin Schrödinger's book, *What is Life?*, inspired many physicists to study biology by suggesting that life processes could be explained in terms of molecular structures and physical laws. Schrödinger's concept of a "code-script" hinted at the existence of genetic material that could transmit biological information.
2. **The Phage Group and the Quest to Understand Heredity**
   - The "Phage Group," led by Max Delbrück, Salvador Luria, and Alfred Hershey, played a crucial role in defining molecular biology's focus on heredity. By studying bacteriophages (viruses that infect bacteria), this group aimed to determine the nature of the genetic material and how it controlled cell functions.
   - The bacteriophage experiments demonstrated that genes could be studied as discrete molecular entities, which reinforced the idea that biological processes could be understood by studying the molecular properties of genes.
3. **The Establishment of Molecular Biology as a Unique Discipline**
   - The interdisciplinary nature of molecular biology attracted researchers from physics, chemistry, and biology, creating a collaborative environment that encouraged innovative thinking. This fusion of disciplines allowed molecular biology to focus on understanding life at the molecular level, distinct from the broader focus of classical biology.

Key Discoveries that Defined Molecular Biology

The emergence of molecular biology was marked by a series of groundbreaking discoveries that set the discipline's direction and provided a molecular framework for studying life.

1. **The Identification of DNA as the Genetic Material**
   - Key experiments in the 1940s and 1950s, including the Avery-MacLeod-McCarty

experiment and the Hershey-Chase experiment, demonstrated that DNA, rather than proteins, was the carrier of genetic information. These findings shifted the focus of genetics from chromosomes and traits to molecular structures and sequences.

2. **Elucidation of the Double Helix Structure of DNA (1953)**
   - The discovery of the DNA double helix by James Watson and Francis Crick was a landmark moment that defined molecular biology as a discipline focused on the molecular basis of heredity. The double-helix model provided insights into how genetic information is stored, replicated, and transmitted, explaining heredity at a molecular level.
   - Rosalind Franklin's contributions through X-ray crystallography and Chargaff's rules on base pairing were critical to this discovery, which highlighted the role of molecular structure in biological function.
3. **The Central Dogma and the Flow of Genetic Information**
   - The central dogma of molecular biology, articulated by Francis Crick in 1958, formalized the understanding of genetic information flow: DNA to RNA to protein. This concept unified molecular biology by linking genetics, transcription, and translation into a single model that described how genes direct cellular functions.
   - The central dogma emphasized that molecular biology aimed to understand life by studying the mechanisms through which genes control proteins, the functional molecules of cells.

Institutionalization of Molecular Biology

The establishment of molecular biology as an independent academic and research discipline was facilitated by the creation of dedicated research centers, academic programs, and funding initiatives.

1. **Research Centers and Institutions**
   - The 1950s and 1960s saw the establishment of prominent molecular biology research centers, including the Medical Research Council (MRC) Laboratory of Molecular Biology in Cambridge, UK, and the Cold Spring Harbor Laboratory in the United States. These institutions became hubs for molecular biology research, attracting scientists from around the world.
   - These centers fostered a culture of collaboration and cross-disciplinary research, promoting the exchange of ideas and accelerating the development of molecular biology techniques and theories.
2. **Academic Programs and Molecular Biology Departments**
   - Molecular biology became a formal academic discipline with the establishment of specialized departments in universities and research institutions. Academic programs in molecular biology offered training in genetics, biochemistry, and cell biology, equipping students with the skills to conduct molecular-level research.
   - The inclusion of molecular biology in academic curricula helped legitimize the field, making it accessible to students and researchers and contributing to the rapid expansion of knowledge in the field.

3. **Funding and Support for Molecular Biology Research**
    - Funding agencies, including the National Institutes of Health (NIH) and private foundations, began to recognize the importance of molecular biology, providing grants for research in areas such as DNA structure, gene expression, and molecular genetics.
    - Increased funding facilitated the development of advanced research tools, such as sequencing technologies and genetic engineering, which enabled scientists to explore molecular biology with unprecedented precision.

Technological Innovations and Methodological Advances

Molecular biology's emergence was supported by technological innovations that provided researchers with tools to investigate cellular processes at the molecular level. These innovations include techniques that remain foundational in molecular biology research today.

1. **X-ray Crystallography**
    - X-ray crystallography, developed by physicists like William Lawrence Bragg, enabled researchers to determine the three-dimensional structures of biological macromolecules. Rosalind Franklin's use of this technique was instrumental in elucidating the structure of DNA.
    - This technique later became essential for studying the structures of proteins, enzymes, and other macromolecules, advancing our understanding of molecular interactions and mechanisms.
2. **Electron Microscopy**
    - Electron microscopy provided the first detailed images of cellular structures and organelles, allowing researchers to visualize cellular components at the molecular level.
    - The ability to observe viruses, ribosomes, and DNA-protein complexes through electron microscopy contributed to the structural understanding of molecular biology.
3. **Development of DNA Sequencing and Cloning Techniques**
    - The invention of DNA sequencing methods by Frederick Sanger and others enabled scientists to determine the exact sequence of nucleotides in DNA. This technology provided a means of studying genes at a molecular level, leading to the sequencing of entire genomes.
    - Cloning techniques, such as restriction enzyme-based cloning, allowed researchers to isolate and manipulate genes, giving rise to recombinant DNA technology and genetic engineering.
4. **Polymerase Chain Reaction (PCR)**
    - The invention of PCR by Kary Mullis in 1983 revolutionized molecular biology by enabling the rapid amplification of specific DNA sequences. PCR has become an indispensable tool for molecular biology research, diagnostics, and forensic science.

Molecular Biology's Influence on Other Disciplines

The emergence of molecular biology has had a profound impact on other fields of biological science, leading to new interdisciplinary areas of research and transforming the study of biology.

1. **Genomics and Bioinformatics**
   - Molecular biology paved the way for genomics, the study of whole genomes, and bioinformatics, which applies computational tools to analyze large-scale genetic data. These fields have enabled the study of genetic variation, population genetics, and personalized medicine.
2. **Systems Biology**
   - Systems biology integrates molecular biology with computational models to study complex interactions within cells and organisms. By understanding how genes, proteins, and pathways interact, systems biology seeks to explain emergent properties of biological systems.
3. **Synthetic Biology**
   - Synthetic biology, an outgrowth of molecular biology, aims to design and engineer new biological functions and systems. Using molecular techniques, synthetic biology seeks to create artificial life forms and molecular machines for applications in medicine, energy, and environmental science.

Key Innovations and Technological Milestones in Molecular Biology

The rapid advancement of molecular biology is built on a foundation of technological innovations and key discoveries that have redefined what is possible in biological research. These innovations have provided scientists with powerful tools to investigate, manipulate, and understand the molecular basis of life. From the discovery of DNA's double helix to gene editing with CRISPR-Cas9, each milestone has expanded our capacity to study genetic material and cellular processes with unprecedented precision and detail.

Discovery of the DNA Double Helix (1953)

1. **The Double Helix Model**
   - The discovery of DNA's double-helix structure by James Watson and Francis Crick in 1953 was a defining moment for molecular biology. Using Rosalind Franklin's X-ray diffraction images and Chargaff's rules, Watson and Crick proposed a model where DNA is structured as two antiparallel strands, each comprising a sequence of nucleotides that pair through hydrogen bonds (A with T, and G with C).
   - This model provided insight into DNA replication and genetic inheritance, as the complementary base-pairing suggested a mechanism for accurate copying of genetic information. The double-helix structure also hinted at the stability and potential for immense data storage in DNA, laying the groundwork for the molecular study of genes and heredity.

2. **Impact of the Discovery**

    - Understanding the double-helix structure was essential for future research, influencing studies on DNA replication, transcription, and translation. It directly led to the central dogma of molecular biology, which describes the flow of genetic information from DNA to RNA to protein.

Development of DNA Sequencing Techniques

1. **Sanger Sequencing (1977)**

    - Frederick Sanger's chain-termination sequencing method, introduced in 1977, allowed scientists to determine the precise sequence of nucleotides in DNA. Sanger sequencing uses modified nucleotides, or dideoxynucleotides, to terminate DNA synthesis at specific points, creating DNA fragments that can be read and assembled into a sequence.
    - This technique was essential for early genome projects, including the Human Genome Project, and has been used to sequence countless genes and genomes, providing critical insights into genetic variation, evolution, and gene function.

2. **Next-Generation Sequencing (NGS)**

    - Next-generation sequencing (NGS) technologies, which emerged in the early 2000s, dramatically improved sequencing speed and reduced costs. Unlike Sanger sequencing, which sequences one DNA fragment at a time, NGS can sequence millions of fragments in parallel.
    - NGS has revolutionized genomics by making whole-genome sequencing accessible for routine research. It enables transcriptomics, epigenomics, and metagenomics studies, opening doors to personalized medicine, microbial ecology, and population genomics.

Polymerase Chain Reaction (PCR) and Its Variants

1. **Polymerase Chain Reaction (PCR) (1983)**

    - The polymerase chain reaction (PCR), invented by Kary Mullis in 1983, is a technique for amplifying specific DNA sequences. PCR uses a thermostable DNA polymerase (such as Taq polymerase), primers, and thermal cycling to generate millions of copies of a DNA segment in a short time.
    - PCR is foundational in molecular biology, enabling applications in diagnostics, forensics, evolutionary studies, and genetic research. It allows scientists to study small DNA samples with high sensitivity, identifying pathogens, mutations, and genetic markers with ease.

2. **Real-Time PCR and Quantitative PCR (qPCR)**

    - Real-time PCR, also known as quantitative PCR (qPCR), enables the quantification of DNA or RNA in real-time by measuring fluorescence emitted during each amplification cycle. qPCR has become an essential tool in diagnostics, gene expression analysis, and viral load quantification.
    - Reverse transcription PCR (RT-PCR) combines PCR with reverse transcription,

allowing RNA to be converted to DNA for quantification. This method is widely used in viral diagnostics, including testing for COVID-19.

Discovery and Development of Restriction Enzymes

1. **Discovery of Restriction Enzymes (1970s)**
   - The discovery of restriction enzymes by Werner Arber, Hamilton Smith, and Daniel Nathans in the 1970s transformed genetic research. These enzymes cut DNA at specific recognition sequences, allowing precise manipulation of DNA.
   - Restriction enzymes laid the foundation for recombinant DNA technology, enabling researchers to cut and splice DNA fragments from different sources. This discovery was pivotal for genetic engineering and cloning, leading to the production of genetically modified organisms (GMOs) and advancements in synthetic biology.
2. **Recombinant DNA Technology**
   - Using restriction enzymes, scientists developed methods to insert DNA fragments into plasmids (small circular DNA molecules) for cloning and expression in bacteria or other hosts. This technology is the backbone of genetic engineering, facilitating the production of insulin, human growth hormone, and other therapeutic proteins.
   - Recombinant DNA technology also enables the creation of genetically modified plants and animals, which has had a significant impact on agriculture, medicine, and industry.

Introduction of Gene Cloning and Expression Systems

1. **Cloning Vectors and Expression Systems**
   - Cloning vectors, such as plasmids, bacteriophages, and artificial chromosomes, are DNA molecules used to insert and replicate foreign DNA within a host cell. By inserting genes of interest into these vectors, scientists can study gene function and produce recombinant proteins.
   - Expression systems allow for the production of proteins from cloned genes, facilitating research on protein function, enzyme activity, and structural biology. Bacterial, yeast, insect, and mammalian cells serve as expression systems, each offering unique benefits for protein production.
2. **Application in Functional Genomics and Biotechnology**
   - Gene cloning and expression systems enable functional genomics studies, where the functions of genes are investigated through overexpression, knockdown, or knockout experiments. This has led to discoveries in gene function, cellular signaling, and metabolic pathways.
   - These systems are also crucial for biotechnology, providing a platform for producing pharmaceuticals, enzymes, and other biomolecules used in research, healthcare, and industry.

The Discovery of CRISPR-Cas9 and Gene Editing

1. **CRISPR-Cas9 Gene Editing (2012)**
   - The discovery of the CRISPR-Cas9 system as a gene-editing tool by Jennifer Doudna and Emmanuelle Charpentier in 2012 marked a new era in molecular biology. Originally an adaptive immune system in bacteria, CRISPR-Cas9 uses RNA to guide the Cas9 enzyme to a specific DNA sequence, where it introduces cuts, allowing for targeted modifications.
   - CRISPR-Cas9 has revolutionized genetics and biotechnology, enabling precise genome editing in diverse organisms. This technology has applications in basic research, disease modeling, and potentially in human gene therapy, opening new avenues for treating genetic disorders.
2. **Development of CRISPR Variants and Applications**
   - Since its discovery, the CRISPR toolkit has expanded to include base editors, prime editors, and epigenetic modifiers, each allowing different types of genome modifications. CRISPR's versatility and efficiency have led to its adoption in agriculture, synthetic biology, and drug development.
   - CRISPR-Cas9 and its variants hold promise for gene therapy applications, offering the potential to correct genetic mutations that cause diseases like sickle cell anemia, cystic fibrosis, and muscular dystrophy.

High-Throughput Technologies and Omics Approaches

1. **Microarrays and High-Throughput Screening**
   - Microarray technology allows for the simultaneous analysis of thousands of genes, enabling researchers to study gene expression patterns, genetic variations, and mutations across entire genomes. This technology has been instrumental in cancer research, pharmacogenomics, and disease diagnostics.
   - High-throughput screening (HTS) combines robotics and automation to test large numbers of compounds or genes, facilitating drug discovery and functional genomics. HTS has accelerated the development of targeted therapies and gene function studies.
2. **Omics Technologies: Genomics, Transcriptomics, Proteomics, and Metabolomics**
   - Omics technologies provide comprehensive data on various biological molecules, enabling a systems-level understanding of biological processes. Genomics (study of genomes), transcriptomics (study of RNA transcripts), proteomics (study of proteins), and metabolomics (study of metabolites) offer insights into complex biological interactions and disease mechanisms.
   - Omics approaches have become essential in precision medicine, allowing researchers to investigate the molecular basis of diseases, identify biomarkers, and develop personalized treatments based on a patient's unique molecular profile.

Bioinformatics and Computational Biology

1. **Bioinformatics Tools and Databases**
   - The integration of molecular biology with computational tools has led to the field of bioinformatics, which involves the storage, retrieval, and analysis of biological data. Bioinformatics databases, such as GenBank and the Protein Data Bank, store DNA, RNA, and protein sequences, providing resources for molecular biology research worldwide.
   - Bioinformatics tools enable sequence alignment, structural prediction, phylogenetic analysis, and data mining, making it possible to interpret complex datasets generated by high-throughput technologies.
2. **Computational Modeling and Systems Biology**
   - Computational modeling and systems biology offer frameworks for studying molecular interactions and cellular processes as integrated networks. By modeling gene regulatory networks, metabolic pathways, and protein-protein interactions, systems biology helps in understanding the emergent properties of biological systems.
   - These computational approaches have applications in drug discovery, synthetic biology, and understanding disease mechanisms at a systems level.

Advanced Imaging Techniques

1. **X-ray Crystallography and Cryo-Electron Microscopy**
   - X-ray crystallography, pioneered by scientists like Rosalind Franklin and John Kendrew, has been essential for determining the three-dimensional structure of macromolecules. It remains a primary technique for studying protein and nucleic acid structures.
   - Cryo-electron microscopy (cryo-EM) is a more recent innovation that allows researchers to visualize biomolecules in their native states at near-atomic resolution without the need for crystallization. Cryo-EM has advanced structural biology by enabling the study of large, dynamic molecular complexes.
2. **Fluorescence Microscopy and Live-Cell Imaging**
   - Fluorescence microscopy, including confocal and super-resolution microscopy, enables scientists to observe the localization and dynamics of molecules within living cells. This technique is critical for studying cellular processes like signaling, trafficking, and gene expression in real-time.
   - Live-cell imaging has revolutionized our understanding of cellular dynamics, allowing scientists to observe how proteins interact, move, and respond to cellular signals in real-time.

2. Central Dogma of Molecular Biology

Certainly! Let's delve deeply into the **Concept and Importance of the Central Dogma** with a focus on sophisticated detail that captures the complexity and significance of this foundational principle in molecular biology.

2.1 Concept and Importance of the Central Dogma

The **Central Dogma of Molecular Biology** encapsulates the flow of genetic information within biological systems, establishing the pathways through which the blueprint of life, encoded in DNA, is transcribed into RNA and ultimately translated into proteins. This concept, first articulated by Francis Crick in 1958, serves as the guiding framework for understanding the molecular mechanisms that underpin cellular function, genetic inheritance, and biological development. Far from a simple one-way process, the central dogma encompasses a complex interplay of molecular processes, checkpoints, and regulatory mechanisms that ensure the accurate transmission and expression of genetic information. The importance of the central dogma cannot be overstated; it is the key to understanding everything from gene expression patterns to evolutionary adaptations, cellular responses, and even modern biotechnological innovations.

Understanding the Central Dogma: A Detailed Examination of DNA → RNA → Protein

The central dogma provides a roadmap for understanding cellular function by establishing a directional flow of information: DNA → RNA → Protein. This process is not merely sequential but highly regulated, ensuring that genetic information is transmitted accurately and in response to the cell's internal and external environments.

1. DNA as the Repository and Blueprint of Genetic Information

At the heart of the central dogma is DNA (deoxyribonucleic acid), the molecular repository of genetic information. DNA's double-helix structure, consisting of two antiparallel strands held together by hydrogen bonds between complementary bases (adenine-thymine and guanine-cytosine), provides an ideal structure for both information storage and replication. Each strand of DNA contains a sequence of nucleotides that, in groups of three, encode specific instructions for building proteins.

1. **Structure and Stability of DNA**
   - The double helix of DNA, with its complementary base pairing, is uniquely suited to serve as a stable blueprint. The hydrogen bonds between base pairs provide enough stability for the structure to be long-lasting, yet they are weak enough to allow for strand separation during replication and transcription.
   - DNA's structure also allows it to be compactly stored within the cell nucleus. In eukaryotes, DNA is further organized into chromatin, where it is wound around histone proteins to form nucleosomes, enabling efficient packaging within the limited nuclear space.
2. **Replication and the Fidelity of Genetic Information**
   - DNA replication is essential for genetic continuity, allowing cells to pass genetic information to daughter cells during cell division. The semiconservative nature of replication, where each new DNA molecule contains one original strand and one newly synthesized strand, ensures high fidelity.
   - DNA polymerases, the enzymes responsible for adding nucleotides during replication, have proofreading abilities that correct errors in real-time, achieving remarkable accuracy in the replication process. This fidelity is vital for preserving the integrity of genetic information across generations.

2. Transcription: Converting DNA into RNA

Transcription is the process by which the information encoded in DNA is transcribed into RNA, specifically messenger RNA (mRNA), which then carries this information from the nucleus (in eukaryotes) to the cytoplasm, where it directs protein synthesis.

1. **The Role of RNA Polymerase and Transcription Factors**
   - Transcription is initiated when RNA polymerase binds to a specific region on DNA known as the promoter. Transcription factors, which are proteins that bind to DNA and influence transcription, help RNA polymerase recognize promoter sequences and initiate transcription.
   - Once bound, RNA polymerase unwinds a segment of the DNA double helix, exposing the template strand, which it reads to synthesize a complementary strand of RNA. Unlike DNA replication, only one strand of DNA is used as a template in transcription, and the resulting RNA strand is single-stranded.
2. **mRNA Processing in Eukaryotes**
   - In eukaryotes, the initial mRNA transcript, known as pre-mRNA, undergoes several modifications before it becomes mature mRNA. These modifications include capping at the 5' end, polyadenylation at the 3' end, and splicing, where non-coding regions (introns) are removed, and coding regions (exons) are joined.
   - Alternative splicing allows for the generation of multiple mRNA variants from a single gene, increasing the diversity of proteins that can be produced. This regulation of mRNA processing is critical for cell differentiation, development, and adaptation to environmental changes.

3. Translation: Synthesizing Proteins from mRNA

Translation is the process by which the sequence of nucleotides in mRNA is interpreted to synthesize a chain of amino acids, forming a protein. This process takes place on the ribosome, a complex molecular machine composed of ribosomal RNA (rRNA) and proteins.

1. **The Genetic Code and Codon Recognition**
   - The genetic code consists of codons, groups of three nucleotides in mRNA, each of which specifies a particular amino acid. The code is degenerate, meaning that multiple codons can code for the same amino acid, providing a buffer against some mutations.
   - Transfer RNA (tRNA) molecules play a key role in translation. Each tRNA has an anticodon that is complementary to a codon on the mRNA and is linked to a specific amino acid. The ribosome facilitates the pairing between codons and anticodons, ensuring that the correct amino acids are added in sequence.
2. **Ribosome Function and Polypeptide Synthesis**
   - The ribosome moves along the mRNA strand, catalyzing the formation of peptide bonds between amino acids as they are delivered by tRNA. This stepwise addition of amino acids forms a polypeptide chain, which will fold into a functional protein.

- Translation consists of three main phases: initiation, where the ribosome assembles at the start codon on the mRNA; elongation, where the polypeptide chain is extended; and termination, where the ribosome reaches a stop codon and releases the newly synthesized protein.

3. **Post-Translational Modifications**

- Once synthesized, many proteins undergo post-translational modifications, such as phosphorylation, glycosylation, or proteolytic cleavage, which influence protein folding, stability, activity, and cellular localization. These modifications allow for further diversification of protein function beyond the genetic code, enabling cells to fine-tune protein activity in response to specific signals or environmental conditions.

The Importance of the Central Dogma in Biological Sciences

The central dogma of molecular biology provides a unifying framework for understanding how genetic information is converted into functional molecules, thereby controlling cellular function, organismal development, and responses to environmental stimuli. The significance of this concept extends across all aspects of biology, from basic research to practical applications.

1. **Understanding Genetic Diseases and Mutations**

- Genetic mutations—alterations in the DNA sequence—can disrupt the flow of information from DNA to protein, leading to diseases. By understanding how specific mutations affect transcription, translation, or protein function, researchers can identify the molecular basis of genetic disorders.
- For example, mutations in genes involved in transcription or splicing can result in aberrant mRNA and dysfunctional proteins, as seen in diseases like muscular dystrophy, cystic fibrosis, and certain cancers.

2. **Drug Development and Targeted Therapies**

- The central dogma provides the basis for identifying molecular targets for drug development. Small molecules, antibodies, and other therapeutic agents can be designed to influence transcription, translation, or protein activity to treat diseases.
- RNA-based therapies, such as RNA interference (RNAi) and antisense oligonucleotides, can specifically target mRNA molecules to downregulate or silence genes, providing new strategies for treating viral infections, genetic disorders, and cancers.

3. **Advances in Biotechnology and Synthetic Biology**

- The principles of the central dogma are foundational in genetic engineering and synthetic biology. By manipulating DNA and RNA, scientists can design synthetic genes, create recombinant proteins, and engineer metabolic pathways in microorganisms for biofuel production, bioremediation, and pharmaceutical synthesis.
- Technologies such as CRISPR-Cas9, which allow precise editing of DNA, are directly based on the central dogma. These tools have revolutionized our ability to alter genetic material with precision, enabling breakthroughs in agriculture, environmental science, and medicine.

4. **Evolutionary Biology and Comparative Genomics**

    - The central dogma offers insights into evolutionary processes by showing how genetic variations are expressed as phenotypic differences among species. By comparing genomes and analyzing the conservation of genes, scientists can reconstruct evolutionary histories and understand how molecular changes contribute to adaptation.
    - Comparative genomics, which involves comparing DNA, RNA, and protein sequences across species, provides insights into gene function, evolutionary conservation, and species divergence, enhancing our understanding of biodiversity and speciation.

Central Dogma: Epigenetics, Non-Coding RNAs, and Alternative Pathways

While the central dogma has provided a foundational understanding of molecular biology, subsequent discoveries have revealed that information flow in cells is more flexible and dynamic than originally thought.

1. **Epigenetic Modifications and Gene Regulation**

    - Epigenetics involves chemical modifications to DNA and histones (e.g., methylation, acetylation) that do not alter the DNA sequence but affect gene expression. Epigenetic marks regulate the accessibility of DNA to transcription factors, thus influencing transcription rates and cellular identity.
    - Epigenetic changes are heritable through cell divisions, meaning they play a crucial role in development and differentiation, and they can be influenced by environmental factors, providing a mechanism for organisms to adapt to changing conditions without altering their genetic code.

2. **Non-Coding RNAs and RNA-Based Regulation**

    - Non-coding RNAs, including microRNAs (miRNAs), small interfering RNAs (siRNAs), and long non-coding RNAs (lncRNAs), play essential regulatory roles in gene expression. These molecules can influence mRNA stability, translation efficiency, and even chromatin structure.
    - The discovery of RNA interference (RNAi) revealed a new layer of gene regulation that allows cells to control protein synthesis post-transcriptionally. This regulatory mechanism has important applications in gene silencing and therapeutic targeting of specific genes.

3. **Alternative Pathways and Exceptions to the Central Dogma**

    - The discovery of reverse transcription in retroviruses, where RNA is converted back into DNA, was an exception to the central dogma and led to the identification of reverse transcriptase. This enzyme has become a valuable tool in molecular biology, enabling the generation of complementary DNA (cDNA) from RNA.
    - Horizontal gene transfer, prevalent in bacteria and certain eukaryotic organisms, challenges the idea of a strictly linear flow of genetic information by allowing genetic material to move between organisms, contributing to genetic diversity and evolution.

2.2 DNA as the Genetic Material

The discovery that DNA is the primary genetic material was a landmark moment in biology, defining our understanding of heredity, cellular function, and the molecular mechanisms that underlie genetic information transfer. DNA, or deoxyribonucleic acid, is not only the carrier of genetic information but also the molecule responsible for directing cellular processes and enabling the continuity of life across generations.

2.2.1 Structure and Properties of DNA

The structure of DNA, first elucidated by James Watson and Francis Crick in 1953, is central to its function as the molecule of heredity. The DNA molecule is a double helix composed of two antiparallel strands, each made of nucleotides that encode genetic instructions. This structure provides both stability and flexibility, making DNA well-suited for storing and transmitting genetic information.

1. Structure of DNA

1. **The Double Helix Model**
   - DNA consists of two long strands of nucleotides twisted into a double helix, with each strand running in opposite directions, or antiparallel, which provides structural stability.
   - The backbone of each strand is composed of alternating sugar (deoxyribose) and phosphate groups, forming a stable phosphate-sugar backbone.
   - Each sugar molecule in the backbone is attached to a nitrogenous base, one of four types: adenine (A), thymine (T), cytosine (C), or guanine (G). The bases of one strand pair with the complementary bases of the other strand (A with T and C with G) via hydrogen bonds, forming base pairs that hold the two strands together.
   - The double-helical structure of DNA is consistent across all known life forms, underscoring its evolutionary importance and stability as a molecule capable of carrying vast amounts of genetic information.
2. **Complementary Base Pairing and Antiparallel Orientation**
   - Complementary base pairing ensures that each DNA strand carries the same information as the other, allowing for precise duplication during replication.
   - The antiparallel orientation (one strand running 5' to 3' and the other 3' to 5') is critical for enzyme binding and replication processes, as many DNA-binding enzymes require specific strand orientation to function.
   - This pairing also protects genetic information from damage and mutation, as any changes to one strand can be corrected by referring to the complementary sequence.
3. **Minor and Major Grooves**
   - DNA's helical structure creates grooves—the major and minor grooves—that run along the length of the molecule. These grooves provide access points for proteins involved in processes such as replication, transcription, and repair.
   - Many DNA-binding proteins recognize specific sequences by binding to these grooves,

enabling precise interactions that are essential for regulating gene expression and DNA metabolism.

2. Properties of DNA as Genetic Material

1. **Stability and Resistance to Chemical Degradation**
   - DNA's double-helical structure, with its phosphate-sugar backbone, is inherently stable and resistant to chemical degradation, which helps it survive under various cellular conditions.
   - The hydrogen bonds between base pairs and hydrophobic interactions between stacked bases within the double helix further enhance DNA's stability. This molecular stability ensures that genetic information remains intact over generations, a fundamental requirement for heredity.
2. **Information Storage Capacity**
   - DNA's sequence of nucleotides provides a virtually limitless capacity for storing genetic information, with each base pair representing a unit of information. Even small DNA molecules can encode complex instructions, as illustrated by viral genomes that encode many functional proteins.
   - In eukaryotic cells, DNA is organized into chromosomes, with vast regions of coding and non-coding sequences that enable complex gene regulation, alternative splicing, and evolutionary adaptations.
3. **Capacity for Accurate Duplication and Mutation**
   - DNA replication is remarkably accurate due to complementary base pairing, which allows each strand to serve as a template for synthesizing a new complementary strand.
   - DNA's ability to mutate is also essential for evolution. Mutations introduce variations in the genetic code, some of which can lead to beneficial adaptations in response to environmental pressures, contributing to the diversity of life.

2.2.2 DNA Replication and Repair Mechanisms

For DNA to function as the hereditary material, it must be accurately replicated in each cell division. Additionally, mechanisms are necessary to detect and repair DNA damage that can compromise genetic integrity. **DNA replication** and **repair mechanisms** are essential processes that maintain the fidelity of genetic information across generations, ensuring both stability and adaptability.

1. DNA Replication

DNA replication is the process by which a cell duplicates its DNA before cell division, resulting in two identical copies of the genome. Replication is semiconservative, meaning that each daughter DNA molecule contains one original strand and one newly synthesized strand.

1. **Initiation of Replication**
   - DNA replication begins at specific sequences called origins of replication, where initiator proteins bind to unwind and open the double helix.

- In eukaryotic cells, replication occurs at multiple origins along the length of each chromosome, while in prokaryotes, a single origin is typically sufficient due to their circular genomes.
- Helicase enzymes separate the two DNA strands by breaking the hydrogen bonds between base pairs, creating a replication fork where the DNA is actively being unwound.

2. **Elongation and Synthesis of New DNA Strands**
   - At each replication fork, DNA polymerase enzymes synthesize new DNA strands by adding complementary nucleotides to each template strand. DNA polymerase requires an RNA primer, synthesized by primase, to initiate DNA synthesis.
   - DNA synthesis occurs in a 5' to 3' direction. This results in continuous synthesis on the leading strand, which runs in a 5' to 3' direction toward the replication fork, and discontinuous synthesis on the lagging strand, which is oriented in the opposite direction.
   - On the lagging strand, short fragments of DNA called Okazaki fragments are synthesized in the 5' to 3' direction and later joined together by DNA ligase to form a continuous strand.
3. **Proofreading and Error Correction by DNA Polymerase**
   - DNA polymerases have proofreading capabilities that detect and correct mismatched nucleotides during DNA synthesis. This proofreading occurs through a 3' to 5' exonuclease activity that removes incorrect nucleotides, followed by insertion of the correct base.
   - This proofreading mechanism ensures high fidelity during DNA replication, resulting in an error rate as low as one mistake per billion nucleotides replicated.
4. **Termination of Replication**
   - In circular genomes, replication terminates when the replication forks meet. In linear chromosomes, however, termination poses unique challenges at the chromosome ends, or telomeres, which shorten with each replication cycle.
   - Telomerase, an enzyme that extends telomeres, helps prevent loss of essential genetic information by adding repetitive nucleotide sequences to the ends of chromosomes, particularly in stem cells and germ cells where frequent replication occurs.

2. DNA Repair Mechanisms

DNA repair mechanisms are essential for correcting damage caused by environmental factors (e.g., UV light, chemicals) and errors that escape proofreading during replication. These repair systems are critical for maintaining genomic stability and preventing mutations that could lead to disease.

1. **Direct Repair Mechanisms**
   - Direct repair is a DNA repair process that corrects specific types of damage without replacing nucleotides. For example, photoreactivation directly reverses UV-induced thymine dimers in prokaryotes and some eukaryotes.
   - Direct repair is efficient and quick but limited to specific types of DNA damage, such as single-base modifications.

2. **Excision Repair Mechanisms**
   - **Base Excision Repair (BER):** BER is a pathway that corrects single-base errors, such as those caused by oxidative damage. In this process, DNA glycosylases recognize and remove damaged bases, creating an abasic site. DNA polymerase then fills in the missing nucleotide, and DNA ligase seals the backbone.
   - **Nucleotide Excision Repair (NER):** NER repairs bulky lesions, such as those caused by UV radiation, which distort the DNA helix. In this process, a stretch of nucleotides surrounding the lesion is removed, and the resulting gap is filled by DNA polymerase and sealed by ligase. NER is particularly important for correcting UV-induced thymine dimers in human cells.
3. **Mismatch Repair (MMR)**
   - Mismatch repair targets replication errors that escape proofreading, such as mismatched base pairs or insertion-deletion loops. In MMR, proteins recognize and bind to the mismatch, remove the incorrect segment, and use the complementary strand as a template for correction.
   - MMR increases replication fidelity and is essential in reducing mutation rates. Defects in mismatch repair genes are linked to certain cancers, such as Lynch syndrome.
4. **Double-Strand Break Repair Mechanisms**
   - **Homologous Recombination (HR):** Homologous recombination is an error-free repair process that uses a sister chromatid as a template to repair double-strand breaks, ensuring accurate repair. This process is active in the late S and G2 phases of the cell cycle, where sister chromatids are available.
   - **Non-Homologous End Joining (NHEJ):** NHEJ is an error-prone repair process that directly joins the ends of double-strand breaks without a template. Although this mechanism can result in small deletions or insertions, it is a crucial repair pathway, particularly in non-dividing cells or in situations where a template is unavailable.
5. **DNA Damage Response and Cell Cycle Checkpoints**
   - In response to DNA damage, cells activate the DNA damage response (DDR), a network of signaling pathways that detect damage, halt cell cycle progression, and initiate repair processes. Key proteins in the DDR, such as p53, play dual roles in repair and apoptosis.
   - Cell cycle checkpoints (G1/S, G2/M) provide additional time for DNA repair before cell division. This mechanism prevents the propagation of mutations, safeguarding genomic integrity.

The Significance of DNA Replication and Repair Mechanisms

The ability of DNA to replicate with high fidelity and repair damage is essential for the preservation of genetic information across generations and for the maintenance of cellular health. Failures in replication accuracy or repair processes lead to mutations, some of which can result in diseases like cancer, genetic disorders, and premature aging syndromes.

Moreover, the study of DNA replication and repair has provided valuable insights into cancer biology, aging, and the evolution of genome stability mechanisms. Many cancer therapies, for instance, target DNA repair pathways to increase cancer cell vulnerability. In regenerative medicine, understanding telomere biology and repair mechanisms opens new possibilities for therapies aimed at aging and age-related diseases.

2.3 RNA and Its Role in Gene Expression

RNA, or ribonucleic acid, is a central molecule in the flow of genetic information within cells, playing crucial roles in gene expression, regulation, and protein synthesis. Unlike DNA, which primarily serves as a stable repository for genetic information, RNA is a versatile molecule that acts both as an intermediary and as an active participant in the cell's molecular machinery. RNA's diverse structures and functions enable it to facilitate and regulate the transfer of information from DNA to proteins, guiding cellular processes in response to internal and external stimuli.

2.3.1 Types of RNA and Their Functions

RNA molecules are highly diverse, both in structure and function, and play multiple roles in gene expression, cellular regulation, and adaptation. Below, we explore the primary types of RNA and their distinct functions in the cell.

1. Messenger RNA (mRNA)

Messenger RNA (mRNA) serves as the blueprint for protein synthesis. Transcribed directly from DNA, mRNA carries the genetic instructions for protein assembly from the cell's nucleus (in eukaryotes) to the ribosomes in the cytoplasm, where translation into protein occurs.

- **Structure and Characteristics of mRNA**: mRNA is single-stranded and varies in length depending on the gene it encodes. It contains a 5' cap and a poly-A tail at the 3' end in eukaryotic cells, which stabilize the molecule and aid in translation.
- **Function**: mRNA's primary function is to encode proteins, with each sequence of three nucleotides (codons) corresponding to a specific amino acid.
- **Regulation of mRNA**: mRNA levels are tightly regulated by mechanisms including transcriptional control, mRNA degradation, and translational regulation, ensuring proteins are synthesized only as needed by the cell.

2. Transfer RNA (tRNA)

Transfer RNA (tRNA) is essential for decoding mRNA sequences during protein synthesis. Each tRNA molecule is linked to a specific amino acid and carries it to the ribosome in response to mRNA codons.

- **Structure of tRNA**: tRNA molecules adopt a cloverleaf structure due to intramolecular base pairing, with an anticodon region that recognizes mRNA codons and an aminoacyl group that binds to an amino acid.
- **Function in Translation**: tRNA molecules match mRNA codons with their corresponding amino acids, facilitating the assembly of amino acids into polypeptides. This role makes tRNA a crucial link between nucleic acid information and protein synthesis.

3. Ribosomal RNA (rRNA)

Ribosomal RNA (rRNA) is a core component of ribosomes, the cellular machinery responsible for synthesizing proteins. rRNA molecules contribute both to the structural integrity of ribosomes and to their catalytic activity in protein synthesis.

- **Structure of rRNA**: rRNA molecules are highly structured and, along with ribosomal proteins, form the large and small subunits of ribosomes. The precise folding of rRNA is critical for ribosome assembly and function.
- **Catalytic Function**: rRNA has peptidyl transferase activity, facilitating peptide bond formation during translation. This catalytic role highlights rRNA as a ribozyme, or RNA molecule with enzymatic function.

4. Small Nuclear RNA (snRNA)

Small nuclear RNAs (snRNAs) are involved in the splicing of pre-mRNA, a key step in mRNA maturation that removes non-coding sequences (introns).

- **Structure and Localization**: snRNAs are small, nuclear-localized RNAs that are part of the spliceosome complex, which performs splicing.
- **Function in Splicing**: snRNAs bind to specific sequences on pre-mRNA and catalyze the removal of introns and the ligation of exons, producing mature mRNA. This role is critical in generating the correct mRNA transcripts for translation.

5. MicroRNA (miRNA) and Small Interfering RNA (siRNA)

MicroRNAs (miRNAs) and small interfering RNAs (siRNAs) are involved in RNA interference (RNAi), a process that regulates gene expression post-transcriptionally by degrading mRNA or blocking its translation.

- **Structure and Processing**: miRNAs and siRNAs are short (typically 20-24 nucleotides) and are processed from precursor molecules by enzymes such as Dicer.
- **Function in Gene Silencing**: miRNAs and siRNAs bind complementary sequences on target mRNAs, leading to degradation or translational inhibition. This process is critical for gene regulation, development, and defense against viral infections.

6. Long Non-Coding RNA (lncRNA)

Long non-coding RNAs (lncRNAs) are RNA molecules longer than 200 nucleotides that do not code for proteins but play regulatory roles in gene expression.

- **Structure and Diversity**: lncRNAs are structurally diverse and can interact with DNA, RNA, and proteins, acting as molecular scaffolds and guides.
- **Regulatory Functions**: lncRNAs are involved in chromatin remodeling, transcriptional regulation, and post-transcriptional processing. Their regulatory roles are vital in development, cellular differentiation, and responses to environmental stimuli.

7. Other Non-Coding RNAs (ncRNAs)

Other non-coding RNAs, such as small nucleolar RNAs (snoRNAs) and PIWI-interacting RNAs (piRNAs), also play roles in cellular regulation.

- **snoRNA**: Small nucleolar RNAs (snoRNAs) are involved in chemical modifications of rRNA, such as methylation and pseudouridylation, essential for ribosome function.
- **piRNA**: PIWI-interacting RNAs (piRNAs) protect the genome from transposable elements by guiding PIWI proteins to silence these sequences, particularly in germ cells.

2.3.2 Transcription Process: From DNA to RNA

Transcription is the process by which a DNA sequence is copied into an RNA molecule. This critical step in the central dogma of molecular biology translates the genetic information encoded in DNA into a form that can be used for protein synthesis and gene regulation.

1. Initiation of Transcription

The transcription process begins at specific regions of DNA called promoters, which signal the start of a gene.

1. **Role of RNA Polymerase**
   - RNA polymerase is the enzyme responsible for synthesizing RNA from a DNA template. In prokaryotes, a single RNA polymerase transcribes all types of RNA, whereas in eukaryotes, different RNA polymerases (Pol I, Pol II, and Pol III) are specialized for rRNA, mRNA, and tRNA synthesis.
   - In eukaryotes, RNA polymerase II is responsible for mRNA synthesis and requires a set of transcription factors to bind to the promoter region, creating a complex that initiates transcription.
2. **Promoter Recognition and Binding**
   - Transcription factors recognize specific sequences within the promoter, such as the TATA box in eukaryotic genes, and recruit RNA polymerase to the transcription start site.
   - The assembly of transcription factors and RNA polymerase at the promoter forms the transcription initiation complex, unwinding the DNA and exposing the template strand.

2. Elongation of the RNA Transcript

During elongation, RNA polymerase moves along the DNA template strand, synthesizing a complementary RNA strand.

1. **RNA Polymerase Activity**
   - RNA polymerase reads the DNA template in the 3' to 5' direction and synthesizes RNA in the 5' to 3' direction, adding ribonucleotides complementary to the DNA template.
   - The enzyme catalyzes the formation of phosphodiester bonds between nucleotides, elongating the RNA strand as it moves along the DNA.

2. **Proofreading and Fidelity**

   - Although RNA polymerase lacks the extensive proofreading capabilities of DNA polymerase, it has a mechanism to correct some errors. However, transcription errors are generally less problematic because RNA is transient and multiple RNA molecules can be synthesized from a single gene.

3. Termination of Transcription

Termination signals the end of transcription, leading to the release of the newly synthesized RNA molecule.

1. **Termination in Prokaryotes**

   - In prokaryotes, transcription can terminate through intrinsic mechanisms, where a specific sequence in the RNA forms a hairpin loop, causing RNA polymerase to detach.

   - Alternatively, a protein called Rho can mediate termination by binding to the RNA and disrupting the transcription complex.

2. **Termination in Eukaryotes**

   - In eukaryotes, termination is more complex and involves cleavage of the pre-mRNA followed by polyadenylation (addition of a poly-A tail).

   - The exact termination signal varies, but it often involves recognition of specific sequences that lead to RNA polymerase detachment and pre-mRNA processing.

2.3.3 RNA Processing and Modifications

Once transcription produces a primary RNA transcript, several modifications are often necessary to convert it into a mature RNA molecule. These modifications ensure RNA stability, facilitate nuclear export, and prepare the RNA for its functional role in the cell.

1. RNA Processing in Eukaryotes

In eukaryotic cells, primary RNA transcripts undergo extensive processing before they become functional.

1. **5' Capping**

   - Shortly after transcription initiation, a modified guanine nucleotide (7-methylguanosine) is added to the 5' end of the nascent RNA. This 5' cap protects the RNA from degradation and aids in ribosome binding during translation.

   - The cap structure also plays a role in nuclear export and is recognized by specific proteins that facilitate translation initiation.

2. **Polyadenylation at the 3' End**

   - At the end of transcription, a string of adenine nucleotides (the poly-A tail) is added to the 3' end of the RNA molecule. This modification increases RNA stability and enhances translation efficiency.

   - The poly-A tail also helps in nuclear export and protects the mRNA from degradation by exonucleases in the cytoplasm.

3. **Splicing of Introns**

   - Eukaryotic genes are often interrupted by non-coding sequences called introns. During splicing, introns are removed, and the remaining exons are joined to produce a contiguous coding sequence.
   - The splicing process is mediated by the spliceosome, a complex of proteins and snRNAs, which recognizes splice sites and catalyzes intron removal. Alternative splicing allows a single gene to produce multiple protein isoforms, increasing proteomic diversity.

2. RNA Editing and Modifications

Beyond the core processing steps, certain RNAs undergo additional modifications that further refine their function.

1. **RNA Editing**

   - RNA editing is a process in which specific nucleotides within an RNA molecule are chemically altered after transcription. Common edits include deamination of adenosine to inosine (A-to-I) and cytidine to uridine (C-to-U).
   - RNA editing diversifies the transcriptome, allowing a single RNA molecule to yield different functional outcomes. This is particularly important in the nervous system, where RNA editing can affect neurotransmitter receptors and ion channels.

2. **RNA Modifications in Non-Coding RNAs**

   - Non-coding RNAs, including tRNA, rRNA, and snRNA, often undergo unique chemical modifications that enhance their stability and function.
   - For example, tRNAs are extensively modified with methylation, pseudouridylation, and thiolation, which contribute to their proper folding and function during translation. rRNAs are similarly modified to support ribosome assembly and stability.

3. **Regulatory Mechanisms through RNA Modifications**

   - Modifications such as N6-methyladenosine (m6A) methylation in mRNA add another layer of regulation to gene expression, influencing mRNA stability, localization, and translation efficiency.
   - These modifications are reversible, allowing cells to dynamically regulate gene expression in response to developmental cues and environmental changes, highlighting the functional versatility of RNA.

2.4 Protein Synthesis and Translation Mechanism

Protein synthesis, or translation, is the process by which the genetic information encoded in mRNA is decoded to produce a polypeptide chain that will fold into a functional protein. This process is fundamental to cellular function, as proteins are the workhorses of the cell, performing a vast array of tasks including catalyzing biochemical reactions, signaling, structural support, and transport. Translation is a highly regulated, multi-step process that ensures that the information encoded within DNA is accurately expressed as specific proteins.

2.4.1 Role of Ribosomes in Translation

The ribosome is a large, complex molecular machine that plays a central role in translation. Composed of ribosomal RNA (rRNA) and proteins, the ribosome is the site where mRNA is decoded into a sequence of amino acids, forming a polypeptide chain that will become a functional protein. Ribosomes exist in both prokaryotic and eukaryotic cells, though with some structural and functional differences.

1. **Structure of Ribosomes**
    - Ribosomes are composed of two subunits: a large subunit and a small subunit. In eukaryotes, the ribosome has a 60S large subunit and a 40S small subunit, which together form the 80S ribosome. In prokaryotes, the 50S large subunit and 30S small subunit form the 70S ribosome.
    - Each subunit is made up of rRNA and ribosomal proteins, with rRNA playing a catalytic role in peptide bond formation. This catalytic activity demonstrates that ribosomes are ribozymes, as they are RNA molecules with enzymatic functions.
2. **Sites of Interaction within the Ribosome**
    - Ribosomes contain three binding sites for transfer RNA (tRNA): the A (aminoacyl) site, the P (peptidyl) site, and the E (exit) site.
    - The **A site** is where the incoming aminoacyl-tRNA, carrying an amino acid, binds according to the codon on the mRNA.
    - The **P site** holds the tRNA carrying the growing polypeptide chain.
    - The **E site** is the exit site where tRNA molecules, after transferring their amino acids, are released from the ribosome.
3. **Ribosome Assembly and Initiation of Translation**
    - Translation begins with the assembly of the ribosome around the start codon on mRNA, typically AUG, which codes for methionine. This step involves various initiation factors that guide the small ribosomal subunit to the mRNA and recruit the large subunit to form a functional ribosome.
    - In prokaryotes, ribosomes recognize the Shine-Dalgarno sequence on mRNA for binding, while in eukaryotes, they recognize the 5' cap structure of mRNA. These recognition processes ensure that translation starts accurately at the beginning of the coding region.
4. **Mechanics of Polypeptide Elongation**
    - During elongation, tRNAs bring amino acids to the ribosome, where they are added to the growing polypeptide chain. Elongation involves the entry of the tRNA at the A site, peptide bond formation at the P site, and the release of the tRNA at the E site.
    - This cycle is repeated for each codon, facilitated by elongation factors that stabilize interactions and maintain the ribosome's movement along the mRNA. This regulated sequence ensures that the amino acids are added in the correct order.

5. **Termination and Ribosome Recycling**
   - Translation ends when the ribosome encounters a stop codon (UAA, UAG, or UGA) on the mRNA. Release factors recognize these stop codons and catalyze the release of the polypeptide from the ribosome.
   - Following termination, the ribosomal subunits dissociate and are recycled for subsequent rounds of translation. This recycling is essential for efficient protein synthesis within the cell.

2.4.2 Genetic Code and Codon Recognition

The genetic code is a set of rules that defines how the nucleotide sequence of mRNA is translated into an amino acid sequence in a protein. This code is universal across nearly all organisms, emphasizing its evolutionary significance, and it relies on the recognition of codons—groups of three nucleotides that specify individual amino acids.

1. **The Triplet Code and Redundancy**
   - The genetic code is composed of triplet codons, each consisting of three nucleotide bases. With four possible bases (adenine, uracil, cytosine, and guanine), there are 64 possible codons, more than enough to code for the 20 amino acids used in protein synthesis.
   - The code is redundant, meaning that multiple codons can specify the same amino acid. For instance, the amino acid leucine is coded by six different codons (UUA, UUG, CUU, CUC, CUA, CUG). This redundancy provides a safeguard against mutations, as certain changes in the DNA sequence may not alter the amino acid sequence of the protein.
2. **Start and Stop Codons**
   - The codon AUG serves as the start codon, signaling the beginning of translation and coding for the amino acid methionine. In eukaryotes, the presence of methionine at the start of most polypeptides is crucial for protein stability and function.
   - Three codons, UAA, UAG, and UGA, act as stop codons, signaling the end of translation. These stop codons do not code for amino acids but instead recruit release factors to terminate protein synthesis.
3. **Wobble Hypothesis and Flexibility in Codon Recognition**
   - The wobble hypothesis, proposed by Francis Crick, explains how certain tRNAs can recognize multiple codons through flexible base pairing at the third position of the codon. This flexibility allows for fewer tRNAs to accommodate the 61 codons that specify amino acids.
   - Wobble pairing occurs due to the unique chemical properties of the nucleotide inosine in tRNA, which can pair with multiple bases. This pairing enhances the efficiency of translation and ensures that the correct amino acids are incorporated, despite the variability in codon sequences.

4. **Reading Frame and Frame Shift Mutations**
   - The reading frame, or the way in which nucleotide triplets are grouped into codons, is crucial for accurate protein synthesis. A shift in the reading frame, caused by insertions or deletions in the mRNA sequence, can lead to a frameshift mutation, drastically altering the protein sequence and often resulting in a nonfunctional protein.
   - Frame shifts underscore the importance of maintaining the integrity of the genetic code, as even small alterations can lead to significant effects on protein structure and function.
5. **Universality of the Genetic Code**
   - The genetic code is nearly universal across organisms, from bacteria to humans, suggesting a shared evolutionary origin. This universality allows for the transfer of genes between species, a principle exploited in genetic engineering and biotechnology.
   - Minor variations in the genetic code exist in certain organelles, such as mitochondria, and in some microorganisms, but these differences are rare and typically involve only a few codons. The universality of the code underscores its efficiency and evolutionary importance.

2.4.3 Post-Translational Modifications

Post-translational modifications (PTMs) are chemical modifications made to a protein after its synthesis on the ribosome. These modifications are crucial for protein maturation, function, localization, and regulation. PTMs allow a single polypeptide to give rise to diverse functional forms, expanding the proteome and enhancing the cell's adaptability.

1. **Phosphorylation**
   - Phosphorylation is the addition of a phosphate group, typically to serine, threonine, or tyrosine residues, catalyzed by enzymes called kinases. This modification plays a pivotal role in cell signaling, as phosphorylation can activate or deactivate enzymes and receptors.
   - Phosphorylation is reversible, with phosphatases removing phosphate groups, providing a dynamic regulatory mechanism for processes like cell division, growth, and apoptosis.
2. **Glycosylation**
   - Glycosylation is the attachment of carbohydrate chains to proteins, primarily occurring on asparagine (N-linked) or serine/threonine (O-linked) residues. This modification is common in membrane and secreted proteins, enhancing stability, cell recognition, and immune response.
   - Glycosylated proteins are often involved in cell-cell interactions and are essential for immune recognition, as in the case of antibodies and cell surface receptors.
3. **Ubiquitination**
   - Ubiquitination is the attachment of ubiquitin, a small regulatory protein, to lysine residues of target proteins. This modification often signals proteins for degradation by the proteasome, a protein complex that breaks down unwanted or damaged proteins.

- Ubiquitination regulates protein turnover, cell cycle progression, and DNA repair. Dysregulation of ubiquitination pathways is implicated in diseases like cancer and neurodegeneration.

4. **Methylation and Acetylation**
   - Methylation involves the addition of a methyl group, often on lysine or arginine residues. Methylation can alter protein-protein interactions, modulate enzyme activity, and regulate gene expression by affecting chromatin structure.
   - Acetylation, especially on histones, plays a central role in gene regulation by modifying chromatin accessibility. Histone acetylation opens chromatin, making DNA accessible for transcription, while deacetylation condenses chromatin and represses gene expression.
5. **Proteolytic Cleavage**
   - Many proteins are synthesized as inactive precursors, or proproteins, which require proteolytic cleavage to become active. This processing is crucial for hormones, enzymes, and signaling molecules, ensuring that they become active only in specific contexts.
   - Proteolytic cleavage is irreversible and often occurs in response to cellular signals or environmental changes, regulating processes like digestion, coagulation, and apoptosis.
6. **Lipidation**
   - Lipidation is the addition of lipid groups to proteins, which targets them to cellular membranes. This modification is common in signaling proteins and receptors that need to be anchored to the cell membrane to function.
   - Lipid-modified proteins play key roles in signaling pathways, such as the Ras protein in cell growth and differentiation, highlighting the importance of lipidation in spatial organization within the cell.
7. **Role of Post-Translational Modifications in Disease**
   - Aberrant PTMs are linked to various diseases. For example, hyperphosphorylation of tau proteins is associated with Alzheimer's disease, while dysregulation in ubiquitination pathways is implicated in certain cancers.
   - Understanding PTMs is therefore crucial in biomedical research, as targeted therapies can modulate these modifications, providing new avenues for treating diseases at the molecular level.

3. Molecular Components of Cells

3.1 Nucleic Acids (DNA and RNA)

Nucleic acids, DNA (deoxyribonucleic acid) and RNA (ribonucleic acid), are fundamental biomolecules that serve as the repositories and transmitters of genetic information in all known life forms. Together, they form the basis of heredity, evolution, and cellular function, enabling the encoding, expression, and transfer of genetic information. DNA and RNA are structurally and functionally distinct, yet they work in tandem to regulate gene expression and protein synthesis.

Structure and Composition of Nucleic Acids

Nucleic acids are polymers composed of nucleotide monomers. Each nucleotide consists of three components: a phosphate group, a pentose sugar, and a nitrogenous base. The structural variations between DNA and RNA arise from differences in these components, which contribute to their unique roles within the cell.

1. **DNA Structure and Composition**
    - DNA is a double-stranded helix composed of two antiparallel strands, each made up of a backbone of alternating phosphate and deoxyribose (a five-carbon sugar) units. The strands are held together by hydrogen bonds between complementary nitrogenous bases.
    - There are four types of nitrogenous bases in DNA: adenine (A), thymine (T), guanine (G), and cytosine (C). Adenine pairs with thymine, and guanine pairs with cytosine, forming the A-T and G-C base pairs.
    - The double-helix structure, proposed by Watson and Crick in 1953, provides stability to the molecule and allows for precise replication. The antiparallel orientation of the strands (one runs 5' to 3' and the other 3' to 5') is essential for the function of DNA-binding enzymes involved in replication and repair.
2. **RNA Structure and Composition**
    - Unlike DNA, RNA is typically single-stranded and contains ribose as its sugar, which includes a hydroxyl group on the 2' carbon. This hydroxyl group makes RNA more chemically reactive and less stable than DNA.
    - RNA also has four bases, but uracil (U) replaces thymine. Thus, adenine pairs with uracil in RNA.
    - Although RNA is generally single-stranded, it can fold into complex secondary and tertiary structures, such as hairpins and loops, through intramolecular base pairing. These structures allow RNA to perform various functions, including catalysis and regulation.

Functions of DNA

DNA's primary function is to store genetic information and ensure its transmission across generations. This genetic code guides the synthesis of proteins, regulates cellular functions, and maintains the continuity of genetic traits.

1. **Information Storage and Heredity**
    - DNA encodes the genetic blueprint of an organism, storing the information required for building proteins, regulating cellular activities, and determining phenotypic traits. The stability and structural integrity of DNA allow it to store genetic information reliably over an organism's lifetime and across generations.
    - Each cell in a multicellular organism typically contains a complete copy of the organism's genome, which enables cells to differentiate and specialize while sharing the same genetic instructions.

2. **DNA Replication**
    - During cell division, DNA undergoes replication to produce identical copies of the genome for each daughter cell. This process is semiconservative, meaning that each new DNA molecule contains one parental strand and one newly synthesized strand, ensuring fidelity and stability in genetic transmission.
    - DNA polymerases, along with other enzymes like helicase, primase, and ligase, coordinate the replication process, ensuring that errors are minimized and corrected. Accurate replication is crucial for cellular integrity and the prevention of mutations.
3. **Gene Expression Regulation**
    - DNA contains regulatory regions, such as promoters, enhancers, and silencers, that control gene expression levels and timing. Transcription factors bind to these regions, modulating the initiation of transcription and influencing cellular response to environmental cues.
    - Epigenetic modifications, including DNA methylation and histone modifications, further regulate gene expression by altering chromatin structure and accessibility. These regulatory mechanisms allow cells to adapt and respond to changes without altering the DNA sequence itself.

Functions of RNA

RNA plays a diverse and dynamic role in gene expression, ranging from serving as a template for protein synthesis to directly participating in regulatory and catalytic processes.

1. **mRNA and Protein Synthesis**
    - Messenger RNA (mRNA) is the intermediary molecule that conveys genetic information from DNA to ribosomes, where proteins are synthesized. During transcription, a DNA template is used to synthesize an mRNA molecule that is complementary to the gene sequence.
    - mRNA is then translated into a specific protein sequence, with each codon (three-nucleotide sequence) specifying a particular amino acid. This mRNA-mediated pathway is essential for the flow of genetic information in the central dogma: DNA → RNA → Protein.
2. **tRNA and Translation**
    - Transfer RNA (tRNA) plays a critical role in translation by linking codons in mRNA with the appropriate amino acids. Each tRNA molecule carries a specific amino acid and contains an anticodon that pairs with a complementary mRNA codon.
    - tRNA facilitates the accurate incorporation of amino acids into the growing polypeptide chain, ensuring that proteins are synthesized according to the genetic instructions. This process highlights tRNA as an essential adaptor molecule in protein synthesis.
3. **rRNA and the Ribosome**
    - Ribosomal RNA (rRNA) is a major structural and functional component of ribosomes,

the cellular machinery responsible for translating mRNA into proteins. rRNA catalyzes the formation of peptide bonds between amino acids, acting as a ribozyme.

- The ribosome's structure, formed by rRNA and ribosomal proteins, ensures precise decoding of mRNA and efficient protein synthesis. The catalytic role of rRNA demonstrates RNA's functional versatility and importance in cellular metabolism.

4. **Regulatory and Catalytic RNAs**

   - Non-coding RNAs, including microRNAs (miRNAs), small interfering RNAs (siRNAs), and long non-coding RNAs (lncRNAs), regulate gene expression post-transcriptionally by influencing mRNA stability, translation, and degradation. These RNA molecules are critical for cellular development, differentiation, and response to environmental changes.
   - Certain RNA molecules also have catalytic activity, such as ribozymes, which can catalyze specific biochemical reactions. For example, self-splicing introns and the ribosome's peptidyl transferase center exemplify RNA's catalytic potential.

The Central Dogma and the Role of Nucleic Acids

The central dogma of molecular biology, which describes the directional flow of genetic information from DNA to RNA to protein, highlights the interplay between DNA and RNA in cellular function. This framework underscores how nucleic acids mediate gene expression, allowing organisms to adapt, survive, and evolve.

1. **Transcription: From DNA to RNA**

   - During transcription, DNA serves as a template for synthesizing RNA. The process is initiated by RNA polymerase, which binds to a promoter region on DNA and unwinds the double helix to access the template strand.
   - RNA polymerase reads the DNA sequence and synthesizes a complementary RNA strand. In eukaryotes, the primary transcript (pre-mRNA) undergoes modifications, such as 5' capping, splicing, and polyadenylation, to become a mature mRNA ready for translation.

2. **Translation: From RNA to Protein**

   - Translation is the process by which ribosomes decode mRNA into a sequence of amino acids, forming a polypeptide chain that folds into a functional protein. Translation relies on tRNA molecules to bring amino acids to the ribosome in accordance with the mRNA codon sequence.
   - This process is facilitated by rRNA, which catalyzes peptide bond formation, emphasizing the active role of RNA in protein synthesis.

3. **Gene Regulation and RNA Interference**

   - RNA molecules regulate gene expression at multiple levels. For example, miRNAs and siRNAs can bind to mRNA molecules and prevent their translation, effectively silencing specific genes. This regulatory mechanism, known as RNA interference, allows cells to fine-tune gene expression in response to internal and external cues.

- The involvement of non-coding RNAs in chromatin remodeling and transcriptional regulation further demonstrates RNA's expansive role beyond its intermediary function, indicating its essential regulatory influence in cellular processes.

Evolutionary Significance of Nucleic Acids

Nucleic acids, particularly RNA, are believed to have played a central role in the early evolution of life. The **RNA world hypothesis** suggests that life may have originated with self-replicating RNA molecules, which possessed both informational and catalytic properties. Over time, DNA and proteins evolved as specialized molecules, with DNA assuming the role of stable information storage and proteins serving as the primary functional molecules.

1. **The RNA World Hypothesis**
   - This hypothesis posits that RNA was the first genetic material due to its dual ability to store information and catalyze reactions. RNA's capacity for self-replication and catalysis may have enabled the earliest life forms to evolve and adapt.
   - Evidence supporting this hypothesis includes the discovery of ribozymes and the catalytic role of rRNA within the ribosome. These findings suggest that RNA could have mediated essential biochemical processes in the absence of proteins.
2. **Transition to DNA-Based Genetics**
   - DNA eventually supplanted RNA as the primary genetic material due to its greater chemical stability, which is attributed to the absence of the 2'-hydroxyl group found in RNA. This structural difference reduces the susceptibility of DNA to hydrolysis, making it better suited for long-term information storage.
   - As DNA became the stable repository of genetic information, RNA adapted to new roles in gene expression, regulation, and catalysis, highlighting the evolutionary adaptability of nucleic acids.
3. **Co-Evolution of DNA, RNA, and Proteins**
   - The co-evolution of DNA, RNA, and proteins has led to a sophisticated division of labor within the cell. DNA serves as a stable genetic archive, RNA acts as the intermediary and regulator, and proteins perform most of the cell's biochemical functions.
   - This molecular specialization enhances cellular efficiency and complexity, enabling organisms to develop intricate regulatory networks, diverse cellular functions, and adaptability to a wide range of environments.

3.2 Proteins: Structure, Function, and Types

Proteins are essential macromolecules that perform a vast array of functions within living organisms, from catalyzing biochemical reactions to providing structural support, facilitating communication, and regulating cellular processes. These diverse functions are made possible by the highly specialized structures of proteins, which are determined by the sequence of amino acids and their spatial arrangements.

Structure of Proteins

Protein structure is highly organized and hierarchical, consisting of four distinct levels—primary, secondary, tertiary, and quaternary. Each level of structure contributes to the protein's overall shape and function, with even minor alterations potentially leading to significant functional changes.

1. **Primary Structure**
   - The primary structure of a protein is its linear sequence of amino acids, linked by peptide bonds to form a polypeptide chain. The amino acid sequence is dictated by the genetic code within DNA, and it ultimately determines the protein's three-dimensional structure and function.
   - The unique sequence of amino acids in a protein is essential, as even a single change can lead to altered function or disease. For example, in sickle cell anemia, a single amino acid substitution in hemoglobin changes its properties, resulting in abnormal red blood cell shapes and impaired oxygen transport.
2. **Secondary Structure**
   - The secondary structure refers to the local folding patterns within a polypeptide chain, stabilized by hydrogen bonds between backbone atoms. The two most common secondary structures are the **alpha helix** and the **beta sheet**.
     - ◊ **Alpha Helix**: This structure is a right-handed coil stabilized by hydrogen bonds between the backbone's carbonyl oxygen and amide hydrogen. Alpha helices are common in fibrous proteins, such as keratin in hair and nails.
     - ◊ **Beta Sheet**: In beta sheets, polypeptide chains lie parallel or antiparallel to each other, forming a pleated structure stabilized by hydrogen bonds between adjacent strands. Beta sheets are common in globular proteins like enzymes and transport proteins.
   - The secondary structure provides local stability and contributes to the protein's overall conformation.
3. **Tertiary Structure**
   - Tertiary structure is the overall three-dimensional shape of a single polypeptide chain. It results from interactions between amino acid side chains, including hydrophobic interactions, hydrogen bonds, ionic bonds, and disulfide bridges (covalent bonds between cysteine residues).
   - This level of structure determines the protein's functional form, creating specific sites for ligand binding, catalytic activity, or interactions with other molecules. For example, the active site of an enzyme is formed by the tertiary structure, allowing it to bind specific substrates.
4. **Quaternary Structure**
   - Quaternary structure is the arrangement of multiple polypeptide chains (subunits)

into a functional protein complex. Not all proteins have quaternary structure; it is only present in multi-subunit proteins.

- Each subunit may have its own distinct function, or multiple subunits may work together to perform a single function. Hemoglobin, for instance, consists of four subunits, each of which binds an oxygen molecule, facilitating efficient oxygen transport in the blood.
- Quaternary structure enables proteins to undergo cooperative interactions and regulatory control, as seen in enzymes and structural proteins.

Functions of Proteins

Proteins are incredibly versatile molecules with functions that are essential to almost every biological process. These functions are often specific to the protein's structure, enabling it to perform roles that range from catalysis to structural support and signaling.

1. **Catalysis (Enzymes)**
    - Enzymes are proteins that catalyze biochemical reactions, speeding up reaction rates without being consumed in the process. They lower the activation energy needed for reactions, enabling cells to carry out complex metabolic processes efficiently.
    - Enzymes have specific active sites where substrates bind, and their activity is often regulated by factors such as pH, temperature, and the presence of inhibitors or activators.
    - Enzymatic functions are central to metabolic pathways, including those involved in energy production, DNA replication, and cellular repair.
2. **Structural Support**
    - Structural proteins provide shape and support to cells and tissues. Collagen, for instance, is a fibrous protein that strengthens connective tissues, including skin, bones, and tendons.
    - Keratin is another structural protein found in hair, nails, and the outer layer of skin, providing resilience and protection.
    - The cytoskeleton, composed of proteins like actin and tubulin, supports cellular shape, facilitates movement, and organizes intracellular transport.
3. **Transport and Storage**
    - Transport proteins, such as hemoglobin, carry essential molecules like oxygen throughout the body. Membrane transport proteins facilitate the movement of ions, nutrients, and other molecules across cell membranes.
    - Storage proteins, such as ferritin, store essential ions and molecules within cells. Ferritin, for example, binds and stores iron, releasing it as needed for processes like hemoglobin synthesis and cellular respiration.
4. **Signal Transduction and Communication**
    - Proteins play key roles in signal transduction, the process by which cells communicate and respond to external stimuli. Receptor proteins, such as G protein-coupled receptors, receive signals from hormones or neurotransmitters and initiate cellular responses.

- Signaling proteins, like kinases, transmit signals through phosphorylation cascades, amplifying cellular responses and regulating processes such as cell division, apoptosis, and metabolism.

5. **Immune Response**

- Proteins are central to the immune system, with antibodies serving as key molecules in recognizing and neutralizing pathogens. Each antibody has a unique binding site that targets specific antigens, marking them for destruction.
- Other immune proteins, such as cytokines, mediate cell signaling within the immune system, coordinating responses to infection and inflammation.

6. **Regulation of Gene Expression**

- Regulatory proteins, such as transcription factors, control the expression of specific genes by binding to DNA and either promoting or inhibiting transcription. This regulation enables cells to respond to changes in their environment and to differentiate into various cell types during development.
- Epigenetic regulatory proteins modify chromatin structure, making DNA accessible or inaccessible to transcriptional machinery, thus influencing gene expression.

Types of Proteins

Proteins are categorized into several types based on their functions and characteristics, with each type contributing uniquely to cellular and organismal function.

1. **Enzymatic Proteins**

- Enzymatic proteins are biological catalysts that facilitate chemical reactions. Each enzyme is highly specific to its substrate and reaction, and enzymes are crucial to pathways like glycolysis, the citric acid cycle, and DNA synthesis.
- Examples include amylase, which breaks down starches into sugars; DNA polymerase, which synthesizes DNA; and proteases, which break down proteins.

2. **Structural Proteins**

- Structural proteins form the physical framework of cells and tissues, providing support, strength, and elasticity.
- Collagen, the most abundant protein in mammals, gives structure to connective tissues, while actin and tubulin form the cytoskeleton, essential for maintaining cell shape and facilitating movement.

3. **Transport Proteins**

- Transport proteins move molecules across cellular membranes and throughout the body. Hemoglobin transports oxygen in the bloodstream, while transporters and channels in cell membranes regulate ion gradients and nutrient uptake.
- Membrane transport proteins, such as sodium-potassium pumps, are essential for maintaining cellular homeostasis and enabling nerve transmission.

4. **Defense Proteins**
    - Defense proteins include antibodies, which are produced by the immune system to recognize and neutralize foreign substances like bacteria and viruses.
    - Antibodies have variable regions that enable them to bind specifically to antigens, forming a critical component of the adaptive immune response. Other immune proteins, like complement proteins, assist in pathogen elimination.
5. **Signaling Proteins**
    - Signaling proteins transmit information between and within cells, coordinating physiological responses. Hormones, such as insulin, are signaling proteins that regulate blood glucose levels, while growth factors stimulate cell division and differentiation.
    - Signal transduction proteins, including kinases and phosphatases, modify other proteins and initiate cellular responses to external signals.
6. **Storage Proteins**
    - Storage proteins store essential ions and molecules, providing a reservoir for cellular and physiological needs. Ferritin stores iron, while myoglobin stores oxygen in muscle cells.
    - Storage proteins allow organisms to maintain supplies of vital nutrients and release them when they are needed for cellular functions.
7. **Motor Proteins**
    - Motor proteins are responsible for movement within cells and tissues. Myosin and actin interact to produce muscle contraction, while kinesin and dynein move cellular cargo along microtubules within cells.
    - These proteins convert chemical energy, typically from ATP, into mechanical work, enabling cellular movement, transport of vesicles, and changes in cell shape.
8. **Regulatory Proteins**
    - Regulatory proteins control gene expression, cell cycle progression, and metabolic pathways. Transcription factors, for instance, bind to DNA to regulate the transcription of specific genes.
    - Regulatory proteins ensure that cellular processes occur at the appropriate times and in response to environmental changes, enabling cells to adapt and function efficiently.

3.3 Lipids and Carbohydrates in Molecular Biology

Lipids and carbohydrates, alongside nucleic acids and proteins, are fundamental macromolecules that play diverse and essential roles in biological systems. Although they are not directly involved in the encoding or transmission of genetic information, lipids and carbohydrates are critical for maintaining cellular structure, storing energy, signaling, and supporting biochemical processes. In molecular biology, these molecules contribute to the structural integrity of cells, facilitate metabolic activities, and mediate cellular communication.

Lipids in Molecular Biology

Lipids are a diverse group of hydrophobic molecules that include fats, oils, phospholipids, steroids, and waxes. Unlike other macromolecules, lipids are not defined by their monomeric units but rather by their insolubility in water and solubility in organic solvents. This hydrophobic nature is due to their long hydrocarbon chains or rings, which allow them to form cellular membranes, store energy, and act as signaling molecules.

1. **Structure of Lipids**
    - The structure of lipids varies considerably depending on the type. Most lipids consist of a glycerol backbone attached to fatty acids, as seen in triglycerides and phospholipids.
    - **Fatty Acids**: Fatty acids are long hydrocarbon chains with a terminal carboxyl group. They can be saturated (no double bonds) or unsaturated (one or more double bonds), affecting the lipid's physical properties. Unsaturated fatty acids introduce kinks in the chain, preventing tight packing and thus increasing membrane fluidity.
    - **Triglycerides**: Triglycerides, or fats, are formed by attaching three fatty acids to a glycerol molecule via ester bonds. They serve as a primary energy storage form in animals, storing more energy per gram than carbohydrates.
    - **Phospholipids**: Phospholipids contain two fatty acids, a glycerol backbone, and a phosphate group. The phosphate group makes the "head" hydrophilic, while the fatty acid "tails" are hydrophobic, creating an amphipathic molecule ideal for forming cellular membranes.
2. **Functions of Lipids in Cellular Structure and Membranes**
    - **Cell Membranes**: Phospholipids are fundamental components of cell membranes, forming a bilayer that separates the intracellular environment from the extracellular space. The hydrophilic heads face outward, while the hydrophobic tails face inward, creating a semi-permeable barrier.
    - **Membrane Fluidity**: The fluidity of cell membranes is influenced by the saturation of fatty acids and the presence of cholesterol. Cholesterol, a lipid with a ring structure, intercalates between phospholipids, stabilizing membrane fluidity and preventing phase transitions at extreme temperatures.
3. **Lipids in Energy Storage**
    - Lipids are the most efficient form of long-term energy storage in organisms. Triglycerides stored in adipose tissue provide a dense energy source that can be mobilized during fasting or intense activity.
    - Fatty acids are broken down through beta-oxidation in the mitochondria, releasing acetyl-CoA, which enters the citric acid cycle for ATP production. This high energy yield makes lipids essential for organisms that require energy reserves for extended periods without food.
4. **Lipids as Signaling Molecules**
    - Lipids play essential roles in cellular communication, particularly through signaling molecules like steroids, eicosanoids, and phosphoinositides.

◊ **Steroid Hormones**: Steroids, such as testosterone, estrogen, and cortisol, are lipids derived from cholesterol. They act as signaling molecules that can cross cell membranes and bind to nuclear receptors, regulating gene expression and influencing growth, metabolism, and immune response.

◊ **Eicosanoids**: Eicosanoids, derived from arachidonic acid, include prostaglandins, thromboxanes, and leukotrienes. These molecules are involved in inflammation, immunity, and blood clotting.

◊ **Phosphoinositides**: Phospholipids like phosphatidylinositol can be phosphorylated to produce phosphoinositides, which serve as secondary messengers in signal transduction pathways, particularly in processes like cell growth and differentiation.

Carbohydrates in Molecular Biology

Carbohydrates are organic molecules composed of carbon, hydrogen, and oxygen, typically in a 1:2:1 ratio. They are a primary source of energy and also serve structural, storage, and signaling functions within cells. Carbohydrates range from simple monosaccharides to complex polysaccharides, with each form adapted to specific roles in cellular and molecular biology.

1. **Structure of Carbohydrates**

   - Carbohydrates are classified based on the number of sugar units they contain: monosaccharides (single units), disaccharides (two units), oligosaccharides (few units), and polysaccharides (many units).
   - **Monosaccharides**: Monosaccharides are simple sugars like glucose, fructose, and galactose. Glucose is a critical energy source, while fructose and galactose contribute to metabolic pathways and synthesis.
   - **Disaccharides**: Disaccharides, such as sucrose (glucose + fructose) and lactose (glucose + galactose), are formed by glycosidic bonds between two monosaccharides and serve as transportable forms of sugar.
   - **Polysaccharides**: Polysaccharides are long chains of monosaccharides linked by glycosidic bonds. They can be either branched or unbranched and are used for energy storage (e.g., glycogen and starch) or structural support (e.g., cellulose and chitin).

2. **Functions of Carbohydrates in Energy Metabolism**

   - Carbohydrates are a primary energy source, particularly glucose, which is metabolized through glycolysis and the citric acid cycle to produce ATP, the energy currency of the cell.
   - **Glycogen**: In animals, excess glucose is stored as glycogen, primarily in the liver and muscle tissues. Glycogen serves as a readily accessible energy reserve that can be mobilized quickly when blood glucose levels drop.
   - **Starch**: In plants, glucose is stored as starch, a polysaccharide that accumulates in chloroplasts and amyloplasts. During periods of low photosynthesis, starch is broken down to release glucose for cellular respiration.

3. **Structural Role of Carbohydrates**

   - Certain polysaccharides provide structural support in cells and tissues. For instance:

     ◊ **Cellulose**: In plants, cellulose forms the cell wall, providing rigidity and protection. Cellulose is a linear polymer of glucose with beta-1,4 linkages, making it resistant to hydrolysis and indigestible to humans.

     ◊ **Chitin**: In fungi and arthropods, chitin forms the structural framework of the cell wall and exoskeleton, respectively. Chitin is a polymer of N-acetylglucosamine, which provides strength and flexibility to the organisms.

4. **Carbohydrates in Cell Recognition and Communication**

   - Carbohydrates are key components of glycoproteins and glycolipids on the cell surface, where they play crucial roles in cell-cell recognition, signaling, and immune responses.
   - **Glycoproteins**: Proteins with attached carbohydrate chains are involved in cell signaling, immune recognition, and cell adhesion. For example, blood type antigens are glycoproteins with specific carbohydrate moieties on the red blood cell surface.
   - **Glycolipids**: Lipids with carbohydrate groups are essential for cell recognition and interaction with extracellular molecules. Glycolipids in the nervous system, such as gangliosides, play a role in nerve cell communication and development.

5. **Role of Carbohydrates in Nucleic Acids**

   - Carbohydrates, specifically ribose in RNA and deoxyribose in DNA, are components of nucleic acids. These sugars form the backbone of nucleic acid molecules, linking with phosphate groups and nitrogenous bases to create DNA and RNA strands.
   - The presence or absence of a hydroxyl group at the 2' carbon in ribose and deoxyribose, respectively, distinguishes RNA from DNA and impacts their stability, with DNA being more stable for long-term genetic storage.

Interplay of Lipids and Carbohydrates in Cellular Processes

Lipids and carbohydrates interact in various cellular processes, particularly in energy metabolism, cellular signaling, and membrane dynamics. These interactions highlight their complementary roles in supporting cellular functions and maintaining cellular integrity.

1. **Energy Storage and Mobilization**

   - While lipids serve as long-term energy storage, carbohydrates provide rapid energy. Glycogen can be quickly mobilized for immediate energy needs, while lipids are broken down more slowly, providing sustained energy over time.
   - Carbohydrate metabolism can directly impact lipid metabolism. For example, when carbohydrate intake is high, excess glucose is converted to fat through lipogenesis. Conversely, low carbohydrate availability stimulates lipolysis, breaking down stored fat for energy.

2. **Membrane Structure and Glycocalyx Formation**

   - The cell membrane is composed of a lipid bilayer with embedded proteins and carbohydrates. Carbohydrates attached to lipids (glycolipids) and proteins (glycoproteins) form a glycocalyx on the cell surface, which is crucial for cell recognition, protection, and signaling.

- The glycocalyx acts as a molecular "fingerprint" that allows cells to recognize one another, which is vital for immune recognition, tissue formation, and cellular communication.

3. **Cell Signaling Pathways**
   - Carbohydrates and lipids play essential roles in cell signaling. For example, glycolipids on the cell surface participate in interactions with extracellular signaling molecules. Additionally, lipid-based molecules like phosphoinositides serve as secondary messengers in intracellular signaling pathways.
   - Lipid rafts, microdomains rich in cholesterol and sphingolipids, organize signaling proteins within the cell membrane, facilitating efficient signal transduction.

3.4 Cell Organelles and Their Molecular Roles

Cell organelles are specialized structures within cells that carry out specific functions essential for cellular survival, growth, and reproduction. Each organelle is equipped with unique molecular machinery and biochemical pathways that contribute to the cell's overall function, from energy production and protein synthesis to waste disposal and genetic information management. The molecular roles of cell organelles are fundamental in molecular biology, as it reveal how cells coordinate their activities and maintain homeostasis.

1. Nucleus: The Command Center of the Cell

The nucleus is the central organelle that houses the cell's genetic material, DNA. It serves as the control center for cellular activities by regulating gene expression and coordinating cell growth, division, and differentiation.

1. **Structure of the Nucleus**
   - The nucleus is surrounded by a double membrane called the nuclear envelope, which contains nuclear pores that regulate the movement of molecules between the nucleus and the cytoplasm.
   - Inside the nucleus, DNA is organized into chromatin, a complex of DNA and histone proteins that condenses into chromosomes during cell division.
2. **Molecular Roles**
   - **Genetic Information Storage**: The nucleus stores the cell's genetic blueprint in the form of DNA, which contains instructions for all cellular functions and protein synthesis.
   - **Gene Expression Regulation**: The nucleus controls gene expression through transcriptional regulation, where genes are transcribed into mRNA in response to cellular signals.
   - **Nucleolus and Ribosome Assembly**: The nucleolus, a dense region within the nucleus, is involved in synthesizing ribosomal RNA (rRNA) and assembling ribosomal subunits, which are exported to the cytoplasm for protein synthesis.

3. **DNA Replication and Repair**

   - The nucleus is the site of DNA replication, where the DNA double helix is duplicated before cell division, ensuring that each daughter cell receives a complete set of genetic information.
   - It is also equipped with DNA repair mechanisms that identify and fix DNA damage, maintaining genomic integrity and preventing mutations.

2. Mitochondria: The Powerhouses of the Cell

Mitochondria are double-membraned organelles responsible for producing the majority of the cell's energy in the form of adenosine triphosphate (ATP) through cellular respiration. They play crucial roles in metabolism and energy regulation.

1. **Structure of Mitochondria**

   - Mitochondria have an outer membrane and a highly folded inner membrane that forms structures called cristae, which increase the surface area for energy production.
   - The matrix, the inner compartment of mitochondria, contains enzymes involved in the citric acid cycle and mitochondrial DNA (mtDNA), which encodes a small number of mitochondrial proteins.

2. **Molecular Roles**

   - **ATP Production**: Mitochondria produce ATP through oxidative phosphorylation, which occurs in the electron transport chain on the inner membrane. This process is driven by the breakdown of glucose and fatty acids.
   - **Metabolism and Biochemical Pathways**: Mitochondria are central to metabolic pathways, including the citric acid cycle (Krebs cycle) and beta-oxidation of fatty acids, providing intermediates for various biosynthetic processes.
   - **Apoptosis Regulation**: Mitochondria play a role in programmed cell death (apoptosis) by releasing cytochrome c, which activates caspases, the enzymes responsible for cell disassembly during apoptosis.

3. **Mitochondrial DNA and Inheritance**

   - Mitochondria have their own DNA, which is inherited maternally. Mitochondrial genes encode components of the electron transport chain, though most mitochondrial proteins are encoded by nuclear DNA and imported into mitochondria.

3. Endoplasmic Reticulum (ER): The Manufacturing and Packaging System

The endoplasmic reticulum is a network of membranous tubules and sacs involved in the synthesis, folding, modification, and transport of proteins and lipids. It is divided into the rough ER (RER) and smooth ER (SER), each with distinct functions.

1. **Rough Endoplasmic Reticulum (RER)**

   - The rough ER is studded with ribosomes on its surface, which synthesize proteins destined for secretion, insertion into cell membranes, or transport to other organelles.

- **Molecular Roles**:
  - ◊ **Protein Synthesis**: Ribosomes on the RER translate mRNA into proteins, which are then translocated into the ER lumen.
  - ◊ **Protein Folding and Quality Control**: The RER contains molecular chaperones that assist in protein folding and quality control, ensuring that only correctly folded proteins proceed to the Golgi apparatus.

2. **Smooth Endoplasmic Reticulum (SER)**
   - The smooth ER lacks ribosomes and is involved in lipid synthesis, detoxification, and calcium storage.
   - **Molecular Roles**:
     - ◊ **Lipid and Steroid Synthesis**: The SER is responsible for synthesizing lipids, including phospholipids and cholesterol, which are essential for membrane formation and repair.
     - ◊ **Detoxification**: Enzymes in the SER metabolize toxins and drugs, converting them into water-soluble compounds that can be excreted.
     - ◊ **Calcium Storage**: The SER stores calcium ions, which are released during cellular signaling events, particularly in muscle cells for contraction.

4. Golgi Apparatus: The Post-Office of the Cell

The Golgi apparatus, or Golgi complex, is an organelle involved in modifying, sorting, and packaging proteins and lipids for transport to their final destinations.

1. **Structure of the Golgi Apparatus**
   - The Golgi apparatus consists of a series of flattened, membrane-bound sacs called cisternae. It has a cis face, where molecules enter, and a trans face, where they exit after processing.
2. **Molecular Roles**
   - **Protein and Lipid Modification**: The Golgi modifies proteins and lipids, often by adding carbohydrate chains (glycosylation) or phosphate groups, which help in targeting molecules to specific locations.
   - **Sorting and Packaging**: The Golgi sorts and packages proteins into vesicles for delivery to the plasma membrane, lysosomes, or other organelles. Vesicles bud from the Golgi and are directed to their destinations by molecular tags.
   - **Production of Lysosomes**: The Golgi is involved in producing lysosomes, membrane-bound vesicles containing digestive enzymes.

5. Lysosomes: The Digestive System of the Cell

Lysosomes are membrane-bound organelles filled with hydrolytic enzymes that break down biomolecules, cellular debris, and foreign particles, maintaining cellular health by recycling and waste removal.

1. **Structure of Lysosomes**
   - Lysosomes are small, spherical organelles containing acidic hydrolytic enzymes capable of digesting proteins, lipids, nucleic acids, and carbohydrates. The acidic environment is maintained by proton pumps on the lysosomal membrane.
2. **Molecular Roles**
   - **Intracellular Digestion**: Lysosomes degrade cellular waste and macromolecules through enzymatic hydrolysis. They digest material taken up by endocytosis, phagocytosis, and autophagy.
   - **Autophagy**: During autophagy, lysosomes fuse with autophagosomes (vesicles containing damaged organelles or proteins) to break down and recycle cellular components, a critical process for cellular maintenance.
   - **Pathogen Defense**: In immune cells, lysosomes help degrade pathogens engulfed during phagocytosis, aiding in the immune response.

6. Peroxisomes: The Detoxification Centers

Peroxisomes are small, membrane-bound organelles that contain enzymes involved in detoxification, fatty acid oxidation, and reactive oxygen species (ROS) management.

1. **Structure of Peroxisomes**
   - Peroxisomes are similar in appearance to lysosomes but contain oxidative enzymes, including catalase, that break down toxic molecules.
2. **Molecular Roles**
   - **Detoxification of Harmful Substances**: Peroxisomes detoxify harmful compounds such as hydrogen peroxide (H_2O_2), which is broken down by catalase to water and oxygen.
   - **Lipid Metabolism**: Peroxisomes play a role in beta-oxidation, breaking down long-chain fatty acids into shorter molecules that can be processed by mitochondria for ATP production.
   - **Biosynthesis of Plasmalogens**: Peroxisomes are involved in the synthesis of plasmalogens, a type of lipid essential for the myelin sheath in nerve cells, underscoring their role in cellular structure and nervous system health.

7. Cytoskeleton: The Cellular Framework

The cytoskeleton is a dynamic network of protein filaments that provides structural support, enables movement, and organizes cellular contents. It is composed of three main types of filaments: microfilaments, microtubules, and intermediate filaments.

1. **Microfilaments (Actin Filaments)**
   - Microfilaments, made of actin, are involved in cell shape maintenance, muscle contraction, and cell motility. Actin polymerization and depolymerization enable cellular movements, such as those in immune cells or during wound healing.

2. **Microtubules**

   - Microtubules, composed of tubulin, are rigid structures that support cell shape, facilitate intracellular transport, and form the mitotic spindle during cell division. Motor proteins like kinesin and dynein travel along microtubules, carrying cellular cargo and facilitating vesicle movement.

3. **Intermediate Filaments**

   - Intermediate filaments provide mechanical strength and resistance to stress. They vary by cell type, with keratin found in epithelial cells and neurofilaments in neurons, helping cells withstand mechanical forces.

8. Plasma Membrane: The Cellular Boundary

The plasma membrane is a selectively permeable barrier that surrounds the cell, regulating the movement of molecules in and out of the cell and enabling communication with the external environment.

1. **Structure of the Plasma Membrane**

   - The plasma membrane is a lipid bilayer composed of phospholipids, cholesterol, and proteins. Embedded proteins and glycoproteins allow selective transport and cell signaling.

2. **Molecular Roles**

   - **Selective Permeability**: The plasma membrane regulates the transport of ions, nutrients, and waste products, maintaining the cell's internal environment.
   - **Signal Reception and Transduction**: Receptor proteins on the membrane bind to signaling molecules, triggering intracellular responses and enabling communication with other cells.
   - **Cell Recognition and Adhesion**: Membrane proteins and carbohydrates enable cell recognition and adhesion, essential for immune response and tissue formation.

4. Molecular Biology Techniques

4.1 DNA Extraction, Purification, and Quantification

DNA extraction, purification, and quantification are foundational techniques in molecular biology, essential for analyzing genetic material in various applications, from clinical diagnostics and forensic investigations to research in genetics and genomics. Each step in this process aims to obtain high-quality DNA in sufficient quantity and purity for downstream analyses, such as PCR, sequencing, and cloning.

DNA Extraction

DNA extraction is the first step in isolating DNA from cells or tissues, enabling researchers to analyze genetic material. This process involves breaking open the cell membrane, releasing DNA into solution, and then separating it from other cellular components. DNA extraction can be performed on a variety of biological samples, including blood, tissue, plants, and bacteria.

1. **Steps in DNA Extraction**
    - **Cell Lysis:** The initial step in DNA extraction is to break down the cell membrane and nuclear envelope (in eukaryotes) to release DNA into solution. Lysis buffers containing detergents like SDS (sodium dodecyl sulfate) disrupt lipid membranes, while mechanical methods (e.g., grinding, vortexing) aid in breaking down tissue samples.
    - **Protein Removal**: Once cells are lysed, enzymes like protease are added to digest proteins, which could interfere with DNA isolation. In some protocols, phenol-chloroform is used to separate proteins from nucleic acids, forming an organic layer for easy removal.
    - **DNA Precipitation**: DNA is precipitated out of solution by adding alcohol (usually ethanol or isopropanol) and salt. Alcohol causes DNA to aggregate and form a visible pellet, which is then washed and re-dissolved in a buffer solution.
    - **RNA Removal**: RNA can contaminate DNA preparations, so an RNase enzyme is often added to degrade RNA, leaving behind pure DNA.
2. **Types of DNA Extraction Methods**
    - **Organic Extraction**: This classic method involves organic solvents like phenol and chloroform to separate DNA from proteins and other contaminants. It is highly effective for high-purity DNA but involves hazardous chemicals.
    - **Inorganic Extraction (Salting Out)**: In this method, salts like ammonium acetate are used to precipitate proteins while DNA remains in solution. This technique is safer than organic extraction and produces high yields.
    - **Silica-Based Extraction**: Common in many commercial DNA kits, silica-based extraction involves binding DNA to a silica membrane in the presence of chaotropic salts, which disrupt cellular structures. DNA is then washed and eluted in a low-salt buffer.
    - **Magnetic Bead-Based Extraction**: Magnetic beads coated with DNA-binding material are used to isolate DNA. When an external magnetic field is applied, the beads (along with the bound DNA) are separated from the solution, allowing for rapid and efficient DNA isolation.
3. **Considerations for DNA Extraction**
    - **Sample Quality**: Fresh samples generally yield higher-quality DNA than degraded samples. Preserving sample integrity by storing tissues or cells at low temperatures can help.
    - **Contamination Prevention**: Contaminants like proteins, RNA, and inhibitors can interfere with downstream applications. Proper use of extraction reagents and thorough washing steps minimize these contaminants.
    - **Yield Requirements**: Different applications require specific DNA quantities. For instance, sequencing typically requires more DNA than PCR. Adjusting sample amount and protocol can help meet these yield requirements.

DNA Purification

Following extraction, DNA purification is necessary to remove residual contaminants that can interfere with downstream applications. Purified DNA is essential for applications requiring high accuracy, such as sequencing and gene cloning.

1. **Purification Methods**
   - **Ethanol Precipitation**: Ethanol or isopropanol is added to precipitate DNA, while contaminants remain in solution. The DNA pellet is then washed with cold ethanol to remove residual salts and dissolved in a suitable buffer.
   - **Silica Column Purification**: This method relies on DNA's ability to bind to silica in the presence of chaotropic salts. After washing away contaminants, DNA is eluted in a low-salt buffer. This method is popular in commercial kits for its simplicity and high purity.
   - **Gel Purification**: In cases where DNA fragments of a specific size are needed, agarose gel electrophoresis is used. DNA bands of the desired size are excised from the gel, and DNA is extracted from the gel matrix, often using a silica-based column or freeze-thaw extraction.
   - **Magnetic Bead Purification**: Magnetic beads coated with silica or other DNA-binding materials are used to bind DNA selectively. This approach is efficient for high-throughput applications and produces high-purity DNA.
2. **Assessing DNA Purity**
   - DNA purity is typically assessed using spectrophotometry, measuring absorbance at wavelengths 260 nm and 280 nm. Pure DNA has an A260/A280 ratio of ~1.8, with lower ratios indicating protein contamination and higher ratios suggesting potential RNA contamination.
   - Contaminants such as phenol, residual proteins, and salts can interfere with DNA quality and downstream applications. Purified DNA should be free of these contaminants to ensure reliability in molecular experiments.
3. **Considerations for DNA Purification**
   - **Application Requirements**: Different applications require different purity levels. For example, PCR may tolerate minor impurities, but sequencing requires highly purified DNA.
   - **Sample Loss**: Each purification step can lead to sample loss, especially in low-yield extractions. Efficient protocols and minimal handling steps can reduce this loss.
   - **Time and Cost**: While silica-based and magnetic bead methods are quick and produce high purity, they are often more expensive. Ethanol precipitation is cost-effective but may require more handling time.

DNA Quantification

Quantification of DNA is an essential step to ensure the concentration is sufficient for downstream applications and to standardize experimental conditions. Quantification methods vary in precision, sensitivity, and ease of use, depending on the requirements of the research or application.

1. **Spectrophotometric Quantification**
   - Spectrophotometry is the most common method for quantifying DNA. DNA absorbs light at a wavelength of 260 nm due to the aromatic rings in its nucleotide bases. The absorbance reading at 260 nm is used to calculate DNA concentration.
   - A reading at 280 nm indicates protein contamination, as proteins absorb at this wavelength. An A260/A280 ratio of around 1.8 is considered pure for DNA. Ratios below this value suggest protein contamination, while values above indicate RNA contamination.
   - **Limitations**: Spectrophotometry is reliable for relatively clean DNA samples but may overestimate concentration if contaminants are present. It is less sensitive for detecting low concentrations.
2. **Fluorometric Quantification**
   - Fluorometry involves using fluorescent dyes that bind specifically to DNA, such as PicoGreen or Qubit, and emit fluorescence when excited by a specific wavelength of light. Fluorometric methods are more sensitive than spectrophotometry and are suitable for low-concentration samples.
   - Fluorescent dyes can distinguish between DNA and RNA, providing more accurate quantification even in mixed samples. Fluorometric quantification is often used for applications requiring high accuracy, such as next-generation sequencing.
   - **Limitations**: Fluorometric methods require specialized equipment (e.g., a fluorometer) and reagents, making them more expensive than spectrophotometry. Additionally, the DNA must be sufficiently pure for reliable results.
3. **Quantitative PCR (qPCR)**
   - Quantitative PCR, or qPCR, is a highly sensitive method for DNA quantification that amplifies DNA and measures its concentration in real-time. By comparing the amplification threshold of a sample to a standard curve, qPCR can quantify DNA with high precision.
   - qPCR is especially useful for detecting low DNA concentrations and measuring specific DNA sequences within a sample. It is commonly used in clinical diagnostics, pathogen detection, and applications requiring high sensitivity.
   - **Limitations:** qPCR is more complex and requires specific reagents and equipment, such as a thermal cycler with fluorescence detection. It is not typically used for general quantification due to its complexity and cost.

Importance and Applications of DNA Extraction, Purification, and Quantification

1. **Clinical Diagnostics**
   - DNA extraction and quantification are fundamental in diagnostics for genetic diseases,

cancer, and infectious diseases. Purified DNA from patient samples enables accurate genetic testing, sequencing, and identification of pathogens.

- In forensic science, DNA quantification ensures that sufficient genetic material is available for analysis, particularly in cases where only minute DNA quantities are present.

2. **Genetic Research and Genomics**
   - DNA extraction and purification are prerequisites for genetic research, allowing scientists to study gene function, genetic variation, and inheritance. Quantified, high-quality DNA is critical for genome-wide association studies (GWAS) and other large-scale genomic analyses.
   - Purified DNA is essential for next-generation sequencing (NGS) technologies, which require high purity and concentration for accurate and efficient sequencing of whole genomes or targeted regions.
3. **Agricultural and Environmental Science**
   - DNA extraction and quantification are used in plant breeding and genetic engineering to identify desirable traits and engineer genetically modified organisms (GMOs). In environmental science, DNA extracted from soil or water samples is used to study biodiversity, monitor ecosystems, and detect invasive species.
   - Metagenomics, which involves analyzing genetic material from environmental samples, relies on precise DNA extraction and quantification for effective community analysis of microbial populations.
4. **Forensic Analysis**
   - Forensic laboratories require highly purified DNA for applications such as DNA fingerprinting, paternity testing, and criminal investigations. Quantifying DNA ensures that samples are sufficient for analysis, which is crucial for solving cases based on trace evidence.
   - In forensic cases, degraded or minute samples must be carefully extracted and purified to obtain usable DNA, with quantification confirming the sample's adequacy for profiling.
5. **Biomedical Research and Biotechnology**
   - Biomedical research depends on DNA extraction and purification to study disease mechanisms, develop therapies, and conduct gene editing using technologies like CRISPR. Quantified DNA is crucial for reproducibility and standardization across experiments.
   - Purified DNA is also necessary for synthetic biology, where precise DNA fragments are manipulated to engineer biological systems, produce pharmaceuticals, or create biofuels.

4.2 Polymerase Chain Reaction (PCR) and Its Variants

Polymerase Chain Reaction (PCR) is a powerful molecular biology technique that allows for the rapid amplification of specific DNA sequences, making it possible to generate millions of copies from a tiny amount of DNA. Since its development by Kary Mullis in 1983, PCR has revolutionized genetic research, diagnostics, forensic science, and many other fields by enabling precise and sensitive detection of DNA. Beyond the basic PCR method, various PCR variants have been developed to address specific research and diagnostic needs, such as quantitative analysis, high specificity, and amplification of RNA.

Basics of Polymerase Chain Reaction (PCR)

PCR amplifies a targeted DNA sequence through a series of repeated heating and cooling cycles, utilizing a thermostable DNA polymerase enzyme to synthesize new DNA strands. Each cycle effectively doubles the number of DNA molecules, resulting in an exponential increase in the target sequence.

1. **Key Components of PCR**
   - **Template DNA**: The DNA sample containing the target sequence to be amplified. Only a small amount of template DNA is required.
   - **Primers**: Short, single-stranded DNA sequences (usually 18-25 nucleotides long) that flank the target region and provide starting points for DNA synthesis. Each PCR reaction requires a pair of primers—one for each strand of the DNA.
   - **DNA Polymerase**: A thermostable enzyme, typically Taq polymerase (isolated from the bacterium *Thermus aquaticus*), that synthesizes new DNA strands by extending the primers. Taq polymerase can withstand the high temperatures used in PCR without denaturing.
   - **dNTPs (deoxynucleotide triphosphates)**: The building blocks (adenine, thymine, cytosine, and guanine) that DNA polymerase incorporates into the growing DNA strand.
   - **Buffer Solution**: Provides the optimal pH and ionic strength for the DNA polymerase enzyme to function efficiently.
2. **PCR Steps and Cycle**
   - **Denaturation**: The reaction mixture is heated to around 94-98°C, causing the double-stranded DNA to separate (denature) into single strands.
   - **Annealing**: The temperature is lowered (typically to 50-65°C) to allow the primers to bind (anneal) to their complementary sequences on the single-stranded DNA template.
   - **Extension/Elongation**: The temperature is raised to the optimal working temperature of Taq polymerase (around 72°C). DNA polymerase extends the primers by adding dNTPs, synthesizing new DNA strands complementary to the template.
   - This cycle is repeated 25-40 times, doubling the target DNA with each cycle and leading to exponential amplification.
3. **Applications of PCR**
   - **Genetic Research**: PCR is widely used to clone DNA, analyze gene expression, and study genetic variations.
   - **Diagnostics**: PCR can detect pathogens, identify genetic mutations, and diagnose diseases by amplifying specific DNA markers.
   - **Forensics**: PCR enables DNA profiling from minute samples, such as blood, saliva, and hair, facilitating identification in criminal investigations.

- **Environmental and Agricultural Science**: PCR is used for detecting genetically modified organisms (GMOs), monitoring microbial populations, and studying biodiversity.

Variants of PCR

Several variants of PCR have been developed to enhance specificity, sensitivity, and functionality in different applications. Below are some of the most widely used PCR variants.

1. Real-Time PCR (qPCR) or Quantitative PCR

Real-Time PCR, or quantitative PCR (qPCR), allows for the quantification of DNA in real-time by measuring the fluorescence emitted during the amplification process. qPCR is especially useful in diagnostics, gene expression analysis, and quantifying DNA in various samples.

1. **Principle**
   - qPCR uses fluorescent dyes or probes (such as SYBR Green or TaqMan probes) that bind to DNA and emit fluorescence in proportion to the amount of DNA synthesized. A qPCR machine detects and measures this fluorescence after each cycle, allowing for real-time monitoring of the amplification.
2. **Types of Detection**
   - **SYBR Green Dye**: SYBR Green binds to double-stranded DNA and emits fluorescence. However, it can bind non-specifically to any double-stranded DNA, which may reduce specificity.
   - **TaqMan Probes**: TaqMan probes are sequence-specific and emit fluorescence only when bound to the target DNA, increasing specificity. They are often used in high-sensitivity applications.
3. **Applications of qPCR**
   - **Gene Expression Analysis**: qPCR quantifies mRNA levels (through reverse transcription qPCR, or RT-qPCR), allowing researchers to measure gene expression under different conditions.
   - **Pathogen Detection and Diagnostics**: qPCR can detect pathogens (e.g., viruses, bacteria) and monitor infection levels in clinical samples by amplifying specific genetic markers.
   - **Genetic Testing**: qPCR identifies specific DNA sequences associated with genetic disorders, providing rapid and quantitative diagnostic insights.

2. Reverse Transcription PCR (RT-PCR)

Reverse Transcription PCR (RT-PCR) is used to amplify RNA sequences by first converting them into complementary DNA (cDNA). RT-PCR is invaluable for studying gene expression, as it enables the amplification of mRNA.

1. **Principle**
   - RT-PCR begins with an enzyme called reverse transcriptase, which synthesizes cDNA from an RNA template. This cDNA then serves as the template for standard PCR amplification.

2. **Applications of RT-PCR**
   - **Gene Expression Studies**: RT-PCR quantifies mRNA levels, enabling researchers to measure gene expression patterns across different tissues or under different conditions.
   - **Viral Detection**: Many viruses have RNA genomes, so RT-PCR is commonly used to detect RNA viruses (e.g., SARS-CoV-2) in clinical samples by amplifying specific viral RNA sequences.
   - **Molecular Biology Research**: RT-PCR is essential in studying RNA-processing events and verifying RNA expression in genetic studies.

3. Multiplex PCR

Multiplex PCR enables the simultaneous amplification of multiple target sequences in a single reaction, saving time and resources.

1. **Principle**
   - Multiplex PCR uses multiple pairs of primers within the same reaction tube, each specific to a different target sequence. By optimizing primer concentrations and annealing conditions, multiple DNA targets can be amplified in one reaction.
2. **Applications of Multiplex PCR**
   - **Pathogen Detection**: In clinical diagnostics, multiplex PCR can detect multiple pathogens simultaneously from a single sample, providing a rapid diagnosis.
   - **Genetic Testing**: Multiplex PCR detects multiple genetic markers or mutations in a single assay, improving efficiency in genetic screening and diagnostics.
   - **Forensics**: Multiplex PCR is used in DNA profiling to amplify multiple loci simultaneously, enhancing the power of forensic analysis.

4. Nested PCR

Nested PCR is designed to increase the specificity and sensitivity of PCR by performing two successive rounds of amplification.

1. **Principle**
   - In the first round of PCR, primers amplify a broad region of DNA, potentially generating non-specific products. In the second round, a new set of primers (nested within the first set) targets a specific region within the first amplicon, increasing specificity and reducing background noise.
2. **Applications of Nested PCR**
   - **Detection of Low-Abundance Targets**: Nested PCR is useful for detecting low-copy-number targets, as it reduces non-specific amplification.
   - **Environmental and Microbial Studies**: Nested PCR is often used in environmental and microbial research, where DNA samples may be highly diluted or contaminated with non-target DNA.

- **Genetic Research**: In gene cloning and mutation analysis, nested PCR enhances the accuracy of target amplification.

5. Hot-Start PCR

Hot-Start PCR improves PCR specificity by preventing non-specific amplification and primer-dimer formation during the initial heating phase.

1. **Principle**
    - In Hot-Start PCR, the DNA polymerase is kept inactive until the reaction reaches a specific temperature (often achieved by chemical modifications or the use of antibodies). This prevents the polymerase from extending primers at lower temperatures, where non-specific binding may occur.
2. **Applications of Hot-Start PCR**
    - **Improved Specificity**: Hot-Start PCR is particularly useful in applications requiring high specificity, such as mutation analysis and genotyping.
    - **Enhanced Accuracy**: By minimizing non-specific products, Hot-Start PCR improves the accuracy of results in clinical diagnostics and research.

6. Digital PCR (dPCR)

Digital PCR (dPCR) is a highly sensitive and precise method for quantifying DNA, allowing absolute quantification without the need for standard curves.

1. **Principle**
    - In dPCR, the sample is partitioned into thousands of individual reactions (such as in droplets or wells), with each partition containing zero, one, or a few copies of the target DNA. Amplification occurs separately in each partition, and the presence or absence of target DNA is detected, enabling quantification based on positive reactions.
2. **Applications of Digital PCR**
    - **Rare Mutation Detection**: dPCR is highly sensitive and can detect rare mutations, making it valuable in cancer research and genetic testing.
    - **Pathogen Detection**: dPCR is used in clinical settings for detecting low-abundance pathogens and quantifying viral loads.
    - **Copy Number Variation (CNV) Analysis**: dPCR can precisely quantify gene copy number, which is useful in genetic studies, pharmacogenomics, and personalized medicine.

7. High-Fidelity PCR

High-Fidelity PCR uses DNA polymerases with proofreading activity to reduce the error rate during DNA synthesis, producing highly accurate and reliable amplicons.

1. **Principle**
    - High-fidelity polymerases, such as Pfu or Q5, have a 3' to 5' exonuclease activity that allows them to proofread and correct errors during DNA synthesis, resulting in fewer mismatches.

2. **Applications of High-Fidelity PCR**
   - **Gene Cloning**: High-fidelity PCR is essential for applications requiring precise DNA sequences, such as gene cloning and recombinant DNA technology.
   - **Sequencing Preparation**: For DNA sequencing, high-fidelity PCR ensures accurate representation of the original sequence, minimizing errors in the data.
   - **Mutagenesis**: In studies involving site-directed mutagenesis, high-fidelity PCR maintains sequence integrity, ensuring that only intended mutations are introduced.

4.3 Gel Electrophoresis and Nucleic Acid Separation

Gel electrophoresis is a fundamental technique in molecular biology used for the separation and analysis of nucleic acids (DNA and RNA) based on their size and charge. By applying an electric field, gel electrophoresis enables researchers to visualize, quantify, and isolate nucleic acid fragments, facilitating applications such as DNA profiling, genetic analysis, cloning, and sequencing. Gel electrophoresis is a versatile tool with various types and methodologies suited to different research needs, providing a clear window into the molecular structure and integrity of nucleic acids.

Principle of Gel Electrophoresis

The basic principle of gel electrophoresis is to separate nucleic acid molecules based on size. Nucleic acids are negatively charged due to their phosphate backbone and thus migrate toward the positive electrode (anode) when an electric current is applied. The gel matrix, typically made of agarose or polyacrylamide, acts as a molecular sieve, with smaller fragments moving more quickly and larger fragments moving more slowly.

1. **Gel Matrix**
   - The gel matrix provides a porous medium that restricts the movement of nucleic acids based on size. Smaller molecules navigate the pores more easily, while larger molecules face greater resistance and move more slowly.
   - **Agarose Gel**: Agarose, a polysaccharide derived from seaweed, is commonly used for DNA separation, especially for fragments larger than 100 base pairs. Agarose concentration can be adjusted to change the resolution, with lower concentrations allowing separation of larger fragments.
   - **Polyacrylamide Gel**: Polyacrylamide is used for separating smaller DNA or RNA fragments, typically under 1000 base pairs. It provides higher resolution for small fragments, making it ideal for applications like DNA sequencing and protein electrophoresis.
2. **Electric Field and Migration**
   - When an electric field is applied, the negatively charged nucleic acids move toward the anode (positive end). The rate of migration is inversely proportional to the size of the molecule, allowing size-based separation.
   - Factors such as gel concentration, buffer composition, voltage, and temperature influence the rate of migration and the resolution of separation.

3. **Visualization and Staining**

    - After electrophoresis, nucleic acids are stained to visualize the bands in the gel. Common dyes include **ethidium bromide**, which intercalates with DNA and fluoresces under UV light, and **SYBR Green**, a safer alternative that provides high sensitivity for both DNA and RNA.
    - The stained gel can be viewed under UV light or a gel documentation system, revealing distinct bands that represent different nucleic acid fragments.

Types of Gel Electrophoresis

Different types of gel electrophoresis are used depending on the size of the nucleic acids, the level of resolution required, and the specific research or diagnostic application. Key types of gel electrophoresis include:

1. **Agarose Gel Electrophoresis**

    - **Application**: Agarose gel electrophoresis is widely used for separating large DNA fragments (typically over 100 base pairs), including genomic DNA, PCR products, and restriction enzyme digests.
    - **Method**: DNA samples are mixed with a loading dye and loaded into wells in the agarose gel. When the current is applied, DNA fragments migrate through the gel based on size, forming distinct bands.
    - **Advantages**: Agarose gel is easy to prepare, non-toxic, and provides sufficient resolution for most DNA analysis applications. Adjusting agarose concentration can fine-tune the gel's resolution for different fragment sizes.

2. **Polyacrylamide Gel Electrophoresis (PAGE)**

    - **Application**: PAGE is used for high-resolution separation of small DNA or RNA fragments, such as those in DNA sequencing or single-strand conformational polymorphism (SSCP) analysis. PAGE is also used in protein electrophoresis.
    - **Method**: Acrylamide and bis-acrylamide form a cross-linked gel matrix, which provides a tighter pore structure than agarose, allowing precise separation of small nucleic acids or proteins.
    - **Advantages**: PAGE offers superior resolution for smaller molecules and provides sharper bands than agarose gel. However, it requires careful handling, as acrylamide is a neurotoxin in its unpolymerized form.

3. **Denaturing Gel Electrophoresis**

    - **Application**: Denaturing gels are used to separate single-stranded DNA or RNA molecules. This technique is commonly used in RNA analysis and DNA sequencing.
    - **Method**: Denaturing agents such as urea or formamide are added to the gel to disrupt hydrogen bonds, ensuring that nucleic acids remain in single-stranded form. Denaturing conditions prevent secondary structures that could affect migration.

- **Advantages**: Denaturing gels provide accurate size separation for single-stranded nucleic acids and allow precise analysis of RNA molecules and single-stranded DNA fragments.

4. **Pulsed-Field Gel Electrophoresis (PFGE)**
    - **Application**: PFGE is designed to separate very large DNA molecules, typically those over 20,000 base pairs, such as whole bacterial genomes or chromosomal DNA.
    - **Method**: PFGE uses alternating electric fields that periodically change direction, allowing large DNA fragments to reorient and move through the gel more effectively.
    - **Advantages**: PFGE provides high-resolution separation of large DNA fragments, which is useful in epidemiological studies, microbial genotyping, and genome mapping.

Steps in Gel Electrophoresis

The general process of gel electrophoresis includes sample preparation, loading, running the gel, staining, and analyzing the results. Below is a step-by-step overview:

1. **Preparation of the Gel**
    - **Agarose or Polyacrylamide Gel Preparation**: The gel is prepared by dissolving agarose or polyacrylamide in buffer, heating to dissolve the powder, and pouring it into a gel casting tray. A comb is placed in the gel to create wells for loading samples. Once the gel has solidified, it is placed in the electrophoresis chamber.
2. **Preparation of DNA Samples**
    - **Sample Loading**: DNA samples are mixed with loading dye, which provides density to the sample (allowing it to sink into the well) and adds visible color to track sample migration. Common loading dyes also contain tracking dyes like bromophenol blue and xylene cyanol.
    - **DNA Ladder**: A DNA ladder (or marker), containing DNA fragments of known sizes, is loaded alongside the samples. The ladder serves as a reference for estimating the sizes of the sample fragments.
3. **Running the Gel**
    - The gel is placed in an electrophoresis chamber filled with running buffer (e.g., TAE or TBE buffer), which conducts the electric current and maintains pH. An electric field is applied, causing the DNA to migrate toward the positive end (anode) of the chamber.
    - **Run Time and Voltage**: The gel is run for a specific time and at a set voltage, both of which depend on the desired resolution and the size of the fragments. Higher voltage speeds up migration but can lead to band distortion.
4. **Staining and Visualization**
    - After electrophoresis, the gel is stained with a fluorescent dye such as ethidium bromide or SYBR Green to visualize DNA. The gel is then exposed to UV light or another light source, causing the dye-bound DNA to fluoresce.

- Bands of DNA appear as distinct lines, each corresponding to a specific fragment size.

5. **Analysis and Interpretation**
   - The DNA ladder is used as a reference to estimate the size of the sample fragments based on their migration distance. Gel documentation systems with imaging software provide quantitative measurements, including band intensity, which can correlate with DNA quantity.

Applications of Gel Electrophoresis in Molecular Biology

Gel electrophoresis is a versatile tool used across various fields of molecular biology, including research, diagnostics, forensics, and environmental science.

1. **DNA Fingerprinting and Forensics**
   - Gel electrophoresis is instrumental in DNA profiling, as it separates DNA fragments produced by restriction enzyme digestion or PCR amplification. In forensics, comparing DNA band patterns allows for individual identification and paternity testing.
2. **Genetic and Molecular Diagnostics**
   - In clinical diagnostics, gel electrophoresis is used to detect genetic mutations, analyze gene expression, and identify pathogens. For example, PCR-amplified fragments of a pathogen's DNA can be visualized to confirm infection.
3. **Cloning and Recombinant DNA Technology**
   - Gel electrophoresis is essential in cloning and recombinant DNA work for verifying the presence and size of inserted DNA fragments. Scientists can isolate specific DNA bands for extraction and further experimentation, such as cloning into vectors.
4. **Genomic and Transcriptomic Studies**
   - Gel electrophoresis allows researchers to assess the quality and quantity of RNA and DNA samples, which is crucial in genomics and transcriptomics. For example, analyzing RNA integrity helps determine sample suitability for RNA sequencing.
5. **Environmental and Evolutionary Biology**
   - In environmental studies, gel electrophoresis can help identify microbial populations and analyze biodiversity through DNA fingerprinting. In evolutionary biology, it is used to study genetic variation and evolutionary relationships.

Limitations of Gel Electrophoresis

While gel electrophoresis is a powerful and widely used tool, it has some limitations:

1. **Resolution Limits**
   - Agarose gel electrophoresis has limited resolution for very small fragments, and polyacrylamide gel is not ideal for very large fragments. Techniques like PFGE are necessary for separating large DNA fragments but require specialized equipment.

2. **Quantitative Limitations**
   - Standard gel electrophoresis provides semi-quantitative results. Accurate quantification typically requires additional methods, such as densitometry, qPCR, or digital PCR.
3. **Sample Processing Time**
   - Gel preparation, staining, and imaging require time, which can slow down workflows in high-throughput applications. Newer capillary electrophoresis methods offer faster analysis but at a higher cost.

4.4 DNA Sequencing: From Sanger to Next-Generation Sequencing

DNA sequencing is the process of determining the precise order of nucleotides within a DNA molecule. Since the development of the first sequencing method by Frederick Sanger in the 1970s, DNA sequencing technology has transformed our understanding of genetics, genomics, and molecular biology. DNA sequencing allows scientists to decode the genetic information that guides biological processes, enabling applications in medical diagnostics, personalized medicine, evolutionary biology, and more.

The Evolution of DNA Sequencing

DNA sequencing has evolved significantly over the past few decades. The journey began with Sanger sequencing, a labor-intensive and relatively low-throughput method. The advent of next-generation sequencing (NGS) revolutionized the field by providing high-throughput, cost-effective, and scalable options, allowing whole genomes to be sequenced rapidly and at unprecedented scales.

Sanger Sequencing (Chain Termination Method)

Sanger sequencing, also known as the chain termination method, was developed by Frederick Sanger in 1977 and became the foundation for DNA sequencing technology. This technique was widely used for over three decades and remains relevant for specific applications due to its high accuracy.

1. **Principle of Sanger Sequencing**
   - Sanger sequencing uses modified nucleotides called **dideoxynucleotide triphosphates (ddNTPs)**, which lack the 3' hydroxyl group required for chain elongation. When a ddNTP is incorporated into a growing DNA strand during synthesis, it terminates the chain, as no further nucleotides can be added.
   - The process involves four separate reactions, each containing one of the four ddNTPs (ddATP, ddTTP, ddCTP, or ddGTP) in addition to the standard deoxynucleotide triphosphates (dNTPs).
   - As DNA polymerase extends the primer, it randomly incorporates ddNTPs, creating fragments of varying lengths, each ending at a specific nucleotide. The fragments are then separated by size using gel or capillary electrophoresis, and the sequence is read based on the terminal ddNTPs.
2. **Methodology**
   - **Primer Annealing**: A primer binds to the DNA template, providing a starting point for DNA synthesis.

- **Chain Termination and Fragment Generation**: In each reaction mixture, DNA polymerase incorporates nucleotides, terminating randomly at each base depending on the ddNTP added.
- **Electrophoresis and Detection**: The resulting fragments are separated by electrophoresis. Early versions used radiolabeled ddNTPs for detection, but modern methods use fluorescently labeled ddNTPs, enabling automated detection.

3. **Applications of Sanger Sequencing**
   - **Small-Scale Sequencing**: Sanger sequencing is still used for sequencing short DNA regions, such as in PCR products, plasmids, and individual genes.
   - **Clinical and Diagnostic Applications**: Due to its accuracy, Sanger sequencing is used in clinical diagnostics for confirming mutations, verifying NGS data, and sequencing specific regions associated with genetic diseases.
   - **Molecular Cloning and Validation**: Sanger sequencing remains valuable for verifying cloned DNA sequences in molecular biology research.
4. **Limitations**
   - Sanger sequencing is relatively low-throughput and expensive for large-scale projects, limiting its use in whole-genome sequencing.
   - The method can only sequence up to 1000 base pairs per reaction, making it impractical for sequencing large genomes quickly.

Next-Generation Sequencing (NGS)

Next-generation sequencing (NGS), also known as high-throughput sequencing, represents a series of advanced sequencing technologies that enable massively parallel sequencing of millions of DNA fragments simultaneously. NGS has transformed genomics by making it feasible to sequence entire genomes quickly and at a fraction of the cost of Sanger sequencing.

1. **Principles of Next-Generation Sequencing**
   - Unlike Sanger sequencing, NGS does not rely on chain termination. Instead, DNA is fragmented, adapters are attached, and millions of fragments are amplified in parallel. Each fragment is sequenced by synthesizing complementary strands in real-time, with the order of nucleotides determined by detecting fluorescent signals or pH changes.
   - NGS technologies employ various sequencing-by-synthesis or sequencing-by-ligation methods, depending on the platform (e.g., Illumina, Ion Torrent, Pacific Biosciences).
2. **Common NGS Platforms and Methods**
   - **Illumina Sequencing (Sequencing by Synthesis)**: Illumina sequencing is based on sequencing by synthesis, where fluorescently labeled nucleotides are added one at a time. Each nucleotide incorporation is detected via fluorescence, allowing sequence determination.

- **Ion Torrent Sequencing**: Ion Torrent detects nucleotide incorporation by measuring changes in pH. Each incorporated nucleotide releases a hydrogen ion, creating a measurable pH change. This method is faster but may be less accurate for homopolymers.
- **Single-Molecule Real-Time (SMRT) Sequencing (Pacific Biosciences)**: SMRT sequencing reads long DNA molecules in real time without amplification. It offers high accuracy and long read lengths, which is useful for complex genome structures and repetitive regions.
- **Nanopore Sequencing (Oxford Nanopore)**: Nanopore sequencing reads long DNA molecules by passing them through a protein nanopore embedded in a membrane. Each nucleotide causes a characteristic disruption in electrical current, allowing real-time base identification. Nanopore sequencing offers ultra-long reads but may have lower accuracy.

3. **Advantages of NGS**
   - **High-Throughput and Scalability**: NGS can sequence millions of DNA fragments simultaneously, making it suitable for whole-genome sequencing, transcriptomics, metagenomics, and large-scale studies.
   - **Cost-Effectiveness**: NGS is more cost-effective than Sanger sequencing for large-scale projects, enabling population-scale studies and personalized genomics.
   - **Fast Turnaround Time**: NGS allows entire genomes to be sequenced in days rather than months, accelerating research and clinical applications.
4. **Applications of NGS**
   - **Whole-Genome Sequencing**: NGS enables the sequencing of entire genomes, providing comprehensive data on genetic variations, structural changes, and mutations.
   - **Transcriptome Analysis (RNA-Seq)**: RNA sequencing with NGS allows researchers to analyze gene expression, alternative splicing, and non-coding RNAs across different conditions.
   - **Targeted Sequencing and Exome Sequencing**: NGS can be used to sequence specific regions or exomes (protein-coding regions), enabling cost-effective analysis of genes associated with diseases.
   - **Microbiome and Metagenomic Studies**: NGS allows for the study of microbial communities in environmental and human samples, revealing insights into diversity, function, and interactions of microbes.
   - **Cancer Genomics and Precision Medicine**: NGS is used to identify mutations, biomarkers, and drug targets in cancer, enabling personalized treatments based on genetic profiles.

Third-Generation Sequencing (TGS) and Beyond

Third-generation sequencing (TGS) platforms, such as Pacific Biosciences and Oxford Nanopore Technologies, have introduced methods that enable real-time, single-molecule sequencing with long read lengths. TGS has the potential to overcome some of the limitations of NGS by providing longer reads, allowing for improved assembly of complex genomes and more accurate detection of structural variants.

1. **Single-Molecule Real-Time (SMRT) Sequencing**
    - **Principle**: SMRT sequencing from Pacific Biosciences captures the incorporation of nucleotides in real-time as DNA polymerase synthesizes a complementary strand. Fluorescently labeled nucleotides are incorporated, and each incorporation is recorded instantly, providing long, continuous reads.
    - **Advantages**: SMRT sequencing allows for long read lengths (up to tens of thousands of base pairs) with high accuracy. It is particularly useful for sequencing repetitive or highly homologous regions and identifying structural variants.
    - **Applications**: SMRT sequencing is used in de novo genome assembly, structural variant detection, and transcriptome analysis of complex genes with multiple isoforms.
2. **Nanopore Sequencing**
    - **Principle**: Nanopore sequencing involves passing a DNA strand through a nanopore embedded in a membrane. Each base creates a unique disruption in the ionic current across the membrane, which is detected and translated into nucleotide sequences in real time.
    - **Advantages**: Nanopore sequencing can generate ultra-long reads (up to hundreds of thousands of base pairs) and enables real-time data acquisition. It can sequence DNA, RNA, and modified nucleotides without PCR amplification.
    - **Applications**: Nanopore sequencing is widely used in rapid pathogen identification, structural variant detection, and field-based sequencing applications, such as monitoring infectious disease outbreaks.

Comparison of Sequencing Technologies

1. **Accuracy and Read Length**
    - **Sanger Sequencing**: High accuracy with short read lengths (up to ~1000 bp).
    - **NGS (Illumina, Ion Torrent)**: High-throughput, short read lengths (typically 50-300 bp), and cost-effective for large-scale projects.
    - **TGS (SMRT, Nanopore)**: Variable accuracy but offers long read lengths, with potential applications in complex genome assembly and structural variant analysis.
2. **Throughput and Cost**
    - **Sanger Sequencing**: Low throughput and higher cost per base for large projects.
    - **NGS**: High throughput with low cost per base, ideal for whole-genome sequencing, RNA-Seq, and population studies.
    - **TGS**: Moderate to high cost per base but advantageous for specific applications requiring long read lengths.

Applications and Impact of DNA Sequencing

1. **Genomics and Personalized Medicine**
   - DNA sequencing enables the identification of genetic variations and mutations associated with diseases, paving the way for precision medicine, where treatments are tailored based on individual genetic profiles.
2. **Evolutionary Biology and Ecology**
   - Sequencing helps in studying evolutionary relationships, genetic diversity, and population genetics. NGS and TGS are widely used to analyze genomes of various species, understand biodiversity, and monitor ecosystems.
3. **Infectious Disease Surveillance**
   - DNA sequencing is critical in identifying pathogens, tracking viral mutations (e.g., COVID-19), and understanding pathogen evolution and transmission dynamics.
4. **Cancer Genomics**
   - Sequencing technologies have transformed cancer research by enabling comprehensive analysis of cancer genomes. This allows the identification of mutations, structural rearrangements, and epigenetic changes, supporting targeted therapy development.
5. **Agriculture and Biotechnology**
   - In agriculture, sequencing is used for crop improvement, identifying genes for disease resistance, and understanding plant genetics. In biotechnology, sequencing aids in developing genetically modified organisms (GMOs) and understanding metabolic pathways.

4.5 RNA Analysis and cDNA Synthesis

RNA analysis and complementary DNA (cDNA) synthesis are essential techniques in molecular biology, allowing scientists to study gene expression, RNA structure, and function. RNA analysis provides insights into transcriptional activity, alternative splicing, and the regulation of gene expression, while cDNA synthesis enables the conversion of RNA into a stable DNA form that can be amplified, sequenced, or used in various downstream applications.

Overview of RNA Analysis

RNA analysis is crucial for studying gene expression, understanding cellular responses to environmental changes, and identifying transcripts of interest. Unlike DNA, RNA is a dynamic molecule that reflects real-time gene expression patterns within cells, making it a valuable indicator of cellular states.

1. **Types of RNA**
   - **mRNA (Messenger RNA)**: mRNA is transcribed from DNA and serves as the template for protein synthesis. Analyzing mRNA levels provides insights into gene expression patterns.
   - **rRNA (Ribosomal RNA)**: rRNA is a component of ribosomes, which catalyze protein synthesis. Although it is highly abundant, rRNA is generally not analyzed in gene expression studies.

- **tRNA (Transfer RNA)**: tRNA delivers amino acids to ribosomes during translation. While not typically analyzed for expression, tRNA abundance and modifications are studied in specialized contexts.
- **Non-Coding RNAs (ncRNAs)**: This category includes various regulatory RNAs, such as microRNAs (miRNAs), small interfering RNAs (siRNAs), and long non-coding RNAs (lncRNAs), which play roles in gene silencing and regulation.

2. **Importance of RNA Analysis**
    - **Gene Expression Profiling**: RNA analysis allows researchers to determine which genes are active and to what extent, under specific conditions or in different cell types.
    - **Alternative Splicing**: RNA analysis helps identify alternatively spliced isoforms, which increase protein diversity and allow cells to respond to environmental and developmental cues.
    - **RNA Processing and Stability**: Studying RNA provides information on post-transcriptional modifications, such as splicing, polyadenylation, and RNA degradation, all of which affect gene expression and stability.
3. **RNA Isolation Techniques**
    - RNA isolation is the first step in RNA analysis and must yield high-quality RNA free of DNA contamination. Methods for RNA isolation include:
        - ◊ **Organic Extraction (e.g., TRIzol)**: TRIzol or phenol-chloroform extraction is a widely used method for isolating RNA, separating RNA from DNA and proteins through phase separation.
        - ◊ **Silica Column-Based Kits**: Commercial kits often use silica-based columns to selectively bind RNA, allowing for quick and efficient RNA isolation.
        - ◊ **Magnetic Bead-Based Isolation**: This method uses magnetic beads coated with RNA-binding material, providing high-throughput and automated RNA isolation.
4. **RNA Quality and Integrity Assessment**
    - The quality of isolated RNA is essential for accurate downstream applications. RNA integrity is often assessed by measuring the 28S and 18S rRNA bands on an agarose gel or using an automated system, such as an Agilent Bioanalyzer, which provides an RNA Integrity Number (RIN) score.
    - The **A260/A280 ratio** is also measured to check for protein contamination, with a ratio around 2.0 indicating pure RNA.

cDNA Synthesis

Complementary DNA (cDNA) synthesis is the process of converting RNA into a DNA form that is more stable and suitable for techniques like PCR, sequencing, and cloning. cDNA synthesis is critical for studying mRNA and gene expression patterns because DNA-based techniques cannot directly amplify or analyze RNA.

1. **Principles of cDNA Synthesis**
   - cDNA is synthesized using an enzyme called **reverse transcriptase**, which transcribes RNA into complementary DNA. Reverse transcriptase, originally discovered in retroviruses, enables researchers to study RNA by converting it into DNA.
   - cDNA synthesis usually focuses on mRNA because it reflects gene expression patterns. Since mRNA lacks introns, cDNA also lacks intronic sequences, providing a representation of the coding regions of genes.
2. **cDNA Synthesis Process**
   - **Primer Selection**: The choice of primer affects which RNA sequences are converted into cDNA:
     - ◊ **Oligo(dT) Primers**: These primers bind to the poly-A tail of eukaryotic mRNA, allowing for selective cDNA synthesis from mRNA. This is particularly useful when studying overall gene expression.
     - ◊ **Random Hexamers**: Short, random sequences that bind throughout the RNA molecule, enabling synthesis of cDNA from all RNA types, including rRNA and fragmented RNA.
     - ◊ **Gene-Specific Primers**: These primers target specific RNA sequences, allowing cDNA synthesis from particular transcripts.
   - **Reverse Transcription Reaction**: The RNA sample, along with primers and reverse transcriptase, is incubated to produce a single-stranded cDNA copy. The reaction is often followed by a step to degrade the original RNA template.
   - **Second-Strand Synthesis (Optional)**: In some applications, a second DNA strand is synthesized to create double-stranded cDNA, which is useful for cloning and certain sequencing methods.
3. **Considerations in cDNA Synthesis**
   - **RNA Quality**: High-quality, intact RNA is essential for successful cDNA synthesis. Degraded RNA will produce incomplete or fragmented cDNA, impacting the accuracy of downstream analyses.
   - **Primer Choice**: The primer selection depends on the RNA source and the goals of the experiment. For example, oligo(dT) primers are preferred for mRNA analysis, while random hexamers are suitable for degraded or fragmented RNA.
   - **Reverse Transcriptase Choice**: Different reverse transcriptases vary in their fidelity, processivity, and ability to transcribe structured RNA. High-fidelity enzymes are used in applications requiring accurate transcript representation.

Applications of RNA Analysis and cDNA Synthesis

RNA analysis and cDNA synthesis are widely used across molecular biology, genetics, medicine, and biotechnology.

1. **Gene Expression Analysis**
   - RNA analysis, particularly with cDNA synthesized from mRNA, enables researchers to quantify gene expression across different tissues, stages of development, or conditions. Techniques such as **quantitative PCR (qPCR)** and **RNA sequencing (RNA-Seq)** provide precise measurements of transcript abundance, enabling comparative gene expression studies.
2. **Transcriptome Profiling**
   - Transcriptome profiling allows scientists to examine the entire set of RNA molecules in a cell or tissue, capturing coding and non-coding RNAs. RNA-Seq has revolutionized transcriptome analysis by enabling high-throughput, quantitative assessment of gene expression, alternative splicing, and novel transcript discovery.
3. **Alternative Splicing and Isoform Analysis**
   - RNA analysis and cDNA synthesis reveal alternative splicing events, producing different mRNA isoforms from the same gene. Studying these isoforms is crucial for understanding functional diversity and regulatory mechanisms in gene expression.
4. **Diagnostics and Biomarker Discovery**
   - RNA analysis plays a critical role in diagnostics, as certain RNA molecules (e.g., mRNA, miRNA) can serve as biomarkers for diseases, including cancer, infectious diseases, and genetic disorders. qPCR and RNA-Seq are used to detect and quantify these biomarkers, providing information for personalized treatment approaches.
5. **Viral Detection and Characterization**
   - Since many viruses contain RNA genomes, cDNA synthesis followed by PCR or sequencing is widely used to detect and characterize viral RNA, which is crucial in epidemiology, diagnostics, and monitoring viral mutations, as seen in the study of SARS-CoV-2.

Techniques in RNA Analysis

RNA analysis encompasses several techniques that allow for the detailed examination of RNA molecules, gene expression levels, and transcript diversity.

1. **Quantitative PCR (qPCR)**
   - qPCR, also known as real-time PCR, is one of the most commonly used methods for quantifying gene expression. In qPCR, cDNA synthesized from RNA serves as a template, and fluorescence is used to measure the amount of DNA produced during each cycle.
   - qPCR provides a precise and sensitive measure of gene expression and is often used to validate RNA-Seq data, detect low-abundance transcripts, and monitor expression changes in response to various treatments or conditions.
2. **RNA Sequencing (RNA-Seq)**
   - RNA-Seq is a high-throughput sequencing method that provides comprehensive

information on transcript abundance, structure, and alternative splicing. cDNA synthesized from RNA is sequenced to reveal the complete transcriptome, allowing for the discovery of novel transcripts and non-coding RNAs.

- RNA-Seq can measure expression levels across thousands of genes simultaneously, providing quantitative data and enabling comparisons between different experimental conditions.

3. **Northern Blotting**
   - Northern blotting is a traditional method for detecting specific RNA sequences. RNA is separated by electrophoresis, transferred to a membrane, and hybridized with a labeled probe complementary to the target RNA. Northern blotting is valuable for studying RNA size, abundance, and alternative splicing events.
4. **In Situ Hybridization (ISH)**
   - ISH allows for the visualization of RNA expression within tissue sections or whole organisms. This technique uses labeled probes that bind to specific RNA sequences, providing spatial information about gene expression patterns in a biological context.
   - ISH is used in developmental biology and pathology to study gene expression within the structural context of tissues.
5. **Microarray Analysis**
   - Microarrays consist of thousands of probes immobilized on a solid surface, each representing a specific RNA sequence. RNA is converted to cDNA, labeled, and hybridized to the microarray. This technique provides a high-throughput means of measuring gene expression for large numbers of genes simultaneously.
   - While RNA-Seq has largely replaced microarrays, they are still useful in certain applications where cost-effectiveness and established platforms are important.

Challenges in RNA Analysis and cDNA Synthesis

1. **RNA Stability and Degradation**
   - RNA is more prone to degradation than DNA due to its single-stranded nature and the ubiquitous presence of RNases, which can degrade RNA rapidly. Working with RNA requires stringent laboratory practices, including the use of RNase-free reagents and tools.
2. **cDNA Synthesis Efficiency**
   - Reverse transcription efficiency can vary based on enzyme fidelity, RNA quality, and primer choice. Low efficiency may lead to incomplete or biased representation of transcripts, affecting downstream analyses.
3. **Data Interpretation in RNA Analysis**
   - Gene expression data can be complex to interpret due to the dynamic nature of RNA expression. Biological variability, experimental conditions, and technical limitations may introduce variability, necessitating rigorous experimental design and statistical analysis.

4.6 Protein Analysis and Western Blotting Techniques

Protein analysis is a core component of molecular biology, biochemistry, and cell biology, allowing researchers to investigate protein expression, structure, function, and interactions. Western blotting, a widely used technique in protein analysis, enables the detection and quantification of specific proteins within complex samples. By using antibodies that specifically bind to target proteins, Western blotting provides insights into protein expression levels, post-translational modifications, and other aspects of protein biology.

Overview of Protein Analysis

Protein analysis encompasses a range of techniques that investigate various properties of proteins, from their expression levels and structure to their interactions with other molecules. Understanding these properties is crucial for studying cellular processes, disease mechanisms, and therapeutic targets.

1. **Purpose of Protein Analysis**
    - **Expression Analysis**: Measuring protein levels across different conditions (e.g., treated vs. untreated samples, healthy vs. diseased tissues) provides insight into cellular responses and regulatory mechanisms.
    - **Structural Analysis**: Determining the three-dimensional structure of proteins is essential for understanding their function and for rational drug design.
    - **Functional Analysis**: Studying the biological activities of proteins, such as enzyme kinetics, binding properties, and interactions, is key to understanding cellular pathways and mechanisms.
    - **Modification Analysis**: Post-translational modifications, including phosphorylation, glycosylation, and ubiquitination, affect protein function, stability, and localization. Identifying these modifications is crucial for understanding protein regulation.
2. **Common Techniques in Protein Analysis**
    - **Protein Purification**: Techniques such as affinity chromatography, ion exchange chromatography, and gel filtration are used to isolate proteins from complex mixtures, allowing for detailed analysis.
    - **SDS-PAGE (Sodium Dodecyl Sulfate Polyacrylamide Gel Electrophoresis)**: SDS-PAGE is commonly used to separate proteins based on their molecular weight. Proteins are denatured with SDS, an anionic detergent, and then separated in a polyacrylamide gel under an electric field.
    - **Mass Spectrometry**: This technique identifies and quantifies proteins and their post-translational modifications by measuring the mass-to-charge ratio of peptide fragments.
    - **ELISA (Enzyme-Linked Immunosorbent Assay)**: ELISA uses antibodies to detect and quantify specific proteins within a sample, often used in diagnostic and research applications.
    - **Western Blotting**: A targeted technique for detecting specific proteins in a complex sample, Western blotting combines electrophoresis with antibody-based detection to provide both qualitative and quantitative information.

Western Blotting: Principle and Methodology

Western blotting, also known as immunoblotting, is a protein analysis technique that detects specific proteins within a sample. The process involves separating proteins by gel electrophoresis, transferring them to a membrane, and probing with antibodies specific to the target protein.

1. **Principle of Western Blotting**
   - Western blotting relies on the principle of antibody-antigen specificity. Following protein separation by electrophoresis, the proteins are transferred to a membrane (typically made of nitrocellulose or polyvinylidene difluoride, PVDF). The membrane is then incubated with an antibody that binds specifically to the protein of interest. Detection is achieved by using a secondary antibody conjugated to a reporter enzyme, enabling visualization of the target protein.
2. **Steps in Western Blotting**
   - **Sample Preparation**: Proteins are extracted from cells or tissues using lysis buffers that often contain detergents, salts, and protease inhibitors to maintain protein integrity.
   - **Gel Electrophoresis (SDS-PAGE)**: The protein mixture is loaded onto an SDS-PAGE gel, where proteins are separated based on size. SDS, a detergent, denatures the proteins and imparts a uniform negative charge, ensuring that migration through the gel is size-dependent.
   - **Transfer to Membrane**: After electrophoresis, proteins are transferred from the gel to a membrane. This transfer process, usually done through electroblotting, enables proteins to bind to the membrane surface for antibody probing.
   - **Blocking**: The membrane is blocked with a protein solution (e.g., bovine serum albumin or non-fat milk) to prevent non-specific antibody binding. Blocking reduces background signal, improving the specificity of detection.
   - **Primary Antibody Incubation**: The membrane is incubated with a primary antibody specific to the target protein. This antibody binds directly to the protein of interest.
   - **Secondary Antibody Incubation**: After washing off unbound primary antibody, the membrane is incubated with a secondary antibody. The secondary antibody is conjugated to a detection enzyme, such as horseradish peroxidase (HRP) or alkaline phosphatase (AP), which amplifies the signal.
   - **Detection**: The target protein is visualized by adding a substrate for the enzyme conjugated to the secondary antibody. The enzyme catalyzes a reaction that produces a detectable signal, often chemiluminescent or colorimetric, indicating the presence of the protein.
3. **Detection Methods in Western Blotting**
   - **Chemiluminescence**: In chemiluminescent detection, the enzyme on the secondary antibody catalyzes a reaction that emits light, which is captured on photographic film or a digital imaging system.
   - **Colorimetric Detection**: In this method, the enzyme catalyzes a color-producing reaction, and bands are visually observed as colored bands on the membrane.

- **Fluorescent Detection**: Fluorescently labeled antibodies allow detection using a fluorescence imaging system. This method provides high sensitivity and allows for multiplexing (detecting multiple proteins on the same membrane).

Applications of Western Blotting

Western blotting has become a standard technique in molecular biology, biochemistry, and medical research due to its specificity and versatility. It is widely used for detecting and analyzing proteins in research, diagnostics, and clinical settings.

1. **Protein Expression Analysis**
   - Western blotting enables researchers to quantify protein expression levels under various conditions. By comparing protein abundance across experimental groups, researchers can determine how treatments, environmental factors, or genetic mutations affect protein expression.
2. **Detection of Post-Translational Modifications**
   - Western blotting is used to study protein modifications, such as phosphorylation, acetylation, and ubiquitination. By using modification-specific antibodies, researchers can assess changes in protein activity, stability, and localization.
3. **Disease Diagnostics**
   - Western blotting is used in the clinical diagnosis of diseases, particularly for identifying specific protein markers or detecting the presence of antibodies against pathogens. For example, Western blotting is a confirmatory test for HIV detection following an initial ELISA test.
   - The technique is also used in diagnosing Lyme disease by detecting antibodies against *Borrelia burgdorferi*.
4. **Validation of Genetic and Proteomic Studies**
   - Western blotting is often used to validate findings from genomic and proteomic studies, such as gene expression analysis and mass spectrometry. It confirms the presence and relative abundance of proteins of interest, ensuring data reliability.
5. **Investigating Cellular Responses and Pathways**
   - By analyzing proteins involved in specific cellular pathways, researchers can investigate signal transduction, apoptosis, cell cycle regulation, and immune responses. Western blotting allows for the targeted analysis of pathway components, providing insights into complex cellular processes.

Advantages and Limitations of Western Blotting

1. **Advantages**
   - **Specificity**: Western blotting offers high specificity, as it relies on antibody-antigen interactions, allowing for the detection of individual proteins in complex mixtures.

- **Sensitivity**: Enhanced by secondary antibodies and enzyme reactions, Western blotting can detect proteins at low concentrations, making it suitable for detecting scarce proteins.
- **Quantitative Potential**: By using densitometry or image analysis software, Western blotting can provide semi-quantitative data on protein expression levels.
- **Versatility**: Western blotting is adaptable for a wide range of proteins and can be customized by using different antibodies, detection methods, and imaging systems.

2. **Limitations**
    - **Time-Consuming**: Western blotting involves multiple steps, including electrophoresis, transfer, and antibody incubation, which can take several hours to days to complete.
    - **Antibody Dependency**: The quality and specificity of antibodies are critical for accurate detection. Poor-quality or non-specific antibodies can lead to false results.
    - **Semi-Quantitative**: While Western blotting provides quantitative data, it is semi-quantitative at best. Precise quantification often requires additional techniques or calibration standards.
    - **Sample Loss and Variability**: Sample loss can occur during the transfer step, and variations in experimental conditions may affect reproducibility. Consistency in protocol and technique is crucial for reliable results.

Complementary Techniques in Protein Analysis

Western blotting is often used alongside other protein analysis techniques to provide a comprehensive understanding of protein characteristics, expression, and interactions.

1. **Mass Spectrometry**
    - Mass spectrometry complements Western blotting by identifying proteins and quantifying them with high accuracy. It also provides detailed information on post-translational modifications.
2. **Immunoprecipitation (IP)**
    - Immunoprecipitation enables the isolation of a specific protein or protein complex from a mixture using antibodies. Western blotting can then be used to analyze the immunoprecipitated proteins, enabling the study of protein-protein interactions.
3. **ELISA (Enzyme-Linked Immunosorbent Assay)**
    - ELISA quantifies proteins in solution, providing a more quantitative approach than Western blotting. It is widely used in diagnostic settings for detecting antigens or antibodies in serum samples.
4. **Co-Immunoprecipitation (Co-IP)**
    - Co-IP allows researchers to study protein-protein interactions by isolating complexes containing the target protein and its binding partners. Western blotting can verify the presence of specific proteins in these complexes.

5. **Flow Cytometry**
    - Flow cytometry enables the quantification of proteins on the cell surface or inside cells by labeling with fluorescently conjugated antibodies. This technique is useful for single-cell analysis, assessing cell populations, and identifying cell subtypes.

4.7 Advanced Molecular Techniques: CRISPR, Gene Editing, and Proteomics

Advances in molecular biology have led to groundbreaking techniques that allow precise manipulation of genetic material and comprehensive analysis of proteins. CRISPR (Clustered Regularly Interspaced Short Palindromic Repeats) and other gene-editing tools have revolutionized gene manipulation, enabling targeted modifications in DNA with high precision. Meanwhile, proteomics, the large-scale study of proteins, has expanded our understanding of protein dynamics, interactions, and functions within biological systems. These techniques have broad applications in research, medicine, agriculture, and biotechnology, pushing the boundaries of genetic and protein analysis.

CRISPR and Gene Editing

CRISPR technology, based on a natural defense mechanism found in bacteria, allows scientists to edit genes with unprecedented precision, efficiency, and flexibility. CRISPR has become one of the most powerful tools in molecular biology and genetics for modifying the genome and studying gene function.

1. **Principle of CRISPR-Cas9**
    - The CRISPR-Cas9 system consists of two main components: the **Cas9 protein** (an RNA-guided DNA endonuclease) and a **guide RNA (gRNA)** that directs Cas9 to a specific DNA sequence. The guide RNA has two regions: a spacer sequence that matches the target DNA sequence and a scaffold sequence that binds to Cas9.
    - Once the Cas9-gRNA complex locates and binds to the target sequence, Cas9 introduces a double-strand break (DSB) in the DNA. The cell repairs this break through **non-homologous end joining (NHEJ)** or **homology-directed repair (HDR)**, resulting in targeted gene disruption, insertion, or correction.
2. **Methodology of CRISPR Gene Editing**
    - **Designing the gRNA**: A gRNA is designed to match the target sequence. Software tools help in designing gRNAs with minimal off-target effects, increasing the precision of CRISPR editing.
    - **Delivery of CRISPR Components**: The CRISPR-Cas9 system can be introduced into cells via plasmids, viral vectors, or ribonucleoprotein (RNP) complexes. Delivery method depends on the cell type and editing purpose.
    - **Gene Editing and Repair**: After the Cas9-induced DSB, the cell's repair machinery attempts to fix the break. NHEJ, an error-prone process, often introduces insertions or deletions (indels), disrupting the gene. HDR, a more accurate repair process, can introduce a desired DNA sequence using a donor template.

3. **Applications of CRISPR in Molecular Biology**
   - **Gene Knockout and Functional Studies**: CRISPR is widely used to knock out specific genes, allowing researchers to study gene function by observing the resulting phenotypic changes.
   - **Disease Modeling**: CRISPR enables the creation of animal and cell models of genetic diseases, facilitating the study of disease mechanisms and the development of new treatments.
   - **Gene Therapy**: CRISPR holds potential for therapeutic applications by correcting disease-causing mutations in genetic disorders such as cystic fibrosis, sickle cell anemia, and muscular dystrophy.
   - **Agriculture and Biotechnology**: CRISPR is used to enhance crop traits, such as resistance to pests and environmental stress, and to improve livestock genetics, making it a valuable tool in agricultural biotechnology.
4. **Limitations and Ethical Considerations**
   - **Off-Target Effects**: CRISPR can sometimes edit unintended sites in the genome, potentially leading to undesired mutations. Improving gRNA design and Cas9 specificity is crucial for minimizing these effects.
   - **Ethical Concerns**: Gene editing, especially in human embryos and germline cells, raises ethical questions. The potential for "designer babies" and unintended ecological impacts are ongoing concerns in the field.

Other Gene Editing Technologies

While CRISPR is the most prominent gene-editing tool, other technologies are also used for precise genetic manipulation.

1. **TALENs (Transcription Activator-Like Effector Nucleases)**
   - TALENs consist of DNA-binding domains derived from transcription activators fused to a nuclease domain. Each TALEN can be designed to recognize specific DNA sequences, enabling precise editing. TALENs are especially useful in editing specific, short sequences but are more complex to design compared to CRISPR.
2. **Zinc Finger Nucleases (ZFNs)**
   - ZFNs are engineered proteins containing DNA-binding zinc finger domains linked to a nuclease. By combining multiple zinc finger domains, researchers can target specific DNA sequences. ZFNs have been used in gene therapy applications but are generally less flexible and more challenging to construct than CRISPR.
3. **Base Editing and Prime Editing**
   - **Base Editing**: Base editing is a precision editing method that converts a single DNA base into another without creating a double-strand break. It is particularly useful for correcting point mutations.

- **Prime Editing**: Prime editing introduces targeted edits with high precision, including small insertions, deletions, and base substitutions. Prime editors combine a reverse transcriptase with Cas9 nickase, guided by a prime editing guide RNA (pegRNA).

Proteomics: The Study of Proteins at a Large Scale

Proteomics is the large-scale study of proteins, the functional molecules in cells. Proteomics analyzes the expression, structure, functions, modifications, and interactions of proteins within a cell or tissue, providing insights into cellular processes and responses to various stimuli.

1. **Principles of Proteomics**
    - Proteomics aims to catalog and quantify all proteins within a biological system under specific conditions. The proteome is highly dynamic and changes in response to cellular signals, environmental changes, disease states, and developmental stages.
    - Since proteins are the primary effectors of biological functions, proteomics complements genomics and transcriptomics by providing a direct measurement of cellular activity.
2. **Types of Proteomics**
    - **Expression Proteomics**: Studies protein expression levels across different conditions, allowing researchers to identify proteins that are up- or down-regulated in response to a stimulus.
    - **Structural Proteomics**: Investigates the three-dimensional structure of proteins and protein complexes to understand how structural variations influence protein function.
    - **Functional Proteomics**: Examines protein interactions, modifications, and activity, providing insights into protein roles within biological pathways and networks.
3. **Proteomics Techniques**
    - **Mass Spectrometry (MS)**: MS is the cornerstone of proteomics. It identifies proteins and post-translational modifications by measuring the mass-to-charge ratio of peptide fragments. Techniques such as tandem MS (MS/MS) improve identification and sequencing of peptides.
    - **Two-Dimensional Gel Electrophoresis (2D-GE)**: 2D-GE separates proteins based on isoelectric point and molecular weight, allowing visualization of protein profiles. It is often followed by MS for protein identification.
    - **Liquid Chromatography-Mass Spectrometry (LC-MS)**: LC-MS combines liquid chromatography with MS to separate proteins based on hydrophobicity before analysis, providing greater resolution and sensitivity.
    - **Protein Microarrays**: Protein microarrays analyze thousands of proteins simultaneously, enabling high-throughput screening of protein interactions, enzyme activity, and biomarker discovery.
4. **Applications of Proteomics**
    - **Biomarker Discovery**: Proteomics identifies protein biomarkers associated with diseases,

aiding in early diagnosis and personalized treatment. Biomarkers can indicate disease presence, progression, or response to therapy.

- **Drug Development**: By studying the interactions between proteins and potential drug candidates, proteomics accelerates drug discovery and helps identify therapeutic targets.
- **Systems Biology**: Proteomics contributes to systems biology by mapping protein networks and pathways, enhancing our understanding of complex cellular processes.
- **Clinical Proteomics**: Proteomic profiling of patient samples helps identify disease mechanisms, monitor treatment effects, and develop individualized treatment strategies.

5. **Challenges in Proteomics**
   - **Complexity and Variability**: The proteome is more complex than the genome due to alternative splicing, post-translational modifications, and protein isoforms. This complexity requires advanced methods for comprehensive analysis.
   - **Data Analysis**: Proteomics generates large datasets that require sophisticated bioinformatics tools for accurate analysis and interpretation.
   - **Sensitivity and Specificity**: Detecting low-abundance proteins and distinguishing them from similar molecules is challenging, necessitating highly sensitive instruments and optimized protocols.

Integration of Gene Editing and Proteomics in Molecular Biology

The integration of gene editing technologies, like CRISPR, with proteomics has opened new avenues for understanding and manipulating cellular functions. By editing genes and then studying the resulting changes in the proteome, researchers can gain insights into gene function and protein interactions, accelerating the discovery of therapeutic targets and the development of novel treatments.

1. **Functional Genomics and Gene-Protein Interaction Studies**
   - By using CRISPR to knock out or modify genes and analyzing the proteomic changes that follow, scientists can map gene functions and protein interactions. This approach allows researchers to link genes to specific cellular pathways and biological processes.
2. **Disease Modeling and Drug Discovery**
   - Gene editing combined with proteomics enables precise modeling of genetic diseases at the molecular level, facilitating the identification of protein biomarkers and therapeutic targets. For instance, editing a gene linked to cancer and analyzing proteomic changes can reveal potential points for therapeutic intervention.
3. **Personalized Medicine**
   - CRISPR and proteomics together can help tailor treatments based on an individual's genetic and proteomic profiles. By understanding how specific gene edits influence protein expression, clinicians can develop more effective, individualized therapies.
4. **Synthetic Biology and Biotechnology**
   - In synthetic biology, CRISPR and proteomics are used to design and construct organisms

with new or enhanced functions. Gene editing can engineer metabolic pathways, while proteomics monitors protein production, stability, and functionality, ensuring the desired outcomes.

5. Applications and Impact of Molecular Biology

5.1 Molecular Biology in Genetics and Genomics

Molecular biology has revolutionized genetics and genomics by providing tools that enable the exploration, manipulation, and understanding of genetic material at a fundamental level. Advances in molecular biology techniques, including DNA sequencing, gene editing, and functional genomics, have allowed scientists to investigate the genetic basis of traits, diseases, and evolutionary relationships. These advancements have far-reaching implications for medicine, agriculture, ecology, and other fields.

The Role of Molecular Biology in Genetics

1. **Understanding Genetic Variation**
   - Molecular biology techniques, such as PCR and DNA sequencing, allow scientists to study genetic variation within and between populations. Genetic variation underpins traits, susceptibility to diseases, and responses to environmental changes.
   - **Applications**: Identifying genetic markers associated with diseases or traits enables precision breeding in agriculture and personalized medicine in human healthcare. Additionally, understanding genetic diversity helps conserve biodiversity by identifying distinct populations and endangered species.
2. **Mapping Genes and Genomes**
   - Molecular biology has enabled comprehensive mapping of genomes through techniques like next-generation sequencing (NGS) and genome-wide association studies (GWAS). These approaches allow scientists to pinpoint the genetic basis of complex traits and understand the structure of entire genomes.
   - **Applications**: Genome mapping provides a framework for identifying disease-associated genes, understanding inheritance patterns, and improving crop traits. In medicine, mapping genetic variants aids in identifying targets for drug development and understanding genetic predispositions.
3. **Functional Genomics**
   - Functional genomics aims to characterize the roles of genes within a genome. Techniques like RNA sequencing (RNA-Seq) and CRISPR-based gene knockouts enable researchers to study gene expression and identify gene functions by observing phenotypic changes following gene modification.
   - **Applications**: Functional genomics helps in understanding gene regulation, cellular processes, and metabolic pathways. In agriculture, it is used to identify genes associated with yield, disease resistance, and environmental tolerance. In human health, it aids in identifying the functions of disease-related genes.

4. **Epigenetics and Gene Regulation**
   - Molecular biology has also provided insights into epigenetics, the study of heritable changes in gene expression without changes in DNA sequence. Techniques such as bisulfite sequencing, chromatin immunoprecipitation sequencing (ChIP-Seq), and RNA-Seq help analyze DNA methylation, histone modifications, and non-coding RNAs that regulate gene expression.
   - **Applications**: Epigenetics plays a role in development, aging, and disease. Understanding epigenetic modifications can lead to therapies for diseases linked to epigenetic dysregulation, such as cancer. In agriculture, manipulating epigenetic markers offers the potential to develop stress-resilient crops.

The Role of Molecular Biology in Genomics

1. **Whole-Genome Sequencing (WGS)**
   - Whole-genome sequencing, enabled by advancements in NGS, provides a complete view of an organism's genetic information. WGS allows for the comprehensive identification of genetic variants, structural variations, and evolutionary relationships within and between species.
   - **Applications**: WGS is used in population genetics to study diversity and evolution. It is also instrumental in identifying mutations in genetic diseases, enabling precision medicine, and facilitating conservation efforts by assessing genetic health in endangered species.
2. **Comparative Genomics**
   - Comparative genomics examines the similarities and differences in genetic material across species, helping to understand evolutionary relationships, gene functions, and adaptation mechanisms.
   - **Applications**: Comparative genomics helps identify conserved genes and regulatory elements, aiding in the study of human diseases by using model organisms. It also supports research on crop and livestock improvement by identifying genes associated with favorable traits.
3. **Genomic Data Analysis and Bioinformatics**
   - The vast amounts of data generated by sequencing projects require bioinformatics tools to analyze, store, and interpret genetic information. Bioinformatics allows researchers to detect patterns, predict gene functions, and uncover complex interactions.
   - **Applications**: In medicine, bioinformatics aids in diagnosing genetic disorders and identifying treatment targets. In agriculture, bioinformatics tools assist in marker-assisted selection, optimizing breeding programs to improve traits such as yield and resistance.
4. **Precision Medicine and Gene Therapy**
   - Molecular biology underpins precision medicine, where genetic information is used to customize treatments based on an individual's genetic profile. Gene therapy techniques, such as CRISPR-Cas9, allow for the correction of mutations that cause genetic disorders.

- **Applications**: Precision medicine is transforming cancer treatment, enabling the development of therapies that target specific genetic mutations. In inherited diseases, gene therapy offers the potential for long-lasting treatment by correcting the underlying genetic cause.

5.2 Role in Entomology and Insect Diagnosis

Molecular biology plays a critical role in entomology, providing tools to study insect genetics, ecology, evolution, and interactions with pathogens. Molecular techniques enable the identification of insect species, understanding of genetic diversity, and development of pest management strategies. These advances are essential for agriculture, environmental conservation, and controlling vector-borne diseases.

Molecular Techniques in Insect Identification and Classification

1. **DNA Barcoding**
    - DNA barcoding uses short genetic sequences from a standardized region of the genome (such as the COI gene in mitochondria) to identify species. DNA barcoding is highly effective for distinguishing closely related species, including cryptic species that are morphologically similar.
    - **Applications**: DNA barcoding is used in ecological monitoring, pest control, and conservation. It helps identify invasive species, assess biodiversity, and manage pest populations by distinguishing between harmful and beneficial insects.
2. **Phylogenetics and Evolutionary Studies**
    - Molecular biology techniques such as whole-genome sequencing and comparative genomics allow researchers to study evolutionary relationships among insect species. Phylogenetic analysis provides insights into species origins, divergence, and adaptive traits.
    - **Applications**: Understanding insect evolution is essential for pest management, as it reveals genetic adaptations to pesticides. In conservation, phylogenetic studies help prioritize species and populations for protection based on their evolutionary uniqueness.

Molecular Diagnostics in Insect-Borne Disease Control

1. **Pathogen Detection in Insects**
    - Molecular diagnostics enable the detection of pathogens within insect vectors. PCR, qPCR, and next-generation sequencing are used to identify viruses, bacteria, and parasites carried by insects that cause diseases in humans, animals, and plants.
    - **Applications**: Molecular diagnostics help monitor the spread of vector-borne diseases, such as malaria, dengue, and Zika, by detecting pathogens in mosquito populations. In agriculture, they aid in controlling plant diseases spread by insect vectors, such as the Asian citrus psyllid in citrus greening.
2. **Genetic Markers for Resistance Detection**
    - Genetic markers help identify resistance genes in insect populations, particularly those

that confer resistance to insecticides. Understanding genetic resistance enables the development of sustainable pest control strategies.

- **Applications**: Insecticide resistance management relies on molecular markers to detect resistance alleles, helping farmers rotate chemicals and implement integrated pest management (IPM) strategies effectively.

3. **Gene Editing for Pest Control**

- Gene-editing technologies, particularly CRISPR, offer potential solutions for pest control by targeting genes essential for reproduction, survival, or susceptibility to pesticides. This approach aims to reduce pest populations without harming non-target species.
- **Applications**: Gene editing is being explored to create gene drives in mosquitoes, which spread genes that reduce malaria transmission. In agriculture, gene editing in pest species could reduce crop damage by disrupting genes linked to feeding and reproduction.

Molecular Biology in Studying Insect-Plant and Insect-Pathogen Interactions

1. **Insect-Plant Interactions**

- Molecular techniques are used to study insect-plant interactions, identifying plant defense mechanisms and insect adaptations. Gene expression analysis helps in understanding how plants respond to herbivory and how insects overcome plant defenses.
- **Applications**: Understanding insect-plant interactions can inform breeding programs to develop pest-resistant crop varieties. Molecular analysis also aids in identifying secondary metabolites that deter pests or attract natural predators.

1. **Insect-Pathogen Interactions**

- Molecular biology techniques help elucidate how insects respond to pathogens and how pathogens evade or manipulate insect immune responses. Transcriptomics and proteomics provide insights into immune responses, pathogen virulence factors, and co-evolution.
- **Applications**: In vector control, understanding insect-pathogen interactions can help disrupt pathogen transmission. Insights into immune responses can aid in developing novel insecticides that compromise insect immunity, reducing vector populations and disease spread.

Application of Molecular Biology in Pest Management and Conservation

1. **Integrated Pest Management (IPM)**

- Molecular techniques enhance IPM strategies by enabling the identification of pests, natural enemies, and beneficial insects. Molecular markers also aid in tracking pest population dynamics and resistance genes.
- **Applications**: In agriculture, molecular tools support IPM by identifying pest infestations early, enabling targeted interventions. Understanding resistance gene flow within pest populations helps refine pesticide applications and reduce the risk of resistance development.

2. **Conservation of Insect Diversity**
   - Molecular biology assists in the conservation of beneficial insects, including pollinators and predators, by assessing genetic diversity and population health. DNA barcoding and genomic analysis contribute to habitat restoration and the protection of endangered species.
   - **Applications**: Conservation programs benefit from molecular tools that assess genetic diversity within insect populations. These insights support habitat restoration, prevent biodiversity loss, and ensure the survival of beneficial insects critical to ecosystems and agriculture.

5.3 Biotechnology and Genetic Engineering

Biotechnology and genetic engineering are pivotal fields in molecular biology that leverage biological systems, cells, and organisms to develop products and technologies for diverse applications in medicine, agriculture, industry, and environmental protection. Genetic engineering, a core component of biotechnology, involves manipulating DNA to modify an organism's traits or capabilities. Advances in genetic engineering techniques, such as recombinant DNA technology, CRISPR gene editing, and synthetic biology, have expanded the possibilities for creating organisms with novel characteristics, enhancing productivity, and solving complex biological problems.

Overview of Biotechnology and Genetic Engineering

1. **Biotechnology: Harnessing Biological Systems**
   - Biotechnology encompasses the use of biological systems or organisms to develop products, processes, or technologies that improve human life. Its applications range from developing pharmaceuticals and diagnostics to producing biofuels and biodegradable materials.
   - **Applications**: Biotechnology plays a critical role in medicine (e.g., creating insulin through recombinant bacteria), agriculture (e.g., pest-resistant crops), and environmental management (e.g., bioremediation using engineered microbes).
2. **Genetic Engineering: Modifying DNA for Desired Traits**
   - Genetic engineering refers to the direct manipulation of an organism's DNA to alter its characteristics. This includes adding, deleting, or editing specific genes to achieve desirable traits or to remove undesirable ones.
   - **Applications**: Genetic engineering has created crops resistant to pests, herbicides, and diseases, improved livestock traits, and developed therapeutic treatments for genetic diseases. Additionally, genetically engineered microorganisms are used to produce pharmaceuticals, biofuels, and enzymes for industrial processes.

Key Techniques in Genetic Engineering

1. **Recombinant DNA Technology**
   - Recombinant DNA technology involves combining DNA from different sources to

create a new DNA molecule, which is then introduced into an organism. This technique enables the insertion of foreign genes into bacteria, plants, or animals, leading to the expression of new proteins and traits.

- **Applications**: Recombinant DNA technology is widely used in producing human insulin, growth hormones, and other therapeutic proteins. In agriculture, recombinant DNA has led to genetically modified (GM) crops with enhanced traits like pest resistance and improved nutritional content.

2. **CRISPR-Cas9 and Gene Editing**
   - CRISPR-Cas9 is a revolutionary gene-editing tool that allows precise modifications of DNA sequences. By using guide RNA to target specific DNA sequences, Cas9 cuts the DNA at the desired location, enabling the insertion, deletion, or modification of genes.
   - **Applications**: CRISPR has potential in gene therapy for correcting genetic disorders, creating transgenic animals for research, and developing crops with improved traits. It has also been used to develop treatments for diseases such as cancer and sickle cell anemia.

3. **Synthetic Biology**
   - Synthetic biology involves designing and constructing new biological parts, devices, and systems or re-designing existing biological systems for useful purposes. This interdisciplinary field combines biology, engineering, and computational sciences to create organisms with novel functionalities.
   - **Applications**: Synthetic biology has led to the creation of microbes capable of producing biofuels, synthetic drugs, and high-value chemicals. It also contributes to biosensor development, enabling the detection of pollutants or pathogens in environmental samples.

4. **Gene Cloning**
   - Gene cloning involves copying a specific gene or DNA sequence, allowing for its analysis, manipulation, and expression in different organisms. It is a fundamental tool in genetic engineering, providing a way to study gene function and produce large quantities of specific proteins.
   - **Applications**: Gene cloning is used in research to investigate gene function, in medicine for producing therapeutic proteins, and in agriculture for developing genetically modified organisms.

Applications of Biotechnology and Genetic Engineering

1. **Medical Applications**
   - **Gene Therapy**: Gene therapy introduces functional genes into patients to treat genetic disorders, such as cystic fibrosis, hemophilia, and muscular dystrophy. It holds potential for treating a range of diseases at the genetic level.
   - **Pharmaceutical Production**: Genetically engineered microbes are used to produce therapeutic proteins, including insulin, vaccines, and monoclonal antibodies, revolutionizing treatments for various diseases.

- **Personalized Medicine**: Biotechnology enables genetic screening to identify individual genetic predispositions to diseases, facilitating the development of personalized treatments.

2. **Environmental Applications**
   - **Bioremediation**: Genetically modified microorganisms are used to break down pollutants, heavy metals, and toxins in soil and water. Bioremediation is effective for cleaning up oil spills, heavy metal contamination, and industrial waste.
   - **Biofuels**: Engineered microbes can produce biofuels from renewable resources, offering a sustainable alternative to fossil fuels. Advances in synthetic biology have enabled the production of bioethanol, biodiesel, and other biofuels.
3. **Industrial Applications**
   - **Enzyme Production**: Industrial biotechnology uses genetically engineered microbes to produce enzymes for applications in detergents, food processing, and bioenergy production.
   - **Synthetic Biology in Chemical Production**: Synthetic biology enables the design of microbes that synthesize valuable chemicals, such as biodegradable plastics, high-value pharmaceuticals, and industrial biochemicals, reducing reliance on non-renewable resources.

Ethical and Regulatory Considerations in Genetic Engineering

The rapid advancement of genetic engineering and biotechnology raises ethical and regulatory concerns, particularly regarding the safety and long-term impacts of genetically modified organisms (GMOs) on health, the environment, and biodiversity. Regulatory frameworks and ethical discussions aim to balance innovation with caution, addressing concerns such as unintended consequences, bioethics, and the equitable distribution of biotechnological benefits.

5.4 Agricultural Applications

Molecular biology, biotechnology, and genetic engineering have had profound impacts on agriculture, addressing challenges such as food security, pest control, environmental sustainability, and crop resilience. Agricultural biotechnology utilizes molecular tools to enhance crop yield, improve nutritional quality, increase resistance to pests and diseases, and reduce environmental impact, leading to more sustainable farming practices.

Crop Improvement Through Genetic Engineering

1. **Genetically Modified (GM) Crops**
   - Genetically modified crops are developed by introducing specific genes that confer desired traits, such as pest resistance, herbicide tolerance, and improved nutritional content.
   - **Examples**: *Bt* crops, such as *Bt* cotton and *Bt* corn, contain genes from *Bacillus thuringiensis* that produce insecticidal proteins, reducing the need for chemical pesticides. Herbicide-tolerant crops, like glyphosate-resistant soybeans, enable farmers to control weeds without harming the crop.

2. **Improving Nutritional Quality**
    - Genetic engineering can enhance the nutritional profile of crops by increasing levels of essential vitamins, minerals, and amino acids.
    - **Examples**: Golden Rice is engineered to produce beta-carotene, a precursor to vitamin A, addressing vitamin A deficiency in regions where rice is a staple food. Biofortification of crops with iron, zinc, and protein improves dietary quality and addresses malnutrition.
3. **Enhanced Crop Resilience**
    - Biotechnology enables the development of crops resilient to environmental stresses, including drought, salinity, and extreme temperatures. Such crops are essential for maintaining productivity in regions affected by climate change.
    - **Examples**: Drought-resistant maize and salt-tolerant rice varieties have been developed through genetic engineering, allowing farmers to cultivate crops in arid or saline soils.

Pest and Disease Management

1. **Biopesticides and Insect-Resistant Crops**
    - Biopesticides derived from natural sources, such as microbial or plant-based compounds, offer environmentally friendly pest control. Genetic engineering produces crops that express biopesticides, such as *Bt* toxin, targeting specific pests.
    - **Examples**: *Bt* crops significantly reduce pest populations without harming beneficial insects, reducing the need for synthetic insecticides and promoting environmental sustainability.
2. **Disease-Resistant Crops**
    - Genetic engineering introduces resistance genes that protect crops from bacterial, fungal, and viral diseases, minimizing yield losses and reducing reliance on chemical treatments.
    - **Examples**: Virus-resistant papaya was developed to combat the papaya ringspot virus, a devastating disease for papaya growers. Similarly, genetically engineered potatoes resist late blight, a fungal disease affecting potato crops worldwide.
3. **Integrated Pest Management (IPM)**
    - Molecular biology tools support Integrated Pest Management (IPM) strategies by providing accurate pest identification, tracking resistance genes, and monitoring pest populations. Genetic markers identify pest resistance genes, enabling targeted pesticide application and reducing chemical usage.

Environmental Sustainability and Resource Efficiency

1. **Reduced Environmental Impact**
    - GM crops reduce the need for chemical pesticides and herbicides, minimizing environmental pollution and preserving beneficial organisms. Reduced chemical usage also helps prevent soil and water contamination.

- **Examples**: *Bt* crops decrease pesticide applications, benefiting biodiversity by protecting non-target insects, pollinators, and soil microorganisms.

2. **Efficient Resource Use**

   - Crops engineered for efficiency in water, nutrient, and fertilizer use reduce the environmental footprint of agriculture. These crops promote sustainable practices and minimize resource wastage.
   - **Examples**: Nitrogen-use efficient crops reduce the need for synthetic fertilizers, decreasing runoff pollution and greenhouse gas emissions from fertilizer use.

Molecular Markers and Precision Breeding

1. **Marker-Assisted Selection (MAS)**

   - Marker-assisted selection uses molecular markers linked to desirable traits to accelerate breeding programs. Unlike traditional breeding, MAS allows breeders to select for traits at the genetic level, improving precision and efficiency.
   - **Applications**: MAS is used to breed crops with enhanced resistance to diseases, pests, and environmental stresses. It is widely applied in crops like wheat, rice, and maize to select for complex traits, such as drought tolerance and disease resistance.

2. **Genomic Selection (GS)**

   - Genomic selection is an advanced breeding technique that uses genomic data to predict an individual's performance, helping breeders select the best candidates for breeding. GS is particularly useful for complex traits influenced by multiple genes.
   - **Applications**: Genomic selection accelerates the breeding of crops and livestock with improved yield, resilience, and nutritional quality, enabling rapid response to changing agricultural demands.

Role of CRISPR and Gene Editing in Agriculture

CRISPR and other gene-editing tools have become powerful assets in agriculture, allowing precise modifications in crops to enhance traits without introducing foreign DNA. This technology, often referred to as "genome editing," can create crops with desirable traits while potentially avoiding some regulatory restrictions associated with GMOs.

1. **Development of Non-GMO Crops**

   - Gene-editing techniques like CRISPR can introduce mutations that confer desired traits without incorporating foreign DNA, producing crops that may be classified as non-GMO in some regions.
   - **Applications**: CRISPR has been used to create crops resistant to fungal diseases, improve shelf life, and enhance nutritional quality, benefiting both producers and consumers.

2. **Rapid Crop Improvement**

   - Gene editing allows for faster improvement of crops by targeting specific genes responsible for desirable traits, bypassing the long timelines associated with traditional breeding.

- **Examples**: CRISPR-edited wheat with reduced gluten content and tomatoes with improved ripening characteristics have been developed, catering to market demands and improving agricultural efficiency.

3. **Gene Editing for Pest and Disease Resistance**
   - CRISPR enables precise editing of genes associated with pest and disease resistance, reducing the need for chemical inputs and enhancing crop durability against biotic stresses.
   - **Applications**: Gene-edited rice resistant to bacterial blight and powdery mildew-resistant wheat demonstrate the potential of gene editing for sustainable agriculture.

5.5 Bioinformatics and Computational Biology

Bioinformatics and computational biology are interdisciplinary fields that apply computational tools and techniques to manage, analyze, and interpret complex biological data. With the explosion of data generated by genomics, proteomics, transcriptomics, and other -omics technologies, bioinformatics and computational biology have become essential for advancing research in genetics, molecular biology, medicine, and environmental science. These fields allow scientists to uncover patterns, make predictions, and gain insights into biological processes on a scale and depth that traditional laboratory techniques alone cannot achieve.

Overview of Bioinformatics and Computational Biology

1. **Bioinformatics**
   - Bioinformatics primarily focuses on developing software, databases, and algorithms to handle large volumes of biological data. It involves storing, retrieving, and analyzing genetic sequences, protein structures, gene expression patterns, and other types of molecular information.
   - **Applications**: Bioinformatics tools are used for sequence alignment, gene prediction, protein structure modeling, and comparative genomics. They enable researchers to visualize and interpret complex data, essential for understanding gene function, evolution, and disease mechanisms.

2. **Computational Biology**
   - Computational biology combines computational methods with biological data to model biological systems, simulate cellular processes, and test hypotheses. While bioinformatics deals mainly with the organization and analysis of data, computational biology is more focused on theoretical modeling, systems biology, and predicting biological behaviors.
   - **Applications**: Computational biology is widely used in systems biology, epidemiology, drug discovery, and ecological modeling. It allows researchers to construct detailed models of cellular pathways, predict responses to drugs, and understand the spread of diseases.

Key Areas in Bioinformatics

1. **Genomics and Transcriptomics**
   - Genomics involves analyzing DNA sequences to understand the genetic makeup and evolutionary relationships of organisms. Transcriptomics focuses on the study of RNA sequences, providing insights into gene expression patterns.
   - **Applications**: Bioinformatics tools are used to assemble genomes, identify mutations, map genes, and analyze gene expression. Techniques like RNA sequencing (RNA-Seq) produce large datasets that are processed using bioinformatics pipelines to quantify gene expression, detect alternative splicing, and identify regulatory networks.
2. **Proteomics**
   - Proteomics involves studying the structure, function, and interactions of proteins. Bioinformatics tools are essential for processing mass spectrometry data, identifying proteins, and analyzing post-translational modifications.
   - **Applications**: In proteomics, bioinformatics tools are used for protein identification, quantification, and structural modeling. These analyses help in understanding cellular processes, identifying disease biomarkers, and developing therapeutic targets.
3. **Structural Biology**
   - Structural bioinformatics focuses on predicting and analyzing the 3D structures of proteins and nucleic acids. Understanding molecular structures is crucial for elucidating the function and interaction of biomolecules.
   - **Applications**: Structural bioinformatics aids in protein folding studies, drug design, and understanding protein-protein interactions. Techniques such as homology modeling, molecular docking, and molecular dynamics simulations are widely used to predict protein structures and identify potential drug-binding sites.
4. **Metabolomics and Systems Biology**
   - Metabolomics studies small molecules, or metabolites, within a biological system, while systems biology integrates various -omics data to understand biological networks and interactions.
   - **Applications**: Bioinformatics tools in metabolomics analyze metabolic pathways and quantify metabolite concentrations, revealing cellular responses to environmental changes. In systems biology, computational models simulate complex networks, such as metabolic pathways and signaling cascades, allowing researchers to predict cellular responses and identify intervention points.
5. **Phylogenetics and Evolutionary Biology**
   - Phylogenetics examines the evolutionary relationships among species. Bioinformatics tools allow scientists to reconstruct phylogenetic trees, analyze sequence divergence, and understand evolutionary processes.
   - **Applications**: Phylogenetic analysis is used in comparative genomics, taxonomy, and

evolutionary studies. By analyzing genetic sequences across species, bioinformatics helps uncover gene conservation, horizontal gene transfer, and speciation events.

Tools and Techniques in Bioinformatics

1. **Sequence Alignment and Analysis**
   - Sequence alignment tools, such as BLAST and ClustalW, compare DNA, RNA, or protein sequences to identify similarities, infer evolutionary relationships, and predict gene functions.
   - **Applications**: Sequence alignment is essential for identifying homologous genes, predicting protein function, and assembling genomes. Multiple sequence alignment also helps in studying conserved motifs and regulatory elements.
2. **Database Management and Data Integration**
   - Bioinformatics relies on databases like GenBank, UniProt, and Ensembl, which store vast amounts of biological data. Efficient data management is crucial for retrieving, curating, and updating information.
   - **Applications**: Biological databases allow researchers to access curated datasets, compare sequences, and perform meta-analyses. Data integration tools combine information from multiple sources, facilitating cross-species comparisons and comprehensive -omics analyses.
3. **Gene Prediction and Annotation**
   - Gene prediction algorithms identify potential gene-coding regions within DNA sequences, while annotation tools provide functional information on genes, such as coding regions, regulatory elements, and pathways.
   - **Applications**: Gene prediction and annotation are essential for genome assembly and analysis, providing insights into gene structure and function. Annotated genomes serve as references for studying genetic variation, evolution, and gene-environment interactions.
4. **Molecular Docking and Drug Design**
   - Molecular docking predicts the interaction between a ligand (e.g., drug molecule) and a target protein. Computational techniques are used to screen compounds, optimize drug candidates, and predict their efficacy and safety.
   - **Applications**: Molecular docking aids in drug discovery by identifying potential binding sites and interactions, accelerating the development of new therapeutics. Combined with high-throughput screening, docking helps prioritize compounds for experimental testing.
5. **Data Mining and Machine Learning**
   - Data mining and machine learning techniques extract patterns from large datasets, enabling predictions and classifications. Machine learning algorithms are increasingly used in genomics, proteomics, and personalized medicine.
   - **Applications**: Machine learning can predict disease outcomes, classify cancer subtypes, and identify biomarkers. In genomics, these techniques help identify mutations associated

with diseases, while in drug discovery, they assist in screening compounds and optimizing leads.

Computational Biology: Modeling and Simulation

1. **Systems Biology Modeling**
   - Systems biology uses computational models to represent complex biological systems, integrating data from multiple sources to simulate cellular processes and predict behavior under different conditions.
   - **Applications**: Systems biology models provide insights into metabolic networks, signaling pathways, and gene regulatory circuits. By simulating perturbations, these models can predict cellular responses to drugs or environmental changes, aiding in drug development and disease treatment strategies.
2. **Population Genetics and Evolutionary Models**
   - Computational biology models help understand population dynamics, genetic drift, and evolutionary selection. Population genetics models analyze genetic variation within populations, revealing insights into migration, adaptation, and evolutionary pressure.
   - **Applications**: Evolutionary models are used in conservation biology, epidemiology, and anthropology. For example, models of viral evolution help predict the spread of infectious diseases, informing vaccine development and public health strategies.
3. **Molecular Dynamics Simulations**
   - Molecular dynamics (MD) simulations study the movement and interactions of atoms and molecules over time. MD simulations help predict molecular behavior and stability, essential for understanding protein folding, binding interactions, and conformational changes.
   - **Applications**: MD simulations are widely used in drug design to predict ligand-protein interactions, in structural biology to study protein dynamics, and in biophysics to investigate molecular mechanisms underlying cellular processes.
4. **Artificial Intelligence in Computational Biology**
   - Artificial intelligence (AI) applications in computational biology include deep learning algorithms that can analyze and predict complex biological patterns. AI is increasingly applied to image analysis, genomics, and drug discovery.
   - **Applications**: AI-driven models assist in interpreting medical images, predicting protein structures, and optimizing drug compounds. In genomics, AI aids in identifying genetic mutations and biomarkers for personalized medicine.

Applications of Bioinformatics and Computational Biology

1. **Precision Medicine and Genomic Medicine**
   - Bioinformatics and computational biology facilitate precision medicine by analyzing genetic data to customize treatments based on an individual's genetic makeup. Tools

such as genome-wide association studies (GWAS) and SNP analysis identify genetic variations linked to diseases.

- **Applications**: Genomic data analysis identifies biomarkers for cancer, cardiovascular diseases, and metabolic disorders, allowing personalized treatment plans. In pharmacogenomics, bioinformatics predicts drug response and toxicity, improving therapeutic outcomes.

2. **Agriculture and Crop Improvement**

- Bioinformatics supports crop breeding by identifying genes associated with desirable traits, such as drought tolerance, disease resistance, and yield. Genomic selection tools enable precise breeding, enhancing agricultural productivity.

- **Applications**: Crop improvement programs use bioinformatics to select for traits that promote resilience and efficiency. Computational biology models simulate crop growth under various conditions, predicting optimal breeding strategies for climate adaptation.

3. **Environmental and Conservation Biology**

- Computational biology and bioinformatics are used to monitor biodiversity, assess ecosystem health, and study environmental impacts on species. Population genomics and ecological modeling support conservation efforts.

- **Applications**: Genetic diversity analysis informs conservation strategies, while environmental DNA (eDNA) monitoring tracks species in natural habitats. Computational tools assess species' vulnerability to climate change, guiding biodiversity conservation.

4. **Drug Discovery and Development**

- Bioinformatics and computational biology accelerate drug discovery by predicting protein targets, screening compounds, and simulating drug interactions. High-throughput screening, molecular docking, and predictive modeling reduce the cost and time of drug development.

- **Applications**: Bioinformatics identifies potential therapeutic targets for diseases, while computational biology simulates drug interactions to optimize efficacy and safety. Machine learning algorithms are used to predict off-target effects and toxicity, improving drug candidate selection.

5. **Evolutionary Biology and Phylogenetics**

- Bioinformatics tools analyze genetic data to reconstruct evolutionary relationships, study species divergence, and trace the origins of traits. Phylogenetic analysis helps understand evolutionary processes and biodiversity.

- **Applications**: Phylogenetic trees reveal evolutionary history and genetic divergence. In infectious disease research, bioinformatics tracks pathogen evolution, enabling public health agencies to monitor outbreaks and develop vaccines.

Conclusion

The historical background and evolution of molecular biology showcase a transformative field that has reshaped scientific understanding and spurred technological progress across diverse domains. From early discoveries elucidating the structure of DNA to the development of powerful molecular techniques, the journey of molecular biology exemplifies human curiosity and the drive to decode the building blocks of life. Central to this evolution is the Central Dogma, which established a framework for understanding genetic flow, while advancements in DNA and protein research underscored the critical molecular components driving cellular functions.

As molecular biology has expanded, it has integrated knowledge from genetics, chemistry, and computational science, fostering interdisciplinary approaches to understanding life at its most fundamental level. The Central Dogma, molecular theories, and advancements in genetic analysis form a core foundation that has facilitated studies on gene regulation, disease mechanisms, and therapeutic interventions. The development of techniques like PCR, DNA sequencing, and CRISPR technology underscores the field's technological evolution, enabling precise genetic manipulation and opening new avenues in genetics, agriculture, and medicine.

The applications of molecular biology are broad, encompassing fields like genetics, biotechnology, and computational biology, and are increasingly critical in addressing global challenges. In medicine, molecular biology aids in diagnosing genetic diseases and developing targeted therapies. In agriculture, it enhances crop resilience and productivity, and in environmental science, it aids in conservation and biodiversity studies. The field's continued progression, supported by innovations in data science and bioinformatics, holds the promise of further transformative impacts, advancing sustainable development and improving human health.

Chapter 2

Analysis of DNA and RNA in Insects

Studying DNA and RNA in insects has significantly advanced our understanding of their biology, ecology, and interactions with their environments. These molecular components are the foundation of genetic information and play critical roles in gene regulation, protein synthesis, and cellular functions. Insects, as one of the most diverse and adaptable groups of organisms, provide an exceptional model for exploring molecular mechanisms and applying genetic technologies. With their pivotal role in ecosystems and agriculture, studying the molecular basis of their traits and behaviors offers practical solutions to manage pests, conserve pollinators, and improve biological control agents.

DNA, the carrier of genetic information, dictates the hereditary traits of organisms through its unique nucleotide sequence. In insects, the decoding of genomic sequences has opened doors to identifying genes responsible for vital functions such as development, reproduction, immunity, and environmental adaptation. Advances in whole-genome sequencing have provided a comprehensive view of the genetic landscape of several insect species, enabling researchers to pinpoint specific genes associated with insecticide resistance, feeding preferences, and pheromone communication. This wealth of genetic information has laid the groundwork for understanding the molecular underpinnings of insect behavior and physiology, offering actionable insights for pest control and ecological conservation.

RNA, the intermediary molecule in the flow of genetic information, serves as a crucial link between DNA and protein synthesis. The study of RNA in insects has revealed the intricacies of transcriptional regulation and post-transcriptional modifications that influence gene expression. RNA molecules, including messenger RNA (mRNA), transfer RNA (tRNA), and ribosomal RNA (rRNA), play specialized roles in translating genetic instructions into functional proteins. Additionally, non-coding RNAs, such as microRNAs and long non-coding RNAs, have emerged as key regulators of gene expression, influencing insect development, immune responses, and adaptation to environmental challenges. By analyzing RNA profiles under various conditions, researchers can gain a dynamic understanding of gene expression patterns and their implications for insect survival and adaptation.

Transcription, the first step in gene expression, involves the synthesis of RNA from a DNA template. This process is mediated by RNA polymerase and regulated by transcription factors that bind to specific DNA sequences, such as promoters and enhancers. In insects, transcriptional regulation

DOI: 10.1201/9781003757405-2

plays a central role in developmental processes, including molting, metamorphosis, and reproductive cycles. For example, the precise expression of juvenile hormone (JH) and ecdysteroid-regulated genes determines the transition between larval, pupal, and adult stages. Disruptions in transcriptional regulation can lead to developmental anomalies, making these pathways attractive targets for pest management strategies.

Translation, the subsequent step, converts the genetic code in mRNA into functional proteins. Ribosomes, composed of rRNA and proteins, orchestrate the translation process by facilitating the binding of tRNA molecules carrying specific amino acids. The synthesis of proteins is vital for maintaining cellular functions, structural integrity, and enzymatic activities in insects. For instance, enzymes involved in detoxifying xenobiotics, such as cytochrome P450s, glutathione S-transferases, and carboxylesterases, are synthesized through tightly regulated translation mechanisms. These proteins enable insects to metabolize insecticides and adapt to toxic environments, contributing to the evolution of resistance in pest populations.

The integration of recombinant DNA technology into insect research has revolutionized the field of molecular entomology. By combining DNA from different sources, scientists can introduce foreign genes into insect genomes to study their function or confer novel traits. This approach has facilitated the development of genetically modified insects for pest control, functional genomics, and disease vector management. One notable application is the creation of sterile insect strains through genetic engineering, which suppresses pest populations by disrupting reproductive cycles. Additionally, recombinant DNA techniques have been instrumental in producing bioinsecticides, such as genetically modified baculoviruses, that specifically target pest species without harming beneficial organisms.

Identifying genes and nucleotide sequences associated with specific traits is a cornerstone of insect molecular biology. Techniques such as polymerase chain reaction (PCR), quantitative PCR (qPCR), and RNA sequencing (RNA-Seq) enable researchers to amplify, quantify, and analyze gene expression patterns in insects. These methods have uncovered genes linked to behaviors such as host-seeking in mosquitoes, pheromone production in moths, and caste differentiation in social insects like bees and ants. Furthermore, whole-genome sequencing and transcriptomic analyses have identified genetic markers for pesticide resistance, providing insights into the molecular basis of resistance mechanisms and informing the design of effective management strategies.

The ability to manipulate genetic information in insects has paved the way for innovative approaches to pest control and ecological preservation. Gene editing technologies, such as CRISPR-Cas9, have enabled precise modifications of insect genomes, allowing researchers to knock out or insert genes to study their function or disrupt essential biological processes. For example, CRISPR has been used to create gene drives in mosquito populations, spreading genes that reduce the transmission of diseases like malaria and dengue. Similarly, genetic modifications in agricultural pests have targeted reproduction, feeding, or resistance mechanisms, reducing crop damage and reliance on chemical pesticides.

RNA interference (RNAi) has emerged as another powerful tool for studying and controlling insect genes. By introducing double-stranded RNA (dsRNA) specific to target genes, RNAi triggers the degradation of complementary mRNA, effectively silencing gene expression. This technique has been used to study gene function in model insects and develop species-specific pest control methods.

RNAi-based bioinsecticides are being explored for their potential to disrupt critical pathways in pests without affecting non-target organisms, offering a sustainable alternative to traditional pesticides.

The identification and characterization of genes associated with insecticide resistance have been a focal point in molecular entomology. Resistance to insecticides poses a significant challenge to pest management, necessitating a deeper understanding of the genetic and biochemical mechanisms underlying resistance. Genes encoding detoxifying enzymes, ion channels, and transporters have been implicated in resistance development. Molecular analyses have revealed mutations in target-site genes, such as the voltage-gated sodium channel and acetylcholinesterase, that reduce insecticide binding and efficacy. Additionally, overexpression of detoxifying enzymes, driven by regulatory changes in resistance-associated genes, contributes to metabolic resistance. These insights have informed resistance management strategies, including the rotation of insecticides with different modes of action and the development of synergists to enhance insecticide potency.

Advancements in molecular tools have also expanded our understanding of insect-host and insect-pathogen interactions. Identifying genes involved in pathogen transmission and immune responses has provided insights into how vectors, such as mosquitoes and aphids, transmit diseases to humans, animals, and plants. For example, genes regulating mosquito salivary gland function and midgut immunity have been studied to develop targeted interventions that block pathogen development or transmission. Similarly, molecular analyses of insect-borne plant viruses have revealed viral and insect factors that facilitate virus acquisition and transmission, enabling the design of resistant crop varieties or vector control strategies.

The application of DNA and RNA analysis in insect taxonomy and phylogeny has transformed the classification and evolutionary study of insects. Molecular markers, such as mitochondrial DNA (mtDNA) sequences, have been used to distinguish cryptic species and trace evolutionary relationships among insect taxa. Phylogenetic studies based on DNA and RNA sequences have provided insights into the origins and diversification of insect lineages, shedding light on the genetic basis of their ecological adaptations. This information is critical for biodiversity conservation, particularly in identifying species at risk of extinction and understanding their ecological roles.

The integration of molecular biology into entomology has not only enhanced our understanding of insect biology but also opened new avenues for sustainable pest management and conservation. By leveraging DNA and RNA analysis, recombinant DNA technology, and gene editing tools, researchers can address pressing challenges in agriculture, public health, and environmental sustainability. The ability to manipulate genetic information in insects offers the potential to control pest populations, protect pollinators, and mitigate the spread of vector-borne diseases, contributing to a more balanced and resilient ecosystem.

2.1 DNA and RNA Analysis in Insects: Transcription and Translation Mechanisms

Transcription and translation are foundational processes that underlie the flow of genetic information from DNA to proteins, which govern every aspect of an organism's biology. In insects, these mechanisms are critical for regulating development, reproduction, and adaptive responses to environmental stimuli. The precise orchestration of transcription and translation not only ensures the survival and function of individual insects but also has profound implications for pest management, ecological balance, and evolutionary adaptation. By examining these processes in detail, researchers gain insights into molecular regulation and can identify targets for interventions such as gene editing, RNA interference (RNAi), and biotechnological applications.

2.1.1 Overview of Transcription in Insects

Transcription is the foundational process in gene expression, during which the genetic code stored in DNA is transcribed into RNA. This step is critical for converting the genetic blueprint into functional molecules that sustain life. In insects, transcription plays a central role in orchestrating biological processes such as development, reproduction, behavior, immune responses, and environmental adaptation. As insects encounter various ecological challenges and complex life cycles, transcriptional regulation ensures that specific genes are activated or repressed at appropriate times and in the required cells, enabling survival and functionality.

Insects are among the most diverse organisms on Earth, with exceptional adaptability to environmental changes. Their ability to evolve resistance to pesticides, adapt to extreme climates, and fine-tune behaviors such as foraging or mating underscores the complexity of their transcriptional networks. This process is not merely a mechanical copying of genetic material but a dynamic, highly regulated interplay of molecular machinery, ensuring precise gene expression. The study of transcription in insects sheds light on how these small creatures manage such extraordinary adaptability and offers pathways for developing pest control strategies, enhancing beneficial insect populations, and mitigating environmental damage.

Key Features of Transcription in Insects

Transcription in insects involves the synthesis of RNA from a DNA template. This process is carried out by RNA polymerase enzymes, which read the DNA sequence and create complementary RNA strands. However, transcription is not a simple task; it is a multi-step process that requires careful regulation. This regulation occurs at several levels, including the recruitment of RNA polymerase, activation or repression of promoter regions, interaction with enhancers and silencers, and modulation by transcription factors. These features ensure that transcription is both accurate and responsive to internal and external signals.

1. **Developmental Control**: Transcription drives key developmental events such as molting and metamorphosis. For example, hormones like ecdysteroids and juvenile hormones regulate transcription factors that activate genes necessary for transitioning from larval to pupal and adult stages.
2. **AdaptiveResponses**: Insects rely on transcriptional changes to adapt to environmental stresses such as temperature extremes, drought, and exposure to pathogens. Genes encoding heat shock proteins (HSPs) are transcribed in response to thermal stress, while immune-related genes are activated during infection.
3. **Fine-Tuned Gene Expression**: Transcriptional regulation ensures that only the necessary genes are expressed, preventing wastage of cellular resources. For instance, enzymes required for detoxifying insecticides are transcribed only in response to exposure, minimizing energy expenditure in their absence.

Key Components of Transcription

1. Promoter Regions

Promoters are DNA sequences located upstream of genes that control the initiation of transcription. They act as docking sites for RNA polymerase and other transcription factors, determining where transcription begins and how frequently it occurs. In insects, promoter regions are specialized to respond to developmental and environmental cues, making them pivotal in transcriptional regulation.

» **Core Promoter Elements**:

- **TATA Box**: A highly conserved sequence that helps position RNA polymerase at the correct starting point. In insects, the TATA box is essential for initiating transcription in genes involved in development and reproduction.
- **Initiator Sequences (Inr)**: Found near the transcription start site, these sequences enhance transcription precision, particularly in genes regulated by hormonal signals.
- **Downstream Promoter Elements (DPE)**: Found in some insect genes, DPEs work with core elements to ensure efficient transcription.

» **Regulation by Promoters**:

Promoters in insects are regulated by enhancers, silencers, and other elements that integrate hormonal and environmental signals. For instance, the promoters of genes involved in detoxification are activated by xenobiotic-responsive transcription factors when an insect encounters toxic substances.

» **Developmental Importance:**

Promoter activity during development is tightly regulated to ensure that genes involved in molting and metamorphosis are expressed at the correct stages. Disruptions in promoter regulation can result in developmental defects.

2. RNA Polymerases

RNA polymerases are specialized enzymes that synthesize RNA from DNA templates. In insects, three types of RNA polymerases work in concert to produce different classes of RNA, each fulfilling distinct cellular roles.

» **RNA PolymeraseI**:

This enzyme transcribes ribosomal RNA (rRNA), the main structural and functional component of ribosomes. Ribosomes are essential for protein synthesis, making RNA Polymerase I activity critical for cellular growth and reproduction.

» **RNA PolymeraseII**:

This enzyme transcribes messenger RNA (mRNA), which encodes the genetic instructions for protein synthesis. It also synthesizes regulatory non-coding RNAs such as microRNAs (miRNAs) and long non-coding RNAs (lncRNAs). RNA Polymerase II is central to gene expression in insects, particularly for genes involved in immunity, stress responses, and behavioral regulation.

» **RNAPolymeraseIII:**

This enzyme produces transfer RNA (tRNA) and small nuclear RNA (snRNA). tRNA molecules are crucial for decoding mRNA into proteins, while snRNAs play roles in RNA splicing and modification. RNA Polymerase III activity is essential for maintaining the protein synthesis machinery.

3. Enhancers and Silencers

Enhancers and silencers are regulatory DNA elements that influence the transcriptional activity of associated genes. They modulate gene expression in response to developmental, hormonal, and environmental cues.

» **Enhancers**: Enhancers increase transcription by recruiting activator proteins and facilitating the assembly of the transcription machinery. In insects, enhancers regulate genes involved in metamorphosis, reproduction, and immune responses. For example, hormone-responsive enhancers activate genes required for molting and tissue remodeling during larval transitions.

» **Silencers**: Silencers suppress transcription by recruiting repressor proteins or altering chromatin structure to reduce DNA accessibility. Silencers are crucial during development to prevent the premature activation of genes required in later stages.

» **Dynamic Regulation**: Enhancers and silencers respond to external stimuli, such as temperature changes or infection, allowing insects to dynamically adjust gene expression. For instance, immune-related enhancers activate antimicrobial peptide genes upon pathogen detection.

4. Transcription Factors

Transcription factors are proteins that bind to specific DNA sequences to regulate transcription. They act as molecular switches, turning genes on or off in response to various signals.

» **General Transcription Factors:**

These are required for the basic assembly of the transcription initiation complex. Examples include TFIID, which binds to the TATA box, and TFIIH, which unwinds DNA to initiate RNA synthesis.

» **SpecificTranscriptionFactors:**

These respond to developmental or environmental signals to regulate specific genes. Examples include:

- **Broad-Complex (BR-C)**: Controls genes involved in molting and metamorphosis, ensuring proper developmental transitions.
- **Kruppel Homolog 1 (Kr-h1)**: Represses adult-specific genes during juvenile stages, maintaining larval characteristics.
- **Relish**: Activates immune genes encoding antimicrobial peptides during bacterial infection.

» **Post-Translational Modifications**:

Transcription factors are often modified by phosphorylation, acetylation, or ubiquitination, affecting their stability and activity. These modifications enable rapid transcriptional responses to environmental stimuli.

Stages of Transcription in Insects

Transcription is a complex multi-step process that involves several well-defined stages: initiation, elongation, and termination. Each stage is tightly regulated to ensure the accurate synthesis of RNA from the DNA template. In insects, the transcriptional process is fine-tuned to meet the demands of their dynamic life cycles and adaptability to environmental changes. The stages of transcription involve specific molecular events, ensuring the precise expression of genes crucial for growth, reproduction, defense mechanisms, and other physiological functions.

1. Initiation

The initiation stage of transcription is critical for determining where RNA synthesis begins. It is during this stage that RNA polymerase is recruited to the DNA template, guided by promoter sequences and assisted by transcription factors. Proper initiation ensures that the RNA polymerase begins transcription at the correct site, producing an accurate RNA transcript.

Key Steps in Initiation

1. **Promoter Recognition**
   - RNA polymerase scans the DNA for promoter sequences, specific regions that serve as starting points for transcription.
   - Promoters in insects typically contain conserved elements such as the TATA box, initiator sequences (Inr), and sometimes downstream promoter elements (DPEs). These sequences guide RNA polymerase to the correct transcription start site.
   - In eukaryotic systems, including insects, promoter regions are flanked by enhancer sequences that amplify transcription efficiency by facilitating the recruitment of activator proteins.
2. **Assembly of the Transcription Initiation Complex**
   - The assembly of the initiation complex begins with the binding of general transcription factors (GTFs) to the promoter.
   - **TFIID** binds first, recognizing the TATA box through its TATA-binding protein (TBP) subunit. This binding serves as a scaffold for the recruitment of additional GTFs, including TFIIA, TFIIB, TFIIE, TFIIF, and TFIIH.
   - **RNA Polymerase II** is recruited to the complex, positioning its active site over the transcription start site. This entire assembly constitutes the pre-initiation complex (PIC).
3. **Promoter Clearance**
   - Once RNA polymerase is correctly positioned, it synthesizes a short RNA strand (approximately 10–15 nucleotides). This step ensures that the polymerase is functioning properly.

- Promoter clearance occurs when RNA polymerase escapes the promoter region and transitions into the elongation phase. This step requires the phosphorylation of the C-terminal domain (CTD) of RNA polymerase II by TFIIH, which activates the polymerase for RNA chain elongation.

Significance of Initiation in Insects

» Precise initiation ensures the accurate expression of genes critical for insect development, such as those regulating molting, tissue remodeling, and reproductive processes.

» Mutations or disruptions in promoter sequences or transcription factors involved in initiation can lead to developmental defects or compromised physiological responses.

2. Elongation

Elongation is the process by which RNA polymerase moves along the DNA template strand, synthesizing a complementary RNA strand. This stage is highly dynamic, requiring the polymerase to overcome structural barriers while maintaining fidelity in RNA synthesis. In insects, elongation plays a central role in generating functional RNA molecules required for cellular and systemic activities.

Key Steps in Elongation

1. **RNA Chain Growth**
   - RNA polymerase reads the DNA template strand in the 3′ to 5′ direction, adding ribonucleotides complementary to the DNA sequence to the growing RNA strand.
   - The RNA chain grows at an average rate of 20–50 nucleotides per second in eukaryotic systems, including insects.
   - **Nucleotide Incorporation**: Each ribonucleotide is selected based on base-pair complementarity (adenine with uracil, cytosine with guanine) and catalyzed by the active site of RNA polymerase.
2. **Transcriptional Pausing**
   - RNA polymerase may pause temporarily during elongation due to DNA sequences, secondary structures in the RNA, or interactions with regulatory proteins.
   - Pausing allows for co-transcriptional processes, such as splicing of pre-mRNA, capping of the 5′ end, or modifications like polyadenylation at the 3′ end.
   - Regulatory checkpoints during pausing also enable fine-tuning of RNA synthesis. For example, in stress conditions, transcriptional pausing ensures that genes required for survival, such as heat shock proteins (HSPs), are transcribed first.
3. **Proofreading and Fidelity**
   - RNA polymerase has intrinsic proofreading capabilities, which help maintain the fidelity of RNA synthesis.
   - Mismatched ribonucleotides are detected and removed, ensuring that the RNA sequence accurately reflects the DNA template.

4. **Histone Modification and Chromatin Dynamics**
   - As RNA polymerase progresses along the DNA, it encounters nucleosomes that must be temporarily displaced to allow transcription.
   - Elongation factors work with chromatin remodelers to modify histones and facilitate RNA polymerase movement. In insects, this dynamic interaction is crucial for transcriptional responses to environmental changes or developmental cues.

Significance of Elongation in Insects

» Efficient elongation ensures the production of functional RNA molecules that drive critical biological processes.

» Pausing mechanisms enable the precise regulation of genes, such as those involved in immune responses to pathogens or stress-induced survival pathways.

» Dysregulation of elongation can lead to incomplete RNA transcripts, disrupting protein synthesis and cellular homeostasis.

3. Termination

Termination is the final stage of transcription, where RNA polymerase halts RNA synthesis and releases the RNA molecule. This stage is critical for ensuring that RNA transcripts are properly processed and ready for downstream activities, such as translation or regulatory functions.

Key Steps in Termination

1. **Recognition of Termination Signals**
   - Termination is guided by specific DNA sequences and RNA signals. In insects, polyadenylation signals such as **AAUAAA** are common markers for termination of mRNA synthesis.
   - These sequences are recognized by cleavage and polyadenylation factors that process the 3′ end of the RNA transcript.
2. **RNA Cleavage and Polyadenylation**
   - Once the termination signal is transcribed, the RNA is cleaved downstream of the polyadenylation site.
   - A poly(A) tail, consisting of adenine nucleotides, is added to the 3′ end of the RNA. This modification stabilizes the RNA molecule and facilitates its export from the nucleus.
3. **Release of RNA Polymerase**
   - Termination factors displace RNA polymerase from the DNA template, completing the transcription process.
   - RNA polymerase is then recycled for subsequent rounds of transcription.

Types of Termination in Insects

» **Poly(A)-Dependent Termination:**

- Common in mRNA synthesis, this type of termination relies on the recognition of polyadenylation signals by cleavage and polyadenylation factors.

» **Non-Poly(A)-Dependent Termination:**

- Found in some non-coding RNAs, termination in these cases is mediated by specific RNA structures, such as stem-loops, that signal RNA polymerase to stop.

Significance of Termination in Insects

» Proper termination ensures that RNA transcripts are complete and correctly processed, enabling them to function effectively.

» Failure in termination can result in aberrant RNA molecules, which may interfere with normal cellular processes or lead to developmental abnormalities.

Biological Significance of Transcription in Insects

Transcription, the initial and pivotal step in gene expression, plays a vital role in regulating the biological processes that enable insects to grow, develop, survive, and reproduce. The precise and timely control of transcription is essential for orchestrating complex physiological processes such as molting, metamorphosis, immunity, and behavioral responses to environmental changes. By governing the expression of genes in a stage-specific and context-dependent manner, transcription provides the molecular foundation for insect adaptability, resilience, and ecological success.

1. Development and Metamorphosis

Development and metamorphosis are hallmark processes in the insect life cycle, involving dramatic changes in physiology, morphology, and behavior. Transcription is central to these transformations, regulating the expression of genes necessary for molting, tissue remodeling, and the formation of adult structures.

» **Stage-SpecificGeneExpression:**

During larval, pupal, and adult transitions, transcription ensures that genes required for each stage are activated, while genes unnecessary for the current stage are repressed. This regulation minimizes energy expenditure and ensures developmental precision.

» **Hormone-ResponsiveTranscriptionFactors:**

Ecdysteroids and juvenile hormones (JH) are key hormonal regulators of insect development. These hormones activate specific transcription factors, such as the Broad-Complex (BR-C) and Krüppel homolog 1 (Kr-h1), which regulate genes responsible for cuticle synthesis, tissue breakdown, and organ differentiation.

- **Broad-Complex (BR-C):** Drives the expression of enzymes involved in molting and cuticle degradation.

- **Kruppel homolog 1 (Kr-h1)**: Maintains larval traits by repressing adult-specific genes during early stages.

» **Coordination of Molting and Metamorphosis**:

Molting is regulated by the transcription of genes encoding chitinases and other enzymes that degrade the old cuticle, while metamorphosis involves the transcription of genes for adult-specific structures such as wings and reproductive organs.

» **Developmental Plasticity**:

In response to environmental cues such as nutrient availability or crowding, transcriptional changes can modify developmental timing or trajectory, allowing insects to optimize their life cycle.

2. Reproduction

Reproductive success in insects is tightly linked to transcriptional regulation. Genes involved in pheromone signaling, gametogenesis, and reproductive organ development are activated in a sex-specific and stage-dependent manner to ensure successful mating and offspring production.

» **Pheromone Production and Detection**:

- Transcriptional regulation controls the production of pheromone biosynthetic enzymes and the expression of pheromone receptors in sensory organs.
- For instance, moths exhibit transcriptional upregulation of pheromone-producing genes during reproductive periods, ensuring effective communication between sexes.

» **Gamete Formation:**

- Transcription drives the expression of genes involved in spermatogenesis and oogenesis.
- In female insects, genes encoding vitellogenin (a precursor of egg yolk protein) are transcribed in response to hormonal signals, ensuring nutrient provision for developing embryos.

» **Reproductive Organ Development**:

- The formation of reproductive structures, such as testes, ovaries, and accessory glands, relies on stage-specific transcriptional programs.
- Hormonal regulation of transcription factors ensures the differentiation of these organs during pupal and adult stages.

» **Mating Behavior and Fertility**:

- Genes involved in courtship behavior and copulation are transcriptionally regulated, enabling insects to optimize reproductive success.
- Transcriptional changes also regulate the post-mating response, such as the activation of egg-laying genes in females.

3. Immunity

Insects lack an adaptive immune system but rely on a robust innate immune response, which is transcriptionally regulated to defend against pathogens such as bacteria, viruses, and fungi. Transcriptional activation of immune-related genes ensures a rapid and effective response to infections.

» **Activation of Antimicrobial Peptide (AMP) Genes**:
  - Upon infection, immune-responsive transcription factors such as Relish (in the Toll and Imd pathways) activate the transcription of AMP genes.
  - AMPs, including defensins, cecropins, and attacins, are synthesized and secreted into the hemolymph to neutralize pathogens.

» **Pathogen-Specific Responses**:
  - » Different transcriptional programs are activated depending on the type of pathogen. For example:
    - Toll pathway activation leads to the transcription of genes targeting fungal and Gram-positive bacterial infections.
    - The Imd pathway induces the transcription of genes targeting Gram-negative bacteria.

» **Regulation of Immune Homeostasis**:
  - Transcriptional feedback loops prevent overactivation of the immune system, which could harm the host. Genes encoding negative regulators, such as inhibitors of NF-κB signaling, are transcribed to maintain immune balance.

» **Epigenetic Memory of Infection**:
  - Insects exposed to repeated infections exhibit transcriptional changes that "prime" the immune system for faster responses, suggesting a form of epigenetically mediated immune memory.

4. Adaptation to Stress

Insects face a wide range of environmental stressors, including temperature fluctuations, desiccation, xenobiotics, and habitat changes. Transcriptional regulation enables insects to adapt to these challenges by activating stress-response genes.

» **Heat Shock Response**:
  - Genes encoding heat shock proteins (HSPs) are rapidly transcribed in response to temperature stress. HSPs act as molecular chaperones, stabilizing proteins and preventing aggregation during heat exposure.
  - Heat shock transcription factor (HSF) regulates this response by binding to heat shock elements (HSEs) in the promoters of HSP genes.

» **Detoxification of Xenobiotics**:
  - Insects exposed to insecticides or other toxins upregulate the transcription of detoxification genes.

- Cytochrome P450 monooxygenases, glutathione S-transferases (GSTs), and carboxylesterases are transcribed to metabolize and neutralize harmful compounds.
- For example, agricultural pests exhibit enhanced transcription of P450 genes, contributing to insecticide resistance.

» **Oxidative Stress Response**:

- Transcriptional activation of antioxidant genes, such as those encoding superoxide dismutase (SOD) and catalase, helps insects mitigate oxidative damage caused by environmental stressors.

» **Diapause and Dormancy**:

- Diapause, a period of arrested development in response to unfavorable conditions, is regulated by transcriptional changes. Genes involved in energy conservation, stress resistance, and metabolic suppression are upregulated during diapause.

5. Behavioral Regulation

Behavioral plasticity in insects, including foraging, mating, migration, and predator avoidance, is underpinned by transcriptional regulation of genes involved in sensory perception, neural function, and circadian rhythms.

» **Sensory Perception**:

- Genes encoding olfactory and gustatory receptors are transcriptionally regulated to adapt to changing environmental conditions.
- For instance, in social insects like bees, transcriptional changes in the antennal lobe regulate foraging behavior based on colony needs.

» **Neural Development and Function**:

- Transcriptional regulation during development ensures the formation of neural circuits required for complex behaviors.
- Genes encoding synaptic proteins and neurotransmitter receptors are transcribed in response to environmental stimuli, allowing insects to modify behavior dynamically.

» **Circadian Rhythms**:

- Insects rely on transcriptional feedback loops involving clock genes such as *period* and *timeless* to maintain circadian rhythms. These rhythms regulate feeding, mating, and migration behaviors.

» **Social Behaviors:**

- In eusocial insects like ants and bees, transcriptional differences between castes (workers, queens, drones) govern role-specific behaviors and physiological traits.
- Epigenetic modifications influence caste-specific gene expression, enabling flexibility in social roles.

2.1.2 RNA Processing in Insects

RNA processing is a critical step in gene expression, bridging the gap between transcription and translation. After transcription, the initial RNA transcript, known as precursor mRNA (pre-mRNA), undergoes a series of intricate modifications to become mature messenger RNA (mRNA). These modifications ensure that the RNA is stable, exportable from the nucleus, and capable of directing efficient protein synthesis. The fundamental steps in RNA processing—capping, splicing, and polyadenylation—are conserved across eukaryotes, including insects. However, insects exhibit unique adaptations and variations in these processes, reflecting the diversity of their life cycles, environmental conditions, and physiological needs.

1. Capping

Capping is the first modification that occurs during RNA processing, involving the addition of a specialized nucleotide structure, called the 5' cap, to the 5' end of the pre-mRNA. This modification occurs early in transcription, often after RNA polymerase II has synthesized about 20 to 30 nucleotides of the RNA strand. The 5' cap is essential for RNA stability, protection from degradation, efficient translation, and proper ribosomal recognition.

1.1. The Composition of the 5' Cap

- The 5' cap consists of a 7-methylguanosine (m7G) linked to the 5' end of the RNA via a unique 5'-to-5' triphosphate bridge.
- In insects, as in other organisms, this cap is further modified with methylation at the ribose of the first and sometimes the second nucleotide, depending on the species and cell type.

1.2. Steps of Capping

The capping process involves three key enzymatic reactions:

1. **Removal of the γ-phosphate:** A triphosphatase enzyme cleaves the terminal γ-phosphate from the 5' triphosphate group of the nascent RNA, leaving a diphosphate.
2. **Addition of guanine:** A guanylyltransferase enzyme attaches guanine monophosphate (GMP) to the diphosphate RNA via a 5'-to-5' triphosphate linkage, forming the basic cap structure.
3. **Methylation:** Methyltransferase enzymes transfer a methyl group to the guanine at the N-7 position, creating the 7-methylguanosine cap (m7G). Additional methyl groups may also be added to the ribose of adjacent nucleotides.

1.3. Functions of the 5' Cap

- **Stability:** The cap protects the RNA from exonuclease activity, preventing degradation by enzymes that target unprotected RNA ends.
- **Ribosome Recruitment:** The cap is recognized by cap-binding proteins (e.g., eIF4E), facilitating the assembly of the translation initiation complex.
- **Nuclear Export:** The cap structure interacts with nuclear export machinery, ensuring efficient transport of mature mRNA to the cytoplasm.

» **Splicing Regulation:** In some insects, the capping process may influence subsequent splicing events by interacting with the spliceosome.

1.4. Variations in Insects

» Insects display variations in the efficiency and timing of capping, potentially linked to differences in transcriptional activity or environmental stress.

» For example, insects that undergo diapause (a state of arrested development) may regulate capping to conserve resources and delay protein synthesis.

2. Splicing

Splicing is the process by which non-coding sequences (introns) are removed from the pre-mRNA, and coding sequences (exons) are joined together to form a continuous open reading frame. This step is crucial for generating functional mRNAs that encode proteins. Insects exhibit unique splicing mechanisms that reflect their diverse gene architectures and functional requirements.

2.1. The Spliceosome

» Splicing is mediated by a large ribonucleoprotein complex called the spliceosome. This complex consists of small nuclear RNAs (snRNAs) and associated proteins that recognize conserved splice sites in the pre-mRNA.

» The spliceosome assembles dynamically on the pre-mRNA, undergoing conformational changes during the splicing reaction.

2.2. Mechanism of Splicing

1. **Recognition of Splice Sites:**
   - The spliceosome recognizes conserved sequences at the 5' splice site (donor site), 3' splice site (acceptor site), and branch point.
   - In insects, these sequences may vary slightly from the consensus sequences observed in other organisms.
2. **Formation of the Lariat Intermediate:**
   - The branch point adenosine attacks the 5' splice site, forming a lariat structure and releasing the 5' exon.
3. **Joining of Exons:**
   - The 3' hydroxyl group of the released exon attacks the 3' splice site, joining the exons and releasing the lariat intron.

2.3. Alternative Splicing in Insects

» Alternative splicing is a common phenomenon in insects, allowing a single gene to produce multiple mRNA isoforms and, consequently, diverse protein products.

» Examples include:

- **Sex Determination in Drosophila:** The *Sex-lethal* (Sxl) gene regulates alternative splicing of the *transformer* (tra) gene, determining male or female development.

- **Developmental Regulation:** Splicing patterns can change during different life stages, such as larval, pupal, or adult stages, to meet specific physiological needs.

2.4. Factors Influencing Splicing

» **Cis-Regulatory Elements:** Sequences within the RNA itself, such as enhancers and silencers, influence splice site selection.

» **Trans-Acting Factors:** Proteins such as SR proteins and heterogeneous nuclear ribonucleoproteins (hnRNPs) modulate splicing by binding to specific RNA sequences.

» **Environmental Factors:** Temperature, diet, and other environmental variables can affect splicing patterns in insects.

3. Polyadenylation

Polyadenylation is the process of adding a poly(A) tail, a string of adenosine residues, to the 3' end of the pre-mRNA. This modification is crucial for mRNA stability, export, and translational efficiency. In insects, polyadenylation follows a tightly regulated sequence of events.

3.1. The Polyadenylation Signal

» Polyadenylation is guided by a conserved sequence motif in the pre-mRNA, typically AAUAAA or a variant, located upstream of the cleavage site.

» Insects may have species-specific variations in these motifs, which can influence the efficiency and outcome of polyadenylation.

3.2. Steps of Polyadenylation

1. **Cleavage of the Pre-mRNA:**
   - The pre-mRNA is cleaved downstream of the polyadenylation signal by a cleavage complex that includes endonucleases and auxiliary factors.
2. **Addition of Adenosines:**
   - Poly(A) polymerase (PAP) adds a series of adenosine residues, typically 100–250 nucleotides in length, to the cleaved 3' end.
3. **Binding of Poly(A)-Binding Proteins (PABPs):**
   - PABPs bind to the poly(A) tail, stabilizing the RNA and promoting translation.

3.3. Functions of the Poly(A) Tail

» **Stability:** The tail protects the mRNA from exonuclease degradation.

» **Nuclear Export:** The poly(A) tail facilitates the export of mature mRNA to the cytoplasm by interacting with nuclear export factors.

» **Translation Regulation:** PABPs enhance translation efficiency by promoting ribosome assembly.

3.4. Polyadenylation Variations in Insects

» Insects exhibit unique patterns of polyadenylation during development and stress responses. For instance:

- During diapause, poly(A) tails may be shortened to suppress translation.
- Some insect mRNAs undergo *alternative polyadenylation*, resulting in transcripts with different 3' untranslated regions (UTRs) that regulate stability and localization.

4. Integration and Regulation of RNA Processing

RNA processing in insects is highly regulated and integrated with transcription and downstream translation. The coordination of capping, splicing, and polyadenylation ensures that mRNA is accurately processed, exported, and translated in response to cellular and environmental cues.

4.1. Coupling with Transcription

» RNA polymerase II coordinates RNA processing by recruiting processing factors to the nascent transcript. This coupling ensures that capping occurs early, splicing proceeds co-transcriptionally, and polyadenylation is precisely timed.

4.2. Role in Development and Adaptation

» RNA processing mechanisms are dynamically regulated during insect development, allowing for stage-specific gene expression.

» Environmental factors such as temperature, light, and food availability influence RNA processing, enabling insects to adapt to changing conditions.

2.1.3 Translation Mechanisms in Insects

Translation is the process by which the genetic information encoded in messenger RNA (mRNA) is decoded to synthesize specific proteins. This critical step of gene expression is fundamental for cellular function, development, and adaptation in insects. The translation process is highly conserved across eukaryotes and involves a series of coordinated steps—initiation, elongation, and termination—facilitated by ribosomes, transfer RNAs (tRNAs), and various protein factors.

1. Overview of Translation in Insects

Translation occurs in the cytoplasm, where ribosomes interact with mRNA to assemble amino acids into a polypeptide chain based on the genetic code. Insects, due to their diverse ecological roles, development stages, and environmental challenges, often exhibit fine-tuned translation mechanisms to regulate protein synthesis. For instance, translation efficiency and regulation can vary between larval and adult stages, during diapause, or under environmental stresses such as temperature fluctuations or nutrient scarcity.

2. Key Components of Translation

Translation is a highly intricate and regulated process requiring the coordinated action of multiple molecular components. In insects, the primary players in translation—ribosomes, transfer RNAs (tRNAs), and translation factors—work synergistically to decode messenger RNA (mRNA) into functional proteins. This section provides an exhaustive exploration of these components, with detailed insights into their structure, functions, and unique adaptations observed in insects.

2.1. Ribosomes

Ribosomes are the molecular factories of protein synthesis. They orchestrate the assembly of amino acids into a polypeptide chain, following the sequence of codons in mRNA. Ribosomes in insects, as in all eukaryotes, are composed of ribosomal RNA (rRNA) and ribosomal proteins that form two distinct subunits.

2.1.1. Structure of Ribosomes

In insects, ribosomes consist of two subunits, which come together during translation to form a functional ribosome:

» **Large Subunit (60S):**

- The 60S subunit is responsible for catalyzing peptide bond formation through the peptidyl transferase center (PTC), which resides in the rRNA core.
- This subunit provides binding sites for the tRNA carrying the growing polypeptide chain and facilitates the exit of the nascent protein.

» **Small Subunit (40S):**

- The 40S subunit interacts with the mRNA and decodes its sequence by aligning it with complementary anticodons on the tRNA.
- This subunit houses the *decoding center*, where the accuracy of codon-anticodon pairing is ensured.

2.1.2. Function of Ribosomes

» **Decoding the Genetic Code:** Ribosomes translate the sequence of nucleotide codons in mRNA into the corresponding sequence of amino acids.

» **Catalysis of Peptide Bond Formation:** The ribosome acts as a ribozyme, with the PTC in the large subunit facilitating the formation of peptide bonds between amino acids.

» **Regulating Translation Efficiency:** Ribosomes in insects can adjust their activity based on developmental cues or environmental conditions to ensure efficient protein synthesis.

2.1.3. Unique Features in Insects

» **Stress Adaptations:**

- In response to environmental stresses such as extreme temperatures or starvation, insects modulate ribosomal activity. For instance, ribosomes may pause translation of non-essential proteins while prioritizing stress-response proteins like heat-shock proteins.
- Some insect species, such as those in diapause, significantly reduce global ribosomal activity to conserve energy during dormant periods.

- **Developmental Regulation:**
  - Ribosomes are dynamically regulated during key developmental transitions in insects. For example, during metamorphosis in *Drosophila melanogaster*, ribosomes selectively translate mRNAs involved in larval tissue degradation and adult structure formation.
- **Ribosomal Heterogeneity:**
  - Variations in ribosomal composition, particularly in rRNA and ribosomal proteins, enable insects to regulate translation in a context-specific manner. This heterogeneity facilitates selective translation of mRNAs required for specific environmental or physiological conditions.

2.2. Transfer RNA (tRNA)

Transfer RNAs (tRNAs) are indispensable components of the translation machinery, serving as adaptors that decode the genetic information in mRNA into amino acids. Each tRNA is uniquely charged with a specific amino acid, ensuring accurate protein synthesis.

2.2.1. Structure of tRNA

tRNAs are highly conserved in structure, with distinct features that enable their adaptor function:

- **Anticodon Loop:** A region that contains a triplet of nucleotides complementary to a codon on the mRNA. This ensures accurate pairing during translation.
- **Amino Acid Attachment Site:** Located at the 3' end of the tRNA molecule, this site binds to its corresponding amino acid via an ester bond.
- **TψC Loop and D Loop:** These regions provide structural stability to the tRNA and facilitate interactions with ribosomes and aminoacyl-tRNA synthetases.

2.2.2. Charging of tRNA

- The charging of tRNAs is performed by a group of enzymes called aminoacyl-tRNA synthetases. This process ensures that each tRNA is loaded with the correct amino acid.
  - **Step 1:** The amino acid is activated by ATP, forming an aminoacyl-adenylate intermediate.
  - **Step 2:** The amino acid is transferred to the tRNA's 3' end, creating an aminoacyl-tRNA complex.

2.2.3. Functions of tRNA

- **Codon Recognition:** tRNAs facilitate codon-anticodon pairing, ensuring that the correct amino acid is added to the growing polypeptide chain.
- **Delivery of Amino Acids:** Charged tRNAs transport specific amino acids to the ribosome, where they are incorporated into the protein.
- **Translational Fidelity:** tRNAs, in conjunction with aminoacyl-tRNA synthetases, maintain high fidelity in translation by ensuring accurate amino acid selection.

2.2.4. Insect-Specific Adaptations

» **tRNA Modifications:**

- Insects often modify tRNAs to enhance their stability and efficiency under environmental stress. For example, methylation of specific nucleotides in tRNA ensures proper function at extreme temperatures.

» **Selective Translation:**

- During diapause or other stress conditions, insects may rely on specific subsets of tRNAs to selectively translate stress-response proteins.

» **Compact Genomes:**

- Some insects have reduced tRNA gene repertoires. Despite this, their translational machinery compensates by employing versatile tRNAs capable of decoding multiple codons (wobble hypothesis).

2.3. Translation Factors

Translation factors are accessory proteins that play essential roles in the three stages of translation: initiation, elongation, and termination. These factors ensure efficiency, accuracy, and regulation of protein synthesis.

2.3.1. Initiation Factors (eIFs)

» **Role in Initiation:**

- Initiation factors facilitate the assembly of the ribosome on the mRNA, ensuring that translation begins at the correct start codon (AUG).

» **Key Examples:**

- **eIF4E:** Binds to the 5' cap of the mRNA, enabling recruitment of the small ribosomal subunit and subsequent scanning for the start codon.
- **eIF2:** Delivers the initiator tRNA$_i$^Met to the ribosome in a GTP-dependent manner.

» **Insect-Specific Insights:**

- In insects, the activity of eIF4E is often regulated during development and stress. For instance, during diapause, translation initiation is selectively suppressed for most mRNAs, while critical mRNAs required for survival are still translated.

2.3.2. Elongation Factors (eEFs)

» **Role in Elongation:**

- Elongation factors assist in the delivery of aminoacyl-tRNAs to the ribosome and facilitate the translocation of the ribosome along the mRNA.

» **Key Examples:**

- **eEF1A:** Responsible for bringing aminoacyl-tRNAs to the ribosome's A site.

- **eEF2:** Facilitates the movement of the ribosome along the mRNA by one codon, ensuring the proper alignment of tRNAs.

» **Insect-Specific Insights**:

- Elongation factors in insects are often regulated to prioritize the synthesis of proteins critical for stress response, such as heat-shock proteins. For example, during nutrient scarcity, elongation rates are reduced to conserve energy.

2.3.3. Release Factors (eRFs)

» **Role in Termination**:

- Release factors recognize stop codons on the mRNA and mediate the release of the completed polypeptide chain.

» **Key Examples:**

- **eRF1:** Recognizes all three stop codons (UAA, UAG, UGA) and catalyzes the hydrolysis of the bond between the polypeptide and tRNA.
- **eRF3:** Works in conjunction with eRF1 to enhance termination efficiency.

» **Insect-Specific Insights:**

- Insects may modulate the activity of release factors during periods of stress or dormancy to conserve resources. This allows for selective translation of essential proteins.

3. Key Steps in Translation

Translation is the process by which the genetic information encoded in mRNA is used to synthesize proteins. It involves three major steps: **initiation**, **elongation**, and **termination**, each of which is essential for proper protein synthesis. Among these, initiation is the most intricate and regulated step, as it establishes the foundation for accurate decoding of the mRNA. In insects, translation mechanisms not only align with general eukaryotic models but also exhibit unique adaptations due to their diverse ecological niches, developmental stages, and physiological responses to environmental stresses.

3.1. Initiation

Initiation is the first and most critical step in translation, as it determines where protein synthesis begins on an mRNA molecule. The assembly of the translation machinery ensures the accurate selection of the start codon (AUG) and the correct positioning of the initiator tRNA. This step is a highly regulated process involving ribosomes, initiation factors, mRNA, and tRNA.

3.1.1. Mechanism of Initiation

The initiation process involves three primary stages: **formation of the pre-initiation complex**, **mRNA binding and scanning**, and **assembly of the full ribosome**. Each stage is characterized by distinct molecular events that prepare the ribosome for efficient translation.

1. Formation of the Pre-Initiation Complex

- » **Overview:**
    - The initiation process begins with the assembly of the **40S small ribosomal subunit**, various initiation factors, and the initiator tRNA$_i$^Met.
- » **This step ensures the ribosome is primed for mRNA binding and codon recognition.**
- » **Molecular Details:**
    - The 40S subunit binds to several initiation factors, including:
        - ◊ **eIF1 and eIF1A:** Stabilize the ribosome in an open conformation, facilitating mRNA entry.
        - ◊ **eIF3:** Acts as a scaffold, coordinating the interaction between ribosomal components and mRNA.
    - The initiator tRNA$_i$^Met, charged with methionine, forms a **ternary complex** with **eIF2** (a GTP-binding protein) and GTP. This complex is recruited to the 40S ribosomal subunit.
    - The pre-initiation complex (PIC) serves as a critical checkpoint, ensuring all components are correctly assembled before progressing to the next stage.

2. mRNA Binding and Scanning

- » **Cap Recognition:**
    - The mRNA is recruited to the PIC through its 5' cap structure, which is recognized by the **eIF4F complex**:
        - ◊ **eIF4E:** A cap-binding protein that anchors the mRNA to the initiation machinery.
        - ◊ **eIF4A:** An RNA helicase that unwinds secondary structures in the 5' untranslated region (UTR) of the mRNA.
        - ◊ **eIF4G:** A scaffolding protein that bridges eIF4E and other initiation factors.
- » **mRNA Scanning:**
    - The 40S ribosomal subunit, along with the attached initiator tRNA$_i$^Met, scans the mRNA in the **5' to 3' direction** to locate the start codon (AUG).
    - The scanning process requires the coordinated activity of eIF4A to resolve mRNA secondary structures and maintain linear progression.
- » **Start Codon Recognition:**
    - When the anticodon of the initiator tRNA$_i$^Met pairs with the AUG codon, a conformational change occurs in the PIC.
    - This event triggers the hydrolysis of GTP bound to eIF2, resulting in the release of initiation factors and stabilization of the codon-anticodon pairing.

3. Assembly of the Full Ribosome

» **Large Subunit Joining:**

- Following start codon recognition, the **60S large ribosomal subunit** joins the 40S subunit to form the functional **80S ribosome**.
- This step is facilitated by **eIF5B**, which ensures proper alignment of the two subunits.

» **Positioning of the Initiator tRNA:**

- The initiator tRNAᵢ^Met is positioned in the **P site** of the ribosome, aligning its methionine with the start codon.

» **Transition to Elongation:**

- With the release of remaining initiation factors, the ribosome transitions to the elongation phase, where the nascent polypeptide chain begins to grow.

3.1.2. Unique Features in Insects

Insects, owing to their diverse life cycles and habitats, exhibit several adaptations in the initiation phase of translation. These adaptations enable insects to modulate protein synthesis in response to developmental cues, environmental stresses, and pathogen interactions.

1. Developmental Regulation

» **Maternal mRNA Control:**

- During early embryogenesis in insects such as *Drosophila melanogaster*, maternal mRNAs stored in the egg are translationally repressed until specific developmental cues activate their translation.
- This regulation is mediated by proteins such as **Smaug** and **Cup**, which interact with eIF4E to suppress translation.

» **Metamorphosis:**

- Insects undergoing complete metamorphosis, such as moths and butterflies, selectively translate mRNAs that encode proteins for tissue remodeling and adult structure formation.
- Initiation factors may be developmentally regulated to prioritize the translation of stage-specific proteins.

2. Environmental Response Mechanisms

» **Cold Adaptation:**

- Insects that survive in cold environments, such as Antarctic midges, exhibit reduced global translation but maintain selective initiation of stress-response mRNAs encoding antifreeze proteins and heat-shock proteins (HSPs).
- During **diapause**, insects suppress cap-dependent translation to conserve energy while maintaining the translation of specific mRNAs via alternative pathways.

» **Heat Shock Response:**

- Heat stress triggers selective translation of HSP mRNAs, which protect cellular proteins from denaturation.
- This process is mediated by IRES elements in the mRNAs, allowing translation to bypass the canonical cap-dependent initiation pathway.

» **Nutritional Stress:**

- Starvation induces a global reduction in translation initiation while preserving the synthesis of essential proteins, such as enzymes involved in lipid metabolism.

3. Viral Pathogen Interactions

» Insects, as vectors for viruses such as dengue and Zika, face unique challenges in maintaining their translational machinery:

- Viruses often hijack host initiation factors, such as eIF4E, to prioritize viral protein synthesis.
- Some insects have evolved mechanisms to selectively inhibit viral mRNA translation without disrupting host protein synthesis.

4. Alternative Initiation Pathways

» **IRES-Driven Translation:**

- Internal ribosome entry sites (IRESs) in the 5' UTR of certain mRNAs enable ribosome assembly independent of the 5' cap structure.
- This mechanism is particularly advantageous during stress conditions or viral infections when cap-dependent translation is suppressed.

» **Cap-Independent Translation:**

- Insect mRNAs with specialized sequences or structures can recruit ribosomes directly to the start codon, bypassing the need for eIF4E activity.

3.1.3. Implications of Initiation Regulation in Insects

The regulation of translation initiation in insects has far-reaching implications for their biology and survival:

1. **Developmental Plasticity:**
   - Precise control of initiation allows insects to synthesize proteins required for growth, metamorphosis, and reproduction at the right time and location.
2. **Stress Tolerance:**
   - Adaptations in initiation mechanisms enable insects to survive extreme environmental conditions, such as freezing temperatures, heatwaves, and nutrient scarcity.
3. **Host-Pathogen Interactions:**
   - The ability to selectively regulate initiation factors allows insects to counteract viral hijacking of their translational machinery.

4. **Pest Management:**
    - Understanding insect-specific initiation pathways can inform the development of targeted pest control strategies, such as inhibitors of critical initiation factors.

3.2 Elongation

Elongation is the phase in translation where the ribosome synthesizes the growing polypeptide chain by sequentially adding amino acids as directed by the mRNA codons. This cyclic process is highly energy-intensive and relies on a coordinated interplay between the ribosome, tRNAs, elongation factors, and mRNA. In insects, elongation exhibits unique regulatory and adaptive features that reflect their diverse habitats, complex life cycles, and environmental challenges. Below, we explore the detailed mechanisms, roles of elongation factors, and the regulatory strategies insects employ during this critical phase of translation.

3.2.1 Mechanism of Elongation

The elongation cycle involves three distinct steps that repeat with each new codon on the mRNA: **codon recognition, peptide bond formation**, and **translocation**. These steps are orchestrated by the ribosome and supported by specific elongation factors to ensure both accuracy and efficiency.

1. Codon Recognition

» **Overview:**
- Codon recognition is the first step in the elongation cycle. The ribosome's **A site** is where incoming aminoacyl-tRNAs are selected based on codon-anticodon complementarity.
- The accuracy of this step is critical for maintaining the integrity of the synthesized protein, as errors in codon recognition can lead to the incorporation of incorrect amino acids.

» **Molecular Process:**
- A charged tRNA (aminoacyl-tRNA) enters the A site of the ribosome.
- The anticodon of the tRNA base-pairs with the corresponding codon on the mRNA. This pairing is stabilized by hydrogen bonds and the structural configuration of the ribosome.
- The incoming tRNA is delivered to the ribosome as part of a ternary complex with **elongation factor eEF1A** and GTP.
- If the codon-anticodon match is correct, the ribosome undergoes a conformational change, triggering GTP hydrolysis by eEF1A.
- eEF1A is released, and the aminoacyl-tRNA is securely positioned in the A site.

» **Fidelity Mechanisms:**
- The ribosome employs a proofreading mechanism to ensure the fidelity of codon recognition:
    - Incorrectly matched tRNAs are ejected before peptide bond formation.
    - This step significantly reduces the error rate in translation, maintaining an accuracy of approximately 1 in 10,000 amino acid incorporations.

» **Biological Significance:**

- Codon recognition ensures that the genetic code in the mRNA is translated accurately, preserving the functional integrity of proteins.

2. Peptide Bond Formation

» **Overview:**

- Peptide bond formation is the chemical reaction that links the amino acids together in the growing polypeptide chain. This step occurs within the ribosome's **peptidyl transferase center (PTC)**, located in the large ribosomal subunit.

» **Molecular Process:**

- The ribosome aligns the carboxyl group of the polypeptide chain attached to the tRNA in the P site with the amino group of the aminoacyl-tRNA in the A site.
- The PTC catalyzes a nucleophilic attack by the amino group on the carboxyl group, forming a peptide bond.
- The growing polypeptide chain is transferred from the tRNA in the P site to the tRNA in the A site.

» **Catalytic Role of rRNA:**

- The PTC is an RNA-based enzymatic center, highlighting the ribosome's function as a ribozyme.
- Ribosomal RNA (rRNA) within the PTC provides the catalytic framework for peptide bond formation.

» **Biological Importance:**

- Peptide bond formation is the central step in protein synthesis, driving the elongation of the polypeptide chain.

3. Translocation

» **Overview:**

- Translocation refers to the movement of the ribosome along the mRNA by one codon after each peptide bond formation. This step resets the ribosome for the next elongation cycle.

» **Molecular Process:**

- After peptide bond formation, the ribosome undergoes a conformational change, facilitated by **elongation factor eEF2**.
- The ribosome shifts the tRNA from the A site to the P site, and the uncharged tRNA in the P site is moved to the **E site** (exit site).
- The tRNA in the E site is released from the ribosome, vacating the site for the next cycle.
- The A site becomes available for the next incoming aminoacyl-tRNA.

» **Energy Requirements:**

- Translocation is powered by GTP hydrolysis by eEF2, making it an energy-dependent step.

» **Biological Importance:**

- Translocation ensures the continuity of translation by properly aligning the ribosome with the next codon on the mRNA.

3.2.2 Role of Elongation Factors

Elongation factors are accessory proteins that play critical roles in ensuring the accuracy, efficiency, and regulation of elongation. These factors are conserved across eukaryotes, including insects, and are essential for smooth progression through the elongation cycle.

1. Elongation Factor eEF1A

» **Function:**

- eEF1A delivers aminoacyl-tRNAs to the A site of the ribosome, ensuring that only tRNAs with the correct anticodon are positioned for codon recognition.

» **Mechanism:**

- eEF1A binds to aminoacyl-tRNAs in a GTP-bound state, forming a ternary complex.
- Once codon recognition occurs, eEF1A hydrolyzes GTP and releases the tRNA into the ribosome's A site.

» **Significance:**

- eEF1A ensures high fidelity in translation by rejecting mismatched tRNAs.

2. Elongation Factor eEF2

» **Function:**

- eEF2 facilitates the translocation of the ribosome along the mRNA.

» **Mechanism:**

- eEF2 binds to the ribosome and hydrolyzes GTP, providing the energy required for the ribosome to shift by one codon.

» **Significance:**

- eEF2 ensures proper alignment of the ribosome for subsequent rounds of elongation.

3.2.3 Regulatory Mechanisms in Insects

Insects regulate the elongation phase of translation to adapt to their unique physiological and environmental contexts. These regulatory mechanisms allow insects to conserve energy, prioritize essential protein synthesis, and respond effectively to stress conditions.

1. Regulation During Environmental Stress

» **Heat Stress:**

- Insects exposed to high temperatures prioritize the translation of heat-shock proteins (HSPs), which protect cellular proteins from denaturation.
- Heat stress reduces global elongation rates but enhances the efficiency of elongation for HSP mRNAs, often through specialized regulatory mechanisms.

» **Cold Stress:**

- Cold-adapted insects, such as Arctic midges, exhibit slower elongation rates to ensure proper folding of proteins under low-temperature conditions.
- These insects may also employ post-translational modifications of elongation factors to optimize their activity in cold environments.

» **Oxidative Stress:**

- Under oxidative stress, insects regulate elongation factors to maintain the synthesis of proteins involved in cellular repair and detoxification.

2. Developmental Regulation

» **Metamorphosis:**

- During metamorphosis, insects selectively translate mRNAs encoding proteins required for tissue remodeling and adult structure formation.
- Elongation factors are often modulated during this stage to support high rates of protein synthesis.

» **Embryogenesis:**

- In early development, maternal mRNAs stored in the egg are selectively translated, requiring precise regulation of elongation to synthesize proteins essential for growth and differentiation.

3. Selective Translation of Essential Proteins

» **Nutritional Stress:**

- Starvation or nutrient limitation triggers a global reduction in elongation rates, conserving energy while prioritizing the synthesis of essential proteins.
- Insects maintain the elongation of mRNAs encoding enzymes involved in lipid metabolism and stress response.

» **Viral Infections:**

- Insects serving as vectors for viruses (e.g., mosquitoes transmitting dengue virus) regulate elongation factors to prevent viral protein synthesis.
- Some insects employ ribosomal heterogeneity or specific elongation factor modifications to selectively inhibit viral translation.

4. Ribosomal Heterogeneity in Elongation Regulation

- Insects exhibit ribosomal heterogeneity, where ribosomes vary in their protein or rRNA composition. This variation allows selective translation of specific mRNAs during elongation.
- For example, during diapause, ribosomes with specialized elongation factor-binding sites may prioritize the translation of stress-response proteins while suppressing general protein synthesis.

3.2.4 Insect-Specific Adaptations in Elongation

» **Extreme Environments:**

- Desert insects modulate elongation to conserve water and energy, reducing global translation rates while synthesizing protective proteins.
- Aquatic insects may adjust elongation mechanisms to cope with hypoxic conditions by prioritizing proteins involved in anaerobic metabolism.

» **Diapause and Dormancy:**

- During diapause, insects conserve energy by downregulating elongation factors. However, selective translation of antifreeze proteins, HSPs, or other stress-related proteins continues.

3.3. Termination

Termination is the final step in translation, marking the end of the protein synthesis process. It occurs when the ribosome encounters a stop codon on the mRNA, signaling the release of the newly synthesized polypeptide chain. This phase ensures that the protein is completed accurately and prepares the ribosome for another round of translation. Termination is a highly regulated process involving stop codon recognition, the activity of release factors, and the dissociation of the translation machinery. In insects, unique regulatory mechanisms in termination contribute to their ability to adapt to various developmental stages and environmental challenges.

3.3.1. Stop Codons

Stop codons are specialized codons that do not code for any amino acids but instead serve as termination signals. In insects, as in all eukaryotes, there are three stop codons: **UAA**, **UAG**, and **UGA**.

» **Recognition by the Ribosome:**

- Stop codons are recognized by protein release factors (e.g., eRF1 and eRF3) rather than by tRNAs.
- The ribosome's decoding center detects the presence of a stop codon, initiating the termination process.

» **Biological Importance of Stop Codons:**

- Stop codons define the boundaries of the protein-coding region, ensuring that translation terminates at the correct location.
- Mutations in stop codons (e.g., nonsense mutations) can lead to premature termination,

resulting in truncated and often nonfunctional proteins. In insects, such mutations may disrupt critical processes like metamorphosis or stress response.

» **Stop Codon Preference in Insects:**

- Certain insects exhibit a bias in the usage of stop codons, reflecting codon optimization in their genomes. For example, *Drosophila melanogaster* may preferentially use UAA as a stop codon due to its higher translation efficiency.

3.3.2. Mechanism of Termination

The termination phase of translation involves three key steps: **recognition of stop codons, release of the polypeptide chain**, and **disassembly of the ribosome**. Each step is mediated by specialized proteins and molecular interactions.

1. Recognition of Stop Codons

» **Role of Release Factors:**

- When a stop codon enters the ribosome's **A site**, it is recognized by release factors:
  - ◊ **eRF1:** Mimics the structure of tRNA and directly binds to the stop codon.
  - ◊ **eRF3:** A GTPase that enhances the efficiency of eRF1 in catalyzing termination.
- eRF1 interacts with the ribosome's decoding center to confirm the presence of a stop codon, initiating the termination process.

» **Mechanism:**

- The release factors induce a conformational change in the ribosome, positioning the catalytic machinery for the next step—cleavage of the peptidyl-tRNA bond.

» **Accuracy Mechanisms:**

- The ribosome ensures that only stop codons trigger termination by recognizing specific structural features of the codon-release factor complex.

2. Release of the Polypeptide

» **Cleavage of the Peptidyl-tRNA Bond:**

- Once the stop codon is recognized, the peptidyl-tRNA bond in the ribosome's P site is hydrolyzed.
- This reaction is catalyzed by the ribosome in conjunction with eRF1, resulting in the release of the completed polypeptide chain.

» **Energy Requirement:**

- GTP hydrolysis by eRF3 provides the energy necessary for this step, ensuring efficient and irreversible release of the polypeptide.

» **Functional Significance:**

- The release of the polypeptide marks the culmination of translation, allowing the protein to fold into its functional conformation and participate in cellular processes.

3. Disassembly of the Ribosome

» **Dissociation of Ribosomal Subunits:**

- After the polypeptide is released, the ribosomal subunits (40S and 60S) dissociate from each other and from the mRNA.
- This process is facilitated by **ribosome recycling factors (RRFs)** and additional proteins like eIF6.

» **Release of tRNA and mRNA**:

- The deacylated tRNA is released from the ribosome's P site.
- The mRNA is also released, making it available for degradation or further rounds of translation, depending on cellular needs.

» **Preparation for Future Translation:**

- The ribosomal subunits are recycled, ready to participate in the initiation of another translation cycle.

3.3.3. Adaptations in Insects

Insects exhibit unique regulatory mechanisms in the termination phase of translation, allowing them to manage protein synthesis efficiently during development, environmental stress, and resource scarcity. These adaptations include modulation of termination efficiency, selective translation during diapause, and strategies to counteract stress-induced damage.

1. Regulation During Development

» **Metamorphosis**:

- During the transition from larva to pupa and eventually to an adult, insects undergo extensive tissue remodeling and organogenesis.
- Precise regulation of termination ensures the synthesis of proteins required for these developmental processes while preventing the accumulation of truncated or faulty proteins.

» **Maternal mRNA Translation:**

- In early embryogenesis, maternal mRNAs are selectively translated, and their termination efficiency may be regulated to optimize protein output during rapid cellular divisions.

2. Adaptations During Diapause

» **Energy Conservation:**

- Diapause is a state of arrested development that insects enter to survive adverse environmental conditions. During this period:
  - ◊ General translation, including termination, is downregulated to conserve energy.
  - ◊ Selective translation of essential proteins, such as stress-response enzymes and antifreeze proteins, continues. Termination factors may be modified to enhance their efficiency for these specific proteins.

» **mRNA Stability:**

- Termination efficiency is tightly linked to mRNA stability, as improper termination can lead to RNA degradation. Insects ensure that critical mRNAs remain intact during diapause for rapid translation upon environmental improvement.

3. Response to Environmental Stress

» **Heat Stress:**

- High temperatures can cause ribosome stalling at stop codons, leading to incomplete protein synthesis. Insects may upregulate heat-shock proteins (HSPs) to stabilize the translation machinery and ensure efficient termination.

» **Cold Stress:**

- In cold-tolerant insects, termination efficiency is adjusted to maintain protein synthesis at low temperatures. This prevents ribosome stalling and ensures proper folding of newly synthesized proteins.

4. Viral Defense Mechanisms

» **Viral Hijacking of Termination Machinery:**

- Many insect-borne viruses, such as those transmitted by mosquitoes, manipulate the host's translation machinery, including termination factors, to prioritize viral protein synthesis.
- Insects counteract this by selectively modifying their termination processes to restrict viral replication while maintaining host protein synthesis.

» **Premature Stop Codon Recognition:**

- Some insects have evolved mechanisms to detect and degrade viral mRNAs with premature stop codons, limiting the production of viral proteins.

5. Resource Scarcity and Nutritional Stress

» **Selective Translation:**

- Under starvation conditions, insects prioritize the termination of mRNAs encoding essential proteins, such as those involved in metabolism and stress tolerance.
- Termination factors may be redistributed or upregulated for specific mRNAs to optimize resource utilization.

6. Evolutionary Insights

» **Stop Codon Usage Bias:**

- Insects often exhibit a bias in the usage of stop codons, with certain codons being more frequently used in highly expressed genes. This bias reflects evolutionary optimization for efficient termination.

- **Termination Factor Variability:**
  - Variations in release factors, both in sequence and expression levels, have been observed across insect species, suggesting evolutionary adaptation to specific ecological niches.

3.3.4. Broader Implications

Understanding the mechanisms and adaptations of translation termination in insects has several practical applications:

- **Pest Control:**
  - Targeting insect-specific termination factors or their regulatory pathways can disrupt protein synthesis, offering a potential strategy for pest management.
- **Vector Control:**
  - Insights into termination regulation can help develop interventions to counteract viral replication in disease vectors, reducing the transmission of diseases like dengue and malaria.
- **Stress Resilience Research:**
  - Studying termination adaptations in insects provides valuable information on how organisms manage protein synthesis under extreme conditions, which can inform agricultural and biotechnological applications.

Applications of Transcription and Translation Studies

Transcription and translation are fundamental biological processes that convert genetic information into functional proteins. Studying these processes in insects has a wide range of practical applications, particularly in fields such as pest management, agricultural biotechnology, and disease control.

1. Identifying Genes Involved in Insecticide Resistance

The evolution of insecticide resistance is a major problem in agriculture and public health. Studies of transcription and translation provide insights into the molecular mechanisms that confer resistance, enabling the development of targeted strategies to mitigate this issue.

- **Mechanisms Unveiled by Transcription Studies:**
  - Transcriptional profiling identifies genes that are upregulated or downregulated in resistant insect populations compared to susceptible ones. Commonly identified genes include:
    - ◊ **Detoxification Enzymes:** Cytochrome P450s, glutathione S-transferases (GSTs), and carboxylesterases, which break down insecticides into less harmful compounds.
    - ◊ **Efflux Transporters:** Genes encoding ATP-binding cassette (ABC) transporters, which actively pump insecticides out of cells.
    - ◊ **Cuticular Proteins:** Genes involved in cuticle thickening, reducing insecticide penetration.

» **Role of Translation Studies:**

- Translation studies focus on post-transcriptional regulation, identifying how resistant insects produce large quantities of detoxification enzymes or other resistance-related proteins.
- Techniques such as ribosome profiling reveal translational efficiencies of key resistance genes.

» **Applications:**

- **Early Detection:** By identifying resistance-associated genes, molecular diagnostics can be developed to detect resistant populations before they cause significant damage.
- **New Insecticide Targets:** Understanding the transcriptional and translational basis of resistance helps in designing new insecticides that bypass or counteract these mechanisms.
- **Resistance Management:** Studies provide insights into deploying insecticides in a way that reduces the selection pressure for resistance genes.

2. Developmental Genes for Pest Management Strategies

Insects undergo complex developmental processes such as metamorphosis, molting, and diapause, regulated by a network of genes at the transcriptional and translational levels. Studying these genes provides valuable opportunities for pest control.

» **Transcriptional Insights into Developmental Genes:**

- Key transcription factors such as *ecdysone receptor (EcR)* and *juvenile hormone (JH)* signaling genes regulate metamorphosis and molting.
- Temporal and spatial transcriptional profiling during insect development helps identify critical periods where pests are most vulnerable.

» **Translational Regulation in Development:**

- Translation studies reveal how environmental factors such as temperature or food availability influence the synthesis of proteins required for development.
- Ribosome profiling can identify stage-specific proteins essential for larval growth, pupation, or adult emergence.

» **Applications**:

» **Targeted Interventions:**

- Disrupting the expression or translation of developmental genes can halt pest growth or reproduction.
- For example, juvenile hormone analogs or antagonists can prevent insects from reaching adulthood, reducing their ability to reproduce.

» **Timing of Pesticide Applications:**

- Knowledge of gene expression during development informs the optimal timing for pesticide applications, ensuring maximum impact on vulnerable life stages.

» **Developmental Disruption:**

- Gene silencing techniques targeting essential developmental genes can provide species-specific pest control without harming beneficial insects.

3. Exploring Molecular Targets for RNA Interference (RNAi) Pest Control Techniques

RNA interference (RNAi) is an innovative and environmentally friendly pest control method that leverages transcription and translation studies to silence specific genes in target insects. RNAi is particularly promising because it is species-specific and can target genes essential for survival or reproduction.

» **Transcriptional Studies for RNAi Target Identification:**

- High-throughput sequencing identifies genes critical for pest survival, such as those involved in metabolism, reproduction, or immunity.
- Gene expression analysis during stress or exposure to pesticides reveals vulnerabilities that can be exploited using RNAi.

» **Translation Studies for Effective RNAi Delivery:**

- Studies of RNA stability and ribosome interactions help optimize RNAi efficiency by identifying:
- The most effective regions of target mRNA for silencing.
- Mechanisms of RNAi resistance, such as increased degradation of double-stranded RNA (dsRNA) in some pests.

» **Applications**:

» **Gene-Specific Pest Control:**

- RNAi can silence genes involved in feeding behavior (e.g., salivary gland genes), reducing crop damage.
- Targeting reproductive genes prevents pests from reproducing, leading to population decline.

» **Overcoming Resistance:**

- RNAi can be used to silence resistance genes, restoring the effectiveness of existing insecticides.

» **Compatibility with Biopesticides:**

- RNAi can be delivered through genetically engineered plants, oral formulations, or microbial biopesticides.

4. Applications in Disease Vector Control

Insects such as mosquitoes, tsetse flies, and sandflies are vectors for diseases like malaria, dengue, and leishmaniasis. Transcription and translation studies in these vectors provide valuable tools for controlling the spread of diseases.

» **Transcriptional Studies in Vectors:**

- Identifying genes involved in pathogen transmission, such as those regulating midgut immunity or salivary gland proteins, reveals potential targets for disrupting disease transmission.
- Gene expression studies identify pathways that could be targeted to reduce vector populations or pathogen replication.

» **Translation Studies in Vector Biology:**

- Translational regulation of genes involved in vector-pathogen interactions can provide insights into how vectors respond to infection.
- For example, understanding the translational regulation of immune proteins in mosquitoes can help develop strategies to enhance their ability to suppress pathogens.

» **Applications:**

» **Genetic Control of Vectors:**

- Transcription and translation studies inform genetic engineering approaches, such as releasing sterile or pathogen-resistant mosquitoes.

» **Interruption of Pathogen Transmission:**

- Targeting vector genes essential for pathogen survival (e.g., specific gut proteins required for malaria parasite development) can block disease transmission.

» **Development of RNAi-Based Vector Control:**

- RNAi techniques can be used to silence vector genes involved in reproduction or host-seeking behavior, reducing vector populations and disease spread.

5. Agricultural Biotechnology and Crop Protection

Transcription and translation studies in insects also contribute to the development of genetically engineered crops and biopesticides that target pest species without harming beneficial organisms.

» **Understanding Pest-Plant Interactions:**

- Transcriptional analysis of pest genes during plant infestation reveals the molecular basis of feeding behavior and plant-host specificity.
- Translation studies identify proteins secreted by pests during feeding that facilitate nutrient extraction or suppress plant defenses.

» **Applications in Biotech Crops:**

- Genetically engineered plants expressing dsRNA for RNAi target pest genes, such as those involved in digestive enzyme production.
- Crops engineered with pest-resistant traits disrupt the transcription or translation of essential pest genes.

6. Insights into Insect Evolution and Adaptation

Transcription and translation studies also offer insights into how insects adapt to changing environments, develop resistance to natural and artificial toxins, and evolve complex behaviors.

» **Applications:**

- Understanding how insects adapt at the molecular level helps predict the emergence of new pest species or resistance mechanisms.
- Insights into **evolutionary conserved pathways can inform the design of broad-spectru**m pest control strategies.

2.2 DNA Recombinant Technology in Insects

Recombinant DNA (rDNA) technology has transformed molecular biology and entomology, offering innovative approaches to studying and manipulating insect physiology, genetics, behavior, and interactions with their environment. By integrating foreign genes into insect genomes, researchers have opened avenues for advancements in pest control, disease management, and the exploration of biological pathways in insects.

2.2.1 Principles of Recombinant DNA Technology

The fundamental goal of recombinant DNA technology is to create genetically modified organisms (GMOs) by incorporating specific genes into their genomes. In the context of insects, this process enables researchers to elucidate gene function, study metabolic pathways, and even engineer insects with desirable traits, such as resistance to pathogens or altered reproductive capabilities. Below are the key components and steps involved:

1. Restriction Enzymes

Restriction enzymes, also known as restriction endonucleases, are indispensable tools in molecular biology, particularly in recombinant DNA (rDNA) technology. These specialized enzymes act as molecular scissors that can cut DNA at specific sequences, enabling precise manipulation of genetic material. Their discovery and subsequent application have revolutionized genetic engineering, making them fundamental components in studying and modifying DNA.

Definition and Function

Restriction enzymes are naturally occurring proteins found in bacteria and archaea. In their native organisms, they serve as part of a defense mechanism against invading viral DNA (bacteriophages). The host organism's DNA is protected from its own restriction enzymes by methylation of the recognition sites, whereas foreign DNA is cleaved and rendered non-functional.

In genetic engineering, restriction enzymes are used to:

» Precisely cut DNA at specific sequences known as **restriction sites**.

» Generate fragments with defined ends that can be easily joined to vectors or other DNA fragments during recombinant DNA construction.

Mechanism of Action

Restriction enzymes scan DNA for their specific recognition sequences, typically 4-8 base pairs in length, and cleave the phosphodiester bonds within or near these sequences. Based on the nature of their cuts, they can generate two types of ends:

1. **Sticky Ends**:
   » Produced by enzymes that make staggered cuts in the DNA, leaving overhanging single-stranded ends.
   » Example:
   - **EcoRI**, which recognizes the sequence 5'-GAATTC-3', cuts between G and A:

     5'---G AATTC---3'

     3'---CTTAA G---5'

This results in two single-stranded overhangs (sticky ends) that are complementary and can easily pair with corresponding sticky ends from another DNA fragment.

2. **Blunt Ends**:
   » Produced by enzymes that cut straight across both strands, leaving no overhangs.
   » Example:
   - **SmaI**, which recognizes the sequence 5'-CCCGGG-3', cuts between C and G:

     5'---CCCGGG---3'

     3'---GGGCCC---5'

Sticky ends are highly advantageous in recombinant DNA technology because their complementary overhangs allow for more efficient and specific ligation. However, blunt ends can also be ligated, albeit less efficiently, often requiring higher concentrations of DNA and ligase.

Examples of Restriction Enzymes

1. **EcoRI (Escherichia coli RY13)**:
   - Recognition Sequence: 5'-GAATTC-3'
   - Cut Pattern: Between G and A, leaving sticky ends.
   - Application: Commonly used for cloning, generating DNA fragments with cohesive ends.
2. **HindIII (Haemophilus influenzae Rd):**
   - Recognition Sequence: 5'-AAGCTT-3'
   - Cut Pattern: Between A and A, leaving sticky ends.
   - Application: Often used in plasmid mapping and constructing recombinant plasmids.
3. **SmaI (Serratia marcescens)**:
   - Recognition Sequence: 5'-CCCGGG-3'

- Cut Pattern: Between C and G, producing blunt ends.
- Application: Useful when sticky ends are not required.

4. **BamHI (Bacillus amyloliquefaciens H):**
   - Recognition Sequence: 5'-GGATCC-3'
   - Cut Pattern: Between G and G, generating sticky ends.
   - Application: Frequently employed in vector construction.

Classification of Restriction Enzymes

Restriction enzymes are classified into four major types based on their structure, cofactor requirements, and cleavage sites relative to their recognition sequences:

1. **Type I:**
   - Recognize specific sequences but cleave DNA at sites distant from the recognition site.
   - Require ATP and other cofactors.
   - Less commonly used in molecular biology due to their non-specific cleavage patterns.
2. **Type II:**
   - Recognize specific sequences and cleave DNA within or close to the recognition site.
   - Require only magnesium ions (Mg^{2+}) as cofactors.
   - Widely used in recombinant DNA technology due to their precision.
3. **Type III:**
   - Recognize specific sequences and cleave DNA a short distance away.
   - Require ATP for activity.
   - Less commonly used than Type II enzymes.
4. **Type IV:**
   - Recognize and cleave methylated or modified DNA.
   - Play roles in epigenetics and DNA modification studies.

Applications in Recombinant DNA Technology

1. **Precise DNA Fragmentation**: Restriction enzymes cut DNA into predictable fragments, enabling researchers to isolate specific genes or DNA regions of interest. For instance, a gene can be excised from its native location using restriction enzymes and then inserted into a vector for cloning.
2. **Creation of Recombinant Molecules**: By cutting both the vector and the DNA fragment of interest with the same restriction enzyme, complementary sticky ends are generated, ensuring seamless integration during ligation.
3. **Plasmid Mapping**: Restriction enzymes are used to generate a map of restriction sites within a plasmid. This helps in verifying the presence and orientation of inserted DNA fragments.

4. **Molecular Cloning**: In molecular cloning, restriction enzymes are essential for:
    - Preparing vectors by opening them at specific sites.
    - Generating DNA inserts with compatible ends.
    - Facilitating ligation to form recombinant DNA.
5. **Gene Editing and Synthetic Biology**: Restriction enzymes are often used in tandem with advanced tools like CRISPR-Cas9 to create tailored DNA constructs for synthetic biology applications.

Advantages of Restriction Enzymes in DNA Manipulation

» **Specificity**: Restriction enzymes recognize precise sequences, allowing for targeted DNA manipulation.

» **Versatility**: Hundreds of restriction enzymes with unique recognition sequences provide flexibility in experimental design.

» **Efficiency**: Their ability to create sticky ends ensures efficient ligation and high cloning success rates.

Limitations and Challenges

» **Limited Recognition Sites**: Some sequences may not be present in the target DNA, restricting enzyme choice.

» **Star Activity**: Under non-optimal conditions (e.g., high glycerol concentration, incorrect buffer), some restriction enzymes may cleave at sequences similar to, but not identical to, their recognition sites.

» **Blunt Ends**: While blunt-end cuts are versatile, they are less efficient for ligation compared to sticky ends.

2. Vectors

Vectors are essential tools in recombinant DNA technology, serving as DNA carriers that introduce the gene of interest into host cells. These vehicles enable the replication, expression, and manipulation of foreign DNA within the target organism. The selection of a vector is critical and depends on the host system (in this case, insect cells) and the specific application, such as gene expression, protein production, or functional genomics studies.

Types of Vectors

1. **Plasmids:**

» **Description**: Plasmids are small, circular DNA molecules naturally found in bacteria. They replicate independently of the chromosomal DNA, making them ideal for carrying foreign genes.

» **Advantages:**

- High replication efficiency in host cells.

- Ease of genetic modification to include desired features.
- Suitable for small DNA fragments (up to ~15 kb).

» **Applications:**

- Used in molecular cloning and gene expression studies.
- Engineered plasmids often contain features like multiple cloning sites (MCS), selectable markers, and strong promoters.
- **Example**: pUC series plasmids, widely used for cloning.

2. **Viruses:**

» **Description**: Viral vectors exploit the natural ability of viruses to infect host cells and deliver genetic material. They are engineered to eliminate pathogenic components while retaining the ability to carry and express foreign genes.

» **Commonly Used Viral Vectors**:

- **Baculoviruses**: Insect-specific viruses widely used for recombinant protein production. They can carry large DNA inserts (up to 38 kb).
- **Retroviruses**: Deliver genes into dividing cells by integrating into the host genome.

» **Applications**:

- Baculoviruses are particularly popular in insect cells for the production of vaccines and therapeutic proteins.
- Efficient for high-level expression of recombinant proteins in cell lines.

3. **Transposons:**

» **Description**: Transposons are mobile genetic elements that integrate into the host genome via a cut-and-paste or copy-and-paste mechanism. They can deliver foreign DNA into insect genomes effectively.

» **Examples:**

- **P-element**: Commonly used in *Drosophila melanogaster* research for gene insertion.
- **piggyBac**: Versatile transposon system capable of delivering large DNA fragments.

» **Applications**:

- Genetic transformation of insects.
- Studying gene function and regulatory elements in insect genomes.

Key Features of Vectors

1. **Origin of Replication (Ori)**:
   - Enables the autonomous replication of the vector within the host cell.
   - A well-designed Ori ensures high plasmid copy number and stable maintenance in the host.

2. **Selectable Markers**:
   - Allow researchers to identify host cells that have successfully taken up the vector.
   - Common markers include:
     - ◊ Antibiotic resistance genes (e.g., ampicillin resistance in bacterial vectors).
     - ◊ Fluorescent markers (e.g., GFP for visual confirmation in eukaryotic cells).
3. **Multiple Cloning Sites (MCS)**:
   - A short DNA region containing multiple restriction enzyme recognition sites.
   - Facilitates the insertion of the gene of interest using a variety of enzymes, providing flexibility in cloning.
4. **Promoters**:
   - Drive the expression of the inserted gene.
   - Examples include:
     - ◊ **T7 Promoter**: High-level expression in bacterial systems.
     - ◊ **Polh Promoter**: Used in baculovirus expression systems for robust protein production.
5. **Reporter Genes**:
   - Allow monitoring of vector activity and gene expression.
   - **Examples: LacZ (for β-galactosidase activity), GFP (green fluorescent pro**tein).

3. Ligation

Ligation is a critical step in the recombinant DNA process, where DNA fragments are joined to create a stable, functional recombinant molecule. This step ensures that the gene of interest is securely inserted into the vector, allowing for replication and expression in the host organism.

Process of Ligation

Ligation involves forming phosphodiester bonds between the 3'-hydroxyl and 5'-phosphate ends of DNA fragments. The enzyme responsible for this process is **DNA ligase**, which catalyzes the reaction in the presence of cofactors like ATP or NAD^+.

1. **Sticky-End Ligation**:
   - Sticky ends are single-stranded overhangs generated by restriction enzymes.
   - Complementary base pairing between sticky ends ensures precise alignment of DNA fragments before ligation.
   - **Advantages**:
     - ◊ High specificity due to complementary pairing.
     - ◊ Efficient and faster ligation process.

2. **Blunt-End Ligation**:
   - Blunt ends lack overhangs, making alignment dependent on proximity and ligase activity.
   - **Challenges**:
     - ◊ Lower efficiency compared to sticky ends.
     - ◊ Requires higher concentrations of DNA and ligase.
   - Specialized ligation protocols and additives (e.g., polyethylene glycol) can improve blunt-end ligation efficiency.

Steps in the Ligation Reaction

1. **Preparation of DNA Fragments**:
   - Both the vector and the insert DNA are digested with compatible restriction enzymes to create sticky or blunt ends.
   - DNA is purified to remove enzymes and contaminants.
2. **Mixing DNA Fragments:**
   - The insert DNA and vector are combined in a molar ratio optimized for ligation (typically 3:1 insert to vector ratio).
3. **Adding DNA Ligase:**
   - T4 DNA ligase is commonly used, requiring ATP as a cofactor.
   - The ligation reaction is incubated at an appropriate temperature (e.g., 16°C for sticky ends or room temperature for blunt ends) for 1–16 hours.
4. **Verification:**
   - The ligation product is analyzed by gel electrophoresis or transformed into host cells for downstream applications.

Factors Affecting Ligation Efficiency

1. **Concentration of DNA Fragments:**
   - Optimal concentrations of vector and insert DNA are crucial for efficient ligation.
   - Excess of one component can lead to undesired side products (e.g., self-ligation of vectors).
2. **Temperature and Incubation Time**:
   - Sticky-end ligation typically occurs at lower temperatures (16°C) to allow proper base pairing.
   - Blunt-end ligation benefits from room temperature incubation.
3. **Cofactors**:
   - ATP or NAD^+ is required for DNA ligase activity.
   - Insufficient cofactors can reduce ligation efficiency.

4. **Quality of DNA**:
   - DNA should be free of inhibitors (e.g., EDTA, salts) that can interfere with ligase activity.
5. **Vector Preparation**:
   - Dephosphorylation of the vector's 5' ends using alkaline phosphatase can prevent self-ligation, increasing the chances of successful insert integration.

Applications of Ligation

1. **Construction of Recombinant DNA Molecules**:
   - Essential for cloning genes into vectors for expression in bacterial, insect, or mammalian systems.
2. **Gene Libraries**:
   - Ligation is used to generate genomic or cDNA libraries by inserting diverse DNA fragments into vectors.
3. **Synthetic Biology**:
   - Enables the assembly of synthetic DNA constructs for creating artificial genetic systems.

4. Transformation and Transfection

The delivery of recombinant DNA into cells or organisms is a pivotal step in genetic engineering. This step ensures the uptake, integration, and eventual expression of the introduced genetic material. The methods used for this process vary depending on the type of target cell—prokaryotic (e.g., bacteria) or eukaryotic (e.g., insect cells). Transformation and transfection are two commonly employed techniques that allow researchers to genetically modify cells for various applications.

A. Transformation

Transformation refers to the process of introducing foreign DNA into bacterial or eukaryotic cells that can take up and maintain the DNA. While transformation is most commonly associated with bacterial systems, certain insect cells can also be transformed. It involves either chemically or physically inducing permeability in the cell membrane to allow DNA to enter.

Methods of Transformation

1. **Electroporation:**
   - **Principle**: High-voltage electric pulses temporarily disrupt the phospholipid bilayer of the cell membrane, creating transient pores. Through these pores, recombinant DNA enters the cell.

» **Procedure:**
   - Cells are suspended in an electroconductive buffer along with the recombinant DNA.
   - A short electrical pulse is applied.
   - The membrane reseals, trapping the DNA inside the cell.

» **Advantages**:

- Efficient for a wide range of cell types, including bacteria and eukaryotes.
- Does not require chemical treatment of cells.

» **Limitations**:

- Requires specialized equipment (electroporator).
- High voltage can damage cells if not optimized.

2. **Chemical Transformation:**

- **Principle**: Certain chemicals, such as calcium chloride, make cell membranes permeable to DNA.

» **Procedure**:

- Cells are treated with a solution containing calcium chloride (or similar compounds), which neutralizes the negative charges on the DNA and cell membrane.
- Heat shock (sudden temperature increase) is applied to create temporary pores, allowing DNA to enter.

» **Advantages**:

- Cost-effective and straightforward.
- Widely used for bacterial systems.

» **Limitations**:

- Less efficient compared to electroporation.
- Mostly limited to certain cell types, especially bacteria.

Applications of Transformation

1. **Bacterial Cloning:**
   - Transformation is commonly used to introduce recombinant plasmids into bacteria for replication and gene expression.
   - Example: Transformation of *E. coli* with plasmids for protein production.
2. **Genetic Studies:**
   - Transformation in insect cells (though rare) can be used to study gene function and pathways.

B. Transfection

Transfection refers specifically to the introduction of recombinant DNA into eukaryotic cells, including insect cells. This technique is crucial for studying gene expression, functional genomics, and protein production. Transfection methods are more diverse than transformation, accommodating the complex structure and physiology of eukaryotic cells.

Methods of Transfection

1. **Lipofection:**
   - **Principle**: DNA is encapsulated in lipid vesicles (liposomes) that fuse with the cell membrane, delivering the DNA into the cytoplasm.
   - **Procedure**:
     - ◊ Recombinant DNA is mixed with lipid reagents that spontaneously form vesicles around the DNA.
     - ◊ The liposome-DNA complex is incubated with cells, allowing fusion with the cell membrane.
   - **Advantages**:
     - ◊ Simple and effective for many eukaryotic cell types, including insect cells.
     - ◊ Non-toxic compared to other methods.
   - **Limitations**:
     - ◊ Efficiency can vary depending on cell type and lipid formulation.
2. **Microinjection:**
   - **Principle**: DNA is directly injected into the cytoplasm or nucleus of cells using a fine glass needle.
   - **Procedure**:
     - ◊ A micromanipulator and needle are used to deliver DNA into specific locations within cells or embryos.
   - **Advantages**:
     - ◊ High precision, ensuring delivery to the desired location.
     - ◊ Useful for modifying insect embryos or studying gene function.
   - **Limitations**:
     - ◊ Labor-intensive and requires specialized equipment.
     - ◊ Low throughput; suitable for small-scale applications.
3. **Baculovirus-Mediated Transfection:**
   - **Principle**: Engineered baculoviruses serve as vectors to deliver recombinant DNA into insect cells.
   - **Procedure**:
     - ◊ Recombinant baculovirus particles are prepared by inserting the gene of interest into the viral genome.
     - ◊ Insect cells are infected with the baculovirus, which delivers the DNA and drives its expression.

- **Advantages**:
  - ◊ Highly efficient for transfection of insect cells.
  - ◊ Allows for high-level recombinant protein production.
- **Limitations**:
  - ◊ Limited to baculovirus-susceptible cell lines or organisms.

4. **Other Methods**:
   - **Calcium Phosphate Precipitation**: DNA forms a precipitate with calcium phosphate, which facilitates its uptake by cells.
   - **Electroporation**: While more commonly used in bacterial systems, it can also transfect eukaryotic cells by creating transient pores in the membrane.
   - **Nanoparticle Delivery**: DNA is bound to nanoparticles, which enhance uptake by cells through endocytosis.

Applications of Transfection

1. **Gene Function Studies:**
   - Transfection enables transient or stable expression of genes in insect cells, allowing researchers to study their roles in biological pathways.
   - Example: Studying gene regulation in *Drosophila melanogaster* cells.
2. **Recombinant Protein Production:**
   - Transfection of insect cell lines, such as Sf9 (derived from *Spodoptera frugiperda*), with recombinant DNA enables large-scale production of proteins.
   - Example: Production of vaccines and therapeutic proteins using the baculovirus expression system.
3. **Gene Editing:**
   - Transfection is a critical step in delivering CRISPR-Cas9 components for genome editing in insect cells.
   - Example: Knockout of specific genes to study their functions.
4. **Developmental Biology:**
   - Transfection of insect embryos (e.g., by microinjection) is used to study gene expression during development.

Comparison of Transformation and Transfection

Aspect	Transformation	Transfection
Target Cells	Bacterial or some eukaryotic cells	Eukaryotic cells (insect cells)
Methods	Electroporation, chemical methods	Lipofection, microinjection, etc.
Efficiency	High in bacteria	Variable, depending on method
Applications	Cloning, plasmid propagation	Protein production, gene studies

Outcomes of Transformation and Transfection

Both transformation and transfection result in the uptake of recombinant DNA by the target cells. Depending on the experimental design, the introduced DNA can either:

» **Integrate into the Host Genome**: Resulting in stable genetic modification and long-term expression.

» **Remain as an Extrachromosomal Element**: Allowing for transient expression without integration.

2.2.2 Advanced Techniques in Recombinant DNA Technology

Recombinant DNA technology has progressed beyond traditional cloning and genetic manipulation methods. The integration of advanced techniques has opened new avenues for precision, scalability, and application in insects. These modern tools are reshaping molecular entomology by providing deeper insights into gene function, offering novel pest control strategies, and engineering disease-resistant insect populations.

1. CRISPR-Cas9 Gene Editing:

Historical Context and Development

The CRISPR-Cas9 system, originally discovered as a bacterial immune defense against viruses, has become a cornerstone of genetic engineering since its adaptation for laboratory use in 2012. CRISPR (Clustered Regularly Interspaced Short Palindromic Repeats) serves as a memory system in bacteria, allowing them to recognize and cleave foreign genetic material. Cas9 is a nuclease enzyme guided by RNA to the target DNA sequence, where it makes precise cuts. By repurposing this system, scientists have achieved unparalleled accuracy in editing insect genomes.

Mechanism of CRISPR-Cas9 in Insects

1. **Designing the Guide RNA (gRNA):**
    - A gRNA sequence is synthesized to complement the target DNA region. This ensures specificity in guiding Cas9 to the correct location in the insect genome.
2. **Cas9-Mediated Cleavage:**
    - The Cas9 enzyme introduces a double-strand break (DSB) at the target site. This DSB is repaired by:
        - ◊ **Non-Homologous End Joining (NHEJ):** An error-prone repair mechanism that may introduce insertions or deletions (indels), disrupting the target gene.
        - ◊ **Homology-Directed Repair (HDR):** A precise repair mechanism using a supplied DNA template, enabling the insertion of new genes.
3. **Delivery of CRISPR Components:**
    - Microinjection into insect embryos is the most common method, ensuring high efficiency in early developmental stages.

Applications in Insects

1. **Gene Knockouts:**
   - CRISPR-Cas9 allows for the creation of loss-of-function mutations in specific genes to study their roles in physiology, behavior, and development.
   - **Example:** Disrupting genes responsible for olfactory receptors in mosquitoes to study host-seeking behavior.
2. **Gene Drives:**
   - A revolutionary application, gene drives propagate a specific genetic trait through a population at rates higher than Mendelian inheritance. This is particularly useful for:
     - ◊ **Pest Control:** Reducing populations of crop pests.
     - ◊ **Disease Vector Management:** Suppressing mosquito populations to combat malaria and dengue.
3. **Pathogen Resistance:**
   - Modifying insect vectors to block the transmission of pathogens.
   - **Example:** Engineering *Anopheles* mosquitoes with genes that kill *Plasmodium* parasites in their gut.

Advantages of CRISPR-Cas9

» **Precision:** Unparalleled accuracy in targeting specific DNA sequences.

» **Versatility:** Applicable to a wide range of insect species.

» **Efficiency:** Relatively easy to design and implement compared to older gene-editing tools like ZFNs or TALENs.

Challenges and Limitations

» **Off-Target Effects:** Unintended edits at non-target sites can lead to undesired mutations.

» **Delivery Challenges:** Effective delivery of CRISPR components into non-model insects remains difficult.

» **Ethical Concerns:** The ecological risks of gene drives, including potential disruption of ecosystems, are significant.

2. RNA Interference (RNAi): A Tool for Targeted Gene Silencing

Introduction to RNAi Technology

RNA interference (RNAi) is a naturally occurring mechanism of post-transcriptional gene silencing first discovered in *Caenorhabditis elegans*. It has since been harnessed as a powerful tool to selectively silence genes in insects and other organisms. The ability to knock down gene expression transiently has made RNAi a preferred method for studying gene function and managing insect pests.

Mechanism of RNAi in Insects

1. **dsRNA Introduction:**
   - Double-stranded RNA (dsRNA) complementary to the target gene is synthesized and introduced into the insect via feeding, injection, or transfection.
2. **Dicer Processing:**
   - The dsRNA is cleaved by the Dicer enzyme into small interfering RNAs (siRNAs).
3. **RISC Assembly:**
   - The RNA-induced silencing complex (RISC) incorporates siRNAs and guides the degradation of the complementary mRNA, preventing protein synthesis.

Applications in Insects

1. **Gene Function Analysis:**
   - RNAi enables researchers to study the roles of specific genes without the need for permanent genetic modifications.
   - **Example:** Silencing detoxification-related genes to study insecticide resistance.
2. **Pest Management:**
   - RNAi-based biopesticides target essential genes in pests, reducing reliance on chemical insecticides.
   - **Example:** *Bt* crops expressing dsRNA to silence genes in pests like *Helicoverpa armigera*.
3. **Disease Control:**
   - RNAi is used to knock down genes in vectors, such as mosquitoes, that are critical for pathogen survival or transmission.
   - **Example:** Silencing genes involved in *Plasmodium* development within *Anopheles* mosquitoes.

Advantages of RNAi

» **Specificity:** Highly selective for the target gene.

» **Reversibility:** Gene silencing is temporary, reducing the risk of long-term ecological impacts.

» **Cost-Effectiveness:** RNAi applications, such as biopesticides, are economically viable for large-scale pest control.

Challenges and Limitations

» **Variable Efficiency:** The effectiveness of RNAi varies across insect species.

» **Delivery Barriers:** Efficient delivery of dsRNA remains a challenge, particularly for pests with robust gut barriers or dsRNA-degrading enzymes.

» **Environmental Stability:** dsRNA molecules can degrade rapidly in environmental conditions.

3. Transgenic Insects: Revolutionizing Insect Biology and Control

Introduction to Transgenic Technology

Transgenic insects are genetically engineered to express foreign genes that confer desirable traits. This involves stable integration of recombinant DNA into the insect genome, typically using tools like transposons, CRISPR-Cas9, or viral vectors. Transgenesis has revolutionized research in developmental biology, pest control, and disease vector management.

Creation of Transgenic Insects

1. **Gene Delivery Systems:**
   - Microinjection of recombinant DNA into early-stage embryos.
   - Use of transposons (e.g., piggyBac) for stable integration into the genome.
2. **Marker Selection:**
   - Selection markers like GFP or antibiotic resistance are used to identify successfully modified insects.
3. **Stable Inheritance:**
   - The foreign gene is stably passed to subsequent generations, enabling long-term studies or applications.

Applications of Transgenic Insects

1. **Functional Genomics:**
   - Transgenic *Drosophila melanogaster* lines expressing GFP are used to study gene expression patterns.
2. **Sterile Insect Technique (SIT):**
   - Engineering sterile males to reduce pest populations.
   - Example: Releasing sterile male *Aedes aegypti* to combat dengue.
3. **Disease Vector Management:**
   - Creating mosquitoes resistant to pathogens like *Plasmodium* or Zika virus.

Advantages of Transgenic Insects

» Long-term expression of desired traits.

» Precise control over genetic modifications.

» Reduced reliance on chemical pest control methods.

Challenges and Limitations

» Ethical concerns regarding ecological impacts.

» Regulatory hurdles for environmental release.

» High cost of development and deployment.

2.2.3 Applications of Recombinant DNA Technology in Insects

Recombinant DNA technology has unlocked a wide range of applications in insect science, offering innovative solutions in agriculture, public health, and biotechnology. By enabling precise genetic modifications, researchers can manipulate insect physiology, behavior, and ecological interactions to address challenges such as pest control, disease management, and functional genomics. The applications can be broadly categorized into pest control, disease vector control, and research and biotechnology.

1. Pest Control

One of the most impactful applications of recombinant DNA technology is its use in pest control. Insects such as mosquitoes, fruit flies, and agricultural pests cause significant damage to crops, livestock, and public health systems. Recombinant DNA technology has introduced transformative methods for managing pest populations in a targeted and environmentally friendly manner.

Gene Drives are among the most revolutionary tools in pest control. Gene drives are designed to spread specific genetic traits rapidly through pest populations, significantly altering their genetic makeup over generations. For example, a gene drive can introduce sterility in insect populations, drastically reducing their numbers. This technique is particularly effective for controlling populations of invasive species or pests resistant to traditional methods such as pesticides. By using tools like CRISPR-Cas9, researchers can engineer gene drives to target specific species while minimizing unintended effects on non-target organisms. Applications of gene drives include reducing populations of *Aedes aegypti*, the mosquito species responsible for spreading diseases like dengue and Zika.

The expression of ***Bt* toxins**, derived from the bacterium *Bacillus thuringiensis*, is another powerful application of recombinant DNA technology. *Bt* toxins are proteins that are lethal to specific insect pests but harmless to humans and most other organisms. Through genetic engineering, *Bt* toxins can be expressed directly in crops, such as cotton, corn, and rice, or in beneficial insects. When pests consume these toxins, the proteins disrupt their digestive systems, leading to mortality. This method significantly reduces the need for chemical pesticides, which can have adverse environmental effects and promote resistance in pests. Additionally, *Bt*-based pest control is species-specific, ensuring minimal impact on beneficial insects like pollinators.

2. Disease Vector Control

Recombinant DNA technology has played a critical role in controlling insect species that act as vectors for diseases. Insects like mosquitoes are responsible for transmitting life-threatening diseases, including malaria, dengue, chikungunya, and Zika virus. Genetic engineering offers innovative strategies to disrupt these transmission cycles, reducing the global burden of vector-borne diseases.

Genetic sterilization is a widely adopted technique that involves releasing genetically modified male insects into the wild. These males are engineered to carry genes that cause sterility or produce non-viable offspring when they mate with wild females. For instance, sterile male mosquitoes produced using recombinant DNA technology can outcompete wild males, leading to a decline in the mosquito population over successive generations. This technique, often referred to as the Sterile Insect Technique (SIT), has been successfully implemented in controlling agricultural pests and shows great promise in vector control.

Another application is **pathogen resistance**, where insect genomes are engineered to block the transmission of pathogens. For example, recombinant DNA technology has been used to develop mosquitoes resistant to *Plasmodium*, the parasite that causes malaria. These genetically modified mosquitoes are incapable of harboring or transmitting the parasite, thereby breaking the cycle of disease transmission. Similar strategies have been explored to engineer resistance against arboviruses like dengue and chikungunya. This approach has the potential to replace existing vector populations with genetically modified ones that cannot sustain or spread diseases.

3. Research and Biotechnology

Recombinant DNA technology has become an indispensable tool in research and biotechnology, providing deeper insights into insect biology and facilitating the production of valuable biomolecules. The creation of transgenic insect models has expanded our understanding of gene function, development, and behavior.

Functional genomics is a major area of application. By generating transgenic insects with specific genetic modifications, researchers can investigate the roles of individual genes in critical biological processes. For instance, *Drosophila melanogaster* serves as a model organism for studying development, neurobiology, and disease mechanisms. Recombinant DNA technology has enabled the creation of *Drosophila* lines expressing fluorescent markers, allowing real-time visualization of gene expression patterns and protein interactions.Another significant application is the use of insect cell systems for **protein production**. Recombinant proteins, including vaccines, antibodies, and enzymes, can be produced on a large scale using insect cells. The baculovirus expression system is particularly popular for this purpose. This system involves using recombinant baculoviruses to infect insect cell lines, leading to the production of high yields of the desired protein. Examples include the production of vaccines for influenza and other infectious diseases. Insect cell systems offer several advantages over traditional systems, including higher post-translational modification fidelity and scalability.

2.2.4 Ethical and Environmental Considerations

While recombinant DNA technology has brought transformative advancements, its use in insects raises several ethical and environmental concerns that must be carefully addressed. These concerns stem from the potential ecological impacts of genetically modified insects, public perceptions, and regulatory challenges.

Biosafety is a primary concern. The release of genetically modified insects into the environment poses risks of unintended ecological consequences. For example, the introduction of gene drives could result in the extinction of target species, disrupting food webs and ecological balance. Non-target effects, where modifications inadvertently affect other species, are another concern. Rigorous risk assessments and controlled releases are essential to mitigate these risks.

The development and release of genetically modified insects also necessitate robust **regulatory frameworks**. Governments and international organizations must establish clear guidelines for the safe use of recombinant DNA technology. These regulations should cover laboratory containment, field testing, and large-scale releases. Ensuring compliance with these frameworks is crucial for maintaining public trust and minimizing environmental risks.

Public acceptance is another critical factor. Misinformation and societal concerns about genetically modified organisms (GMOs) can hinder the adoption of these technologies. Transparent communication, public engagement, and education are necessary to address misconceptions and highlight the benefits of recombinant DNA technology. Ethical considerations must also account for the potential socioeconomic impacts, particularly in regions heavily reliant on traditional pest control methods.

2.2.5 Future Perspectives

The future of recombinant DNA technology in insects holds immense promise, with ongoing advancements in gene editing, synthetic biology, and computational approaches likely to expand its applications further. As precision tools and knowledge improve, the potential to tackle pressing global challenges in agriculture, medicine, and environmental conservation grows.

Precision gene editing using advanced CRISPR-Cas technologies will play a central role in future research. Tools like base editing and prime editing are expected to allow single-nucleotide changes and precise gene insertions without causing double-strand breaks. These advancements will enhance the safety and efficacy of genetic modifications in insects, reducing off-target effects and improving delivery systems.

Synthetic biology is poised to revolutionize insect genetic engineering. This field involves designing and constructing entirely synthetic genomes or genetic circuits tailored for specific applications. For instance, synthetic genomes could be engineered to create insects with enhanced traits, such as increased resistance to pathogens or improved compatibility with biocontrol programs. The integration of synthetic biology with recombinant DNA technology will enable the creation of programmable insect systems with unprecedented functionality.

Integration with AI and big data is another area of potential growth. Computational tools and machine learning algorithms can analyze vast datasets to predict gene functions, design optimal genetic constructs, and model ecological impacts. This integration will streamline the development of recombinant DNA technologies and improve decision-making in their application.As these advancements continue, recombinant DNA technology in insects will increasingly contribute to sustainable pest management, improved public health, and fundamental scientific discoveries. Addressing ethical and environmental concerns while fostering innovation will be key to realizing the full potential of these transformative technologies.

2.3 Identification of Genes/Nucleotide Sequences for Characters of Interest

The identification of genes and nucleotide sequences associated with specific traits in insects is a pivotal aspect of molecular entomology. This field has seen remarkable advancements due to the integration of genomic, transcriptomic, and proteomic tools. By unraveling the genetic basis of traits such as resistance, behavior, and reproduction, researchers can design targeted strategies for pest control, pollinator protection, and species conservation.

2.3.1 Techniques for Gene and Sequence Identification

To study genes and nucleotide sequences of interest in insects, researchers use an array of molecular techniques that vary in scope, resolution, and application. These methods allow for the comprehensive analysis of insect genomes, identification of gene expression patterns, and detection of genetic variants associated with specific traits.

Whole-Genome Sequencing (WGS):

Whole-genome sequencing decodes the entire DNA sequence of an organism, providing a complete genetic blueprint. This technique has revolutionized molecular entomology, enabling researchers to identify genes that underpin vital insect traits such as pesticide resistance, reproduction, and feeding. For example, sequencing the genome of *Aedes aegypti* has revealed the genetic basis of its capacity to transmit arboviruses such as dengue and Zika. Similarly, WGS of agricultural pests like the fall armyworm (*Spodoptera frugiperda*) has uncovered genes associated with resistance to *Bt* toxins and other insecticides.WGS also facilitates comparative genomics, where the genomes of different insect species or populations are compared to identify conserved and divergent genetic elements. This comparative approach has been instrumental in studying adaptive evolution, such as the emergence of resistance traits in pest populations. Advanced sequencing platforms like Illumina, PacBio, and Oxford Nanopore have further enhanced the accuracy and efficiency of WGS, allowing for the assembly of highly contiguous genomes.

Transcriptomics (RNA-Seq):

While WGS provides a static picture of an organism's genetic potential, transcriptomics reveals dynamic gene expression patterns. RNA sequencing (RNA-Seq) profiles all the RNA molecules present in a cell or tissue under specific conditions, providing insights into which genes are active and their relative expression levels. In insects, transcriptomics has been used to identify genes involved in key processes such as molting, host-seeking behavior, and response to environmental stressors like pesticides.

For instance, transcriptomic analysis of resistant and susceptible populations of the cotton bollworm (*Helicoverpa armigera*) has highlighted the upregulation of detoxifying enzyme genes, such as cytochrome P450s and glutathione S-transferases (GSTs), in resistant individuals. Additionally, RNA-Seq can reveal alternative splicing events, non-coding RNAs, and regulatory networks, all of which contribute to trait diversity in insects.

Polymerase Chain Reaction (PCR):

PCR is a foundational technique in molecular biology that amplifies specific DNA sequences, enabling the detection and characterization of genes of interest. In insect studies, PCR is commonly used to confirm the presence of genes associated with resistance, reproduction, or behavior. For example, PCR can detect kdr (knockdown resistance) mutations in the voltage-gated sodium channel gene, which confer resistance to pyrethroids in mosquitoes.Moreover, PCR is essential for genotyping individual insects in field studies, allowing researchers to monitor the spread of resistance alleles or track population dynamics. Its high specificity, speed, and cost-effectiveness make PCR a cornerstone technique in both basic and applied insect research.

Quantitative PCR (qPCR):

Quantitative PCR (qPCR) expands upon traditional PCR by quantifying gene expression levels in real-time. This technique is particularly valuable for studying the roles of specific genes in physiological and behavioral processes. For example, qPCR can measure the expression of genes involved in pheromone production in moths, shedding light on the molecular basis of mating behaviors. Similarly, qPCR is used to analyze the upregulation of metabolic enzyme genes in pesticide-resistant insects, providing insights into the mechanisms of detoxification.The ability of qPCR to generate quantitative data makes it indispensable for validating transcriptomic findings and conducting functional studies on candidate genes.

Genome-Wide Association Studies (GWAS):

GWAS involves analyzing genetic polymorphisms across populations to identify variants associated with specific traits. This method relies on the genotyping of thousands to millions of genetic markers, such as single-nucleotide polymorphisms (SNPs), in a population with known phenotypic variation. GWAS has been widely used in insects to study traits like insecticide resistance, thermal tolerance, and behavioral adaptations.For example, GWAS has uncovered SNPs in detoxification-related genes associated with resistance to neonicotinoid insecticides in honeybees. In agricultural pests, GWAS has identified genetic variants linked to host plant preferences, providing insights into pest-host coevolution and aiding the development of targeted pest management strategies.

2.3.2 Markers for Traits of Interest

Genetic markers are specific sequences within the genome that are associated with particular traits. Identifying these markers provides a means to study the genetic basis of traits and monitor their prevalence in natural populations. In insects, markers linked to resistance, behavior, and development have wide-ranging applications in pest control and ecological studies.

Resistance Genes:

Resistance genes encode proteins that help insects survive exposure to pesticides. Common resistance mechanisms include the detoxification of insecticides, target-site insensitivity, and enhanced cuticular barriers. Markers associated with resistance genes include mutations in the acetylcholinesterase (AChE) gene, conferring organophosphate and carbamate resistance, and in the voltage-gated sodium channel gene, associated with pyrethroid resistance.Additionally, upregulated expression of detoxifying enzymes, such as cytochrome P450s, GSTs, and carboxylesterases, is a hallmark of metabolic resistance. Identifying and monitoring these markers allows researchers to develop strategies to mitigate resistance, such as designing new insecticides or restoring susceptibility through gene silencing.

Behavioral Genes:

Behavioral traits, such as host-seeking, mating, and pheromone signaling, are governed by specific genetic pathways. For instance, genes involved in the synthesis and detection of sex pheromones are critical for reproduction in moths and other insects. Identifying these genes enables the development of targeted pest management strategies, such as disrupting mating communication using synthetic pheromones.Similarly, genes regulating olfactory receptors in mosquitoes are key to understanding host-seeking behavior. By targeting these genes, researchers can develop interventions that reduce human-vector contact, thereby controlling the spread of vector-borne diseases.

Developmental Genes:

Genes involved in development regulate processes such as molting, metamorphosis, and reproduction. For example, the ecdysone receptor (EcR) gene is crucial for molting and metamorphosis in insects. Mutations or disruptions in these genes can arrest development, making them potential targets for pest control. Mapping developmental genes also provides insights into life cycle regulation and the evolution of complex traits.

2.3.3 Applications in Insect Molecular Biology

The identification and characterization of genes in insects have far-reaching implications in pest control, pollinator protection, and evolutionary studies.

Pest Control:

Recombinant DNA technology allows for the silencing or editing of resistance genes to restore insect susceptibility to insecticides. For example, RNA interference (RNAi) can be used to silence detoxifying enzyme genes, rendering resistant insect populations vulnerable to chemical treatments. Additionally, CRISPR-Cas9-based gene drives can spread traits like sterility or susceptibility through pest populations, providing a sustainable solution for population suppression.

Gene Drives:

Gene drives leverage genetic engineering to bias inheritance patterns, enabling the rapid spread of specific traits in natural populations. This technology is particularly effective for controlling vector-borne diseases and invasive species. For example, gene drives have been proposed to suppress *Aedes aegypti* populations by introducing sterility or disrupting genes essential for survival.

Pollinator Protection:

Identifying genes linked to resilience in pollinators, such as honeybees and bumblebees, is crucial for safeguarding their populations against environmental stressors like pesticides and pathogens. Genetic markers associated with detoxification pathways, immune response, and thermal tolerance provide insights into the molecular mechanisms underlying pollinator health. This knowledge informs conservation strategies and the development of bee-friendly agricultural practices.

Taxonomy and Phylogeny:

Genetic sequences serve as valuable tools for species identification and evolutionary studies. DNA barcoding, based on the mitochondrial cytochrome oxidase I (COI) gene, is widely used to differentiate insect species. Phylogenetic analyses of genetic sequences also shed light on the evolutionary relationships among insect taxa, helping researchers understand speciation, adaptation, and diversification.

Chapter 3

Genetic Improvement of Natural Enemies

3.1 Introduction

The genetic improvement of natural enemies represents a transformative approach to enhancing biological control in agriculture, aiming to bolster the efficiency, specificity, and adaptability of organisms that naturally regulate pest populations. Natural enemies, including parasitoids, predators, and pathogens, are integral to maintaining ecological balance and minimizing the reliance on chemical pesticides. However, their effectiveness can be limited by factors such as environmental conditions, host specificity, and the rapid adaptation of pest populations. Genetic improvement strategies have emerged as a solution to address these challenges, utilizing advanced biotechnological tools to optimize the traits of natural enemies for pest management.

The application of genetic engineering techniques has enabled scientists to modify and enhance various biological control agents, ranging from beneficial microorganisms to macroscopic predators. Insect cell lines have become indispensable tools in this process, providing platforms for studying genetic modifications and enabling the mass production of genetically enhanced pathogens and parasites. These cell lines facilitate the development of recombinant baculoviruses, which are natural insect pathogens engineered to deliver enhanced pest control by incorporating genes for increased virulence or faster action. Similarly, the bacterium *Bacillus thuringiensis* (*Bt*), renowned for its insecticidal properties, has been genetically manipulated to produce novel toxins and incorporated into transgenic crops, providing season-long protection against targeted pests. This approach reduces the need for chemical interventions and supports sustainable agriculture by preserving beneficial organisms and minimizing environmental impacts.

Entomopathogenic fungi, such as *Beauveria bassiana* and *Metarhizium anisopliae*, are another critical component of this effort, serving as natural enemies capable of infecting and killing insect pests. Advances in genetic engineering have allowed researchers to enhance the efficacy of these fungi by increasing their virulence, broadening their host range, and improving their resistance to environmental stressors like UV radiation and high temperatures. These modifications not only make entomopathogenic fungi more effective but also improve their viability as commercial biopesticides. Despite the promising potential of genetically improved natural enemies, challenges such as regulatory

DOI: 10.1201/9781003757405-3

approval, public perception, and the need for resistance management must be addressed to ensure their long-term success and acceptance in agricultural systems.

By integrating cutting-edge genetic tools with traditional biological control strategies, the genetic improvement of natural enemies offers a pathway to more effective, sustainable pest management solutions. This approach not only aligns with the goals of reducing chemical pesticide use and environmental degradation but also contributes to the broader effort of achieving global food security in the face of increasing pest pressures and climatic challenges. Through continued research and collaboration, the genetic improvement of natural enemies holds the potential to revolutionize integrated pest management, ushering in a new era of precision and sustainability in agriculture.

3.2 Cell Lines for Genetic Improvement

Insect cell lines are foundational tools in modern genetics and biotechnology, serving as platforms for understanding molecular mechanisms, developing recombinant technologies, and implementing biological control strategies. These cell lines enable scientists to investigate intricate genetic pathways, test the efficacy of gene editing tools, and produce biologically active compounds such as recombinant proteins and vaccines. In addition to their research applications, insect cell lines play a crucial role in industrial biotechnology, particularly in fields such as agriculture, medicine, and environmental science.

3.2.1 Development of Insect Cell Lines

The establishment of insect cell lines involves isolating cells from various tissues of insects and adapting them to grow under in vitro conditions. The goal is to create a stable and immortalized population of cells that can be propagated indefinitely, providing researchers with a consistent and reproducible experimental platform. Developing insect cell lines is a complex process that requires careful optimization of growth conditions, nutrient media, and cell handling techniques.

Origins of Insect Cell Lines

Insect cell lines are typically derived from tissues with high regenerative or proliferative capacity. Commonly targeted tissues include:

1. **Embryonic Tissues**: Embryonic cells are a rich source for establishing cell lines due to their pluripotency and ability to adapt to artificial culture conditions. For example, embryonic-derived lines can differentiate into diverse cell types, making them suitable for various research applications.
2. **Ovarian Tissues**: Ovarian tissues are among the most frequently used sources for insect cell lines. These cells are metabolically active and often easier to culture compared to other tissues. Cell lines such as Sf9 and Sf21, derived from the ovarian tissue of *Spodoptera frugiperda*, are widely used in biotechnology.
3. **Hemocytes**: Hemocytes, the insect equivalent of blood cells, are commonly used in immune system studies. These cells are particularly useful for understanding host-pathogen interactions and cellular defense mechanisms.
4. **Midgut Epithelium**: Cells from the midgut are useful for studying digestion, nutrient absorption, and interactions with gut pathogens. These cell lines have applications in pest management, especially in testing the effects of insecticides and toxins.

Process of Establishment

Establishing insect cell lines is a multistage process requiring precision and patience:

1. **Dissection and Tissue Isolation:**Target tissues are carefully dissected from donor insects under sterile conditions. For example, ovarian tissue from *Spodoptera frugiperda* larvae or embryos is commonly used. The dissection must minimize contamination and maximize the yield of viable cells.
2. **Dissociation and Seeding**: Isolated tissues are dissociated into single cells or small clusters using enzymatic treatments (e.g., trypsin or collagenase) or mechanical disruption. These cells are then seeded into culture vessels containing nutrient-rich growth media.
3. **Primary Culture Maintenance**: During the primary culture stage, cells adapt to the artificial in vitro environment. This step often involves optimizing factors such as temperature, pH, and nutrient concentrations. Most insect cells thrive at temperatures around 27°C.
4. **Subculturing and Selection**: Cells that show consistent growth and proliferation are subcultured into new vessels, allowing researchers to select for stable and robust populations. It may take several months to establish a stable cell line.
5. **Immortalization**: Some cell lines naturally achieve immortality, while others require genetic modifications or specific growth conditions to sustain long-term proliferation.
6. **Cryopreservation**: Once a cell line is established, it is cryopreserved in liquid nitrogen for long-term storage. This ensures the availability of consistent cells for future experiments.

Commonly Used Insect Cell Lines

1. **Sf9 and Sf21 (Spodoptera frugiperda)**:
   - Derived from the ovarian tissue of Spodoptera frugiperda, these lines are among the most versatile and widely used in research and industry.
   - Applications include recombinant protein production, baculovirus studies, and high-throughput screening of compounds.
2. **High Five (Tn5B1-4)**:
   - Originating from *Trichoplusiani* (cabbage looper), High Five cells are known for their high productivity in recombinant protein production. They are often preferred over Sf9 cells due to their superior glycosylation capabilities.
3. **BmN (Bombyx mori Nucleopolyhedrovirus)**:
   - Derived from the silkworm *Bombyx mori*, this cell line is particularly useful for studying baculoviruses and producing silk-related proteins.
4. **Aedes albopictus (C6/36)**:
   - Derived from the mosquito *Aedes albopictus*, this line is widely used in arbovirus research, including studies on dengue and Zika virus.

Optimization of Culture Conditions

The success of insect cell lines depends heavily on the optimization of culture conditions, including:

- **Growth Media**: Insect cells are commonly cultured in specialized media such as Grace's medium or Sf-900 serum-free medium. These media are supplemented with nutrients, vitamins, and amino acids tailored to the metabolic needs of insect cells.
- **Serum-Free Adaptation**: Many insect cell lines have been adapted to grow in serum-free conditions, which reduces costs and eliminates variability associated with fetal bovine serum (FBS).
- **Environmental Factors**: Optimal temperature (typically 27–28°C), pH (around 6.2–6.4), and oxygen levels are crucial for maintaining healthy cell growth.

3.2.2 Applications of Insect Cell Lines

Insect cell lines have found applications across a wide spectrum of scientific disciplines and industrial sectors. Their versatility, ease of culture, and unique biological characteristics make them ideal for various purposes.

Recombinant Protein Production

One of the most prominent applications of insect cell lines is the production of recombinant proteins using the baculovirus expression vector system (BEVS). This system takes advantage of the ability of insect cells to produce complex, biologically active proteins with post-translational modifications similar to those in higher eukaryotes. Key applications include:

- **Vaccine Production**: Insect cell lines are used to produce vaccines for diseases such as influenza, HPV, and COVID-19. For example, the Flublok influenza vaccine is manufactured using BEVS and Sf9 cells.
- **Therapeutic Proteins**: Recombinant enzymes, growth factors, and antibodies are produced in insect cells for use in medicine.
- **Industrial Enzymes**: Enzymes used in detergents, food processing, and bioremediation are often produced using insect cell systems.

Virus Propagation

Insect cell lines are widely used to propagate viruses, particularly baculoviruses and arboviruses. This is crucial for:

- **Biocontrol**: Baculoviruses are employed as environmentally friendly biopesticides for controlling agricultural pests. Insect cell lines facilitate the mass production of baculovirus formulations.
- **ArbovirusResearch**: Mosquito-derived cell lines such as C6/36 are essential for studying viruses like dengue, chikungunya, and Zika. These studies help identify viral mechanisms and evaluate antiviral compounds.

Gene Function Studies

Insect cell lines provide a controlled environment for studying gene function through techniques such as RNA interference (RNAi), CRISPR-Cas9, and overexpression studies. Applications include:

» Investigating the roles of detoxification enzymes in pesticide resistance.
» Understanding genetic pathways involved in insect development and reproduction.
» Identifying genes responsible for immune responses to pathogens.

Pesticide and Drug Screening

Insect cell lines are invaluable for evaluating the efficacy and toxicity of pesticides, insecticides, and pharmaceutical compounds. By exposing cultured cells to these substances, researchers can assess their impact on cellular metabolism, viability, and genetic integrity.

Host-Pathogen Interaction Studies

Insect cell lines serve as models for studying host-pathogen interactions, providing insights into the molecular mechanisms of infection. These studies are critical for developing strategies to combat insect-borne diseases and improve biocontrol methods.

3.2.3 Advantages of Using Insect Cell Lines

The use of insect cell lines offers several advantages that make them indispensable in genetic research and industrial biotechnology:

1. **Ease of Culture:**
    - Insect cell lines grow well at ambient temperatures (25–28°C), eliminating the need for expensive incubators.
2. **Cost-Effectiveness:**
    - Growth media for insect cells are less expensive than those required for mammalian cells, reducing production costs.
3. **Scalability:**
    - Insect cell cultures can be easily scaled up for industrial applications, such as recombinant protein production.
4. **Ethical Considerations:**
    - In vitro studies using cell lines reduce the reliance on live insects, aligning with ethical research practices.
5. **Post-Translational Modifications:**
    - Insect cells can perform complex post-translational modifications, making them suitable for producing functional proteins.

3.2.4 Challenges in Developing and Using Insect Cell Lines

Despite their many advantages, insect cell lines face several challenges:

1. **Limited Availability**:
    - While cell lines for model insects like *Spodoptera frugiperda* are well-established, there is a lack of cell lines for non-model and ecologically important species.

2. **Genomic Instability**:
   - Prolonged culturing can lead to genetic drift, chromosomal abnormalities, and loss of specific traits.
3. **Media Requirements**:
   - Developing species-specific media for non-model insects can be costly and time-consuming.
4. **Adaptation Difficulties:**
   - Not all insect tissues adapt well to in vitro conditions, limiting the establishment of new cell lines.

3.2.5 Future Perspectives

The future of insect cell lines is bright, with emerging technologies promising to expand their utility:

1. **Stem Cell-Derived Lines**:
   - Advances in stem cell research may enable the development of cell lines with enhanced differentiation potential.
2. **Synthetic Biology**:
   - Integrating synthetic biology approaches with insect cell systems could allow the creation of custom cell lines optimized for specific applications.
3. **High-Throughput Screening**:
   - Automation and robotics will enable large-scale screening of compounds, accelerating research in pest control and drug development.
4. **CRISPR-Cas9 Integration**:
   - Genome editing technologies will enhance the genetic tractability of insect cell lines, allowing for precise modifications tailored to research needs.
5. **Expansion to Non-Model Species**:
   - Efforts to establish cell lines from diverse insect species will broaden the scope of research and industrial applications.

3.3 Genetic Engineering in Baculoviruses

Baculoviruses, naturally occurring pathogens specific to insects, have been studied extensively for their potential applications in pest management, biotechnology, and molecular biology. Genetic engineering has enhanced their versatility, transforming these viruses into powerful tools for environmentally friendly pest control, recombinant protein production, and vaccine development.

3.3.1 Baculoviruses as Biocontrol Agents

Baculoviruses, particularly nucleopolyhedroviruses (NPVs) and granuloviruses (GVs), are highly effective natural biocontrol agents against insect pests. Their unique ability to target specific insect species without harming non-target organisms makes them a cornerstone of integrated pest management (IPM) and sustainable agriculture.

Specificity to Target Pests

One of the defining characteristics of baculoviruses is their species-specific nature. Unlike chemical pesticides, which often affect a broad range of organisms, baculoviruses primarily infect pests within the orders *Lepidoptera*, *Diptera*, and *Hymenoptera*. For example, the nucleopolyhedrovirus (NPV) strain infecting the cotton bollworm (*Helicoverpaarmigera*) targets this pest specifically, sparing beneficial pollinators like bees and predatory insects that contribute to ecosystem health.This specificity arises from the baculovirus infection process. Viral particles must first be ingested by the host insect. Once inside the midgut, the protective polyhedral occlusion body dissolves in the alkaline environment, releasing infectious virions.These virions invade the midgut epithelial cells and subsequently spread to other tissues, causing a systemic infection that leads to liquefaction of the host body. This targeted mode of action ensures that only the pest species is affected, leaving other organisms unharmed.

Eco-Friendly Attributes

Baculoviruses offer a safer alternative to traditional chemical pesticides, which often persist in the environment and accumulate in non-target species. In contrast, baculoviruses degrade rapidly under natural conditions, particularly when exposed to ultraviolet (UV) light. This biodegradability reduces environmental contamination and the risk of toxic residues in crops, soil, and water. Moreover, baculoviruses do not pose risks to human health, making them suitable for use in agricultural systems near human settlements.

Compatibility with Integrated Pest Management (IPM)

Baculoviruses fit seamlessly into IPM frameworks, where they can be used alongside other control methods such as biological predators, crop rotation, and resistant crop varieties. For instance, the simultaneous application of baculoviruses and predatory insects like ladybugs enhances pest suppression while minimizing ecological disruption. Unlike chemical pesticides that often kill beneficial organisms, baculoviruses allow natural enemies of pests to thrive, further contributing to pest population control.

Applications in Major Crops

Baculoviruses have been successfully used to manage pests in various agricultural systems. Examples include:

» **Cotton Bollworm Control**: The *Helicoverpaarmigera* NPV (HearNPV) has been deployed in cotton fields to suppress bollworm populations, reducing the reliance on synthetic insecticides.

» **Fruit and Orchard Pest Management**: The Cydia pomonellagranulovirus (CpGV) has been instrumental in controlling codling moth infestations in apple and pear orchards, significantly reducing crop losses.

» **Vegetable Crop Protection**: Baculoviruses targeting pests like the cabbage looper (*Trichoplusiani*) have improved yields in vegetables such as cabbage, broccoli, and lettuce.

Challenges of Natural Baculoviruses

Despite their many advantages, natural baculoviruses face limitations that hinder their widespread adoption:

» **Slow Speed of Kill**: Infected insects may continue feeding on crops for several days before succumbing to the virus, resulting in significant damage during the infection period.

» **UV Sensitivity**: Baculoviruses degrade quickly under sunlight, necessitating frequent applications or the addition of UV protectants.

» **Production Costs**: Large-scale production of natural baculoviruses can be costlier than chemical pesticides, though advancements in fermentation technology are addressing this issue.

Efforts to overcome these challenges have focused on genetic engineering, which has significantly enhanced the efficacy and commercial viability of baculoviruses.

3.3.2 Genetic Modifications

Genetic engineering has revolutionized the use of baculoviruses as pest control agents, addressing key limitations such as slow speed of kill, narrow host range, and environmental stability. By incorporating foreign genes and modifying existing ones, researchers have transformed baculoviruses into more effective and versatile biocontrol agents.

Enhancing Virulence

One of the primary goals of genetic engineering in baculoviruses is to enhance their lethality against insect pests. This is often achieved by introducing genes encoding insecticidal proteins. For example, genes from *Bacillus thuringiensis* (*Bt*), which produce toxins that disrupt the midgut cells of insects, have been successfully integrated into baculovirus genomes. Recombinant baculoviruses expressing *Bt* toxins not only kill pests more rapidly but also reduce the likelihood of resistance development due to the combined effects of viral infection and toxin action.In addition to *Bt* toxins, other insecticidal proteins, such as those derived from arachnids (spiders and scorpions), have been explored. These neurotoxins target the nervous systems of pests, causing paralysis and death. The inclusion of such genes broadens the spectrum of action and enhances the effectiveness of baculoviruses against resistant pest populations.

Accelerating Speed of Kill

Natural baculoviruses often take several days to kill their hosts, allowing pests to cause extensive damage before succumbing. Genetic modifications have addressed this limitation by incorporating genes that disrupt key physiological processes in the host. For instance:

» **Juvenile Hormone Esterase (JHE)**: This enzyme degrades juvenile hormone, a critical regulator of insect development and reproduction. Recombinant baculoviruses expressing JHE induce premature molting or developmental arrest, leading to faster pest mortality.

» **Apoptosis-Inducing Genes**: Certain engineered baculoviruses express genes that trigger programmed cell death in infected tissues, shortening the time to kill. These modifications not only reduce crop damage but also enhance the competitive advantage of baculoviruses over chemical pesticides.

Enhancing Target Specificity

Genetic engineering has also improved the precision of baculoviruses in targeting specific pest species. By incorporating host-specific promoters—genetic elements that regulate gene expression in a species-specific manner—researchers ensure that toxic proteins are produced only in the intended pest species. For example, midgut-specific promoters can drive the expression of insecticidal genes in the gut cells of target pests, leaving non-target organisms unaffected.

Improving Environmental Stability

The susceptibility of baculoviruses to UV degradation has been a significant limitation in field applications. Genetic modifications aimed at improving environmental stability include:

» **UV-Protective Proteins**: Genes encoding proteins that shield viral DNA from UV light have been introduced, extending the efficacy of baculoviruses under sunlight.

» **Enhanced Longevity**: By modifying viral genes that regulate persistence in the environment, scientists have created recombinant baculoviruses with prolonged activity on plant surfaces.

Other Modifications

» **Fluorescent Markers**: The addition of genes encoding fluorescent proteins (e.g., GFP) allows researchers to track the spread and infection dynamics of recombinant baculoviruses in field settings.

» **Self-Limiting Traits**: Certain recombinant baculoviruses are designed with self-limiting mechanisms, ensuring that they do not persist indefinitely in the environment, addressing ecological concerns.

These genetic enhancements have made baculoviruses more effective, versatile, and commercially viable for large-scale pest control applications.

3.3.3 Production and Application

The production and field application of genetically engineered baculoviruses requires scalable, cost-effective methods and carefully designed strategies to maximize their impact as biocontrol agents. Advances in insect cell culture and formulation technology have enabled the large-scale production of recombinant baculoviruses for commercial use.

Production in Insect Cell Lines

Insect cell lines, such as Sf9 and High Five, are widely used for producing recombinant baculoviruses. These cell lines provide a controlled environment for viral replication and protein expression, ensuring consistent and high-yield production. The production process involves several steps:

1. **Virus Amplification**: A small quantity of recombinant baculovirus is used to infect insect cell cultures. The virus replicates within the cells, producing large numbers of viral particles.
2. **Harvesting**: After the infection cycle, viral particles are harvested from the culture medium and purified to remove cellular debris and contaminants.

3. **Formulation**: Purified viral particles are formulated with additives such as UV protectants, stabilizers, and wetting agents to enhance their efficacy and shelf life.

The scalability of insect cell cultures enables the production of recombinant baculoviruses in bioreactors, meeting the demands of large-scale agricultural applications.

Field Application

Recombinant baculoviruses are applied to crops through spraying, similar to chemical pesticides. Upon application, the viral particles adhere to plant surfaces, where they are ingested by feeding pests. Field studies have demonstrated the effectiveness of recombinant baculoviruses in managing pests such as:

» **Cotton Bollworm (*Helicoverpa armigera*)**: Recombinant baculoviruses expressing *Bt* toxins have significantly reduced bollworm populations, minimizing crop damage and pesticide use.

» **Cabbage Looper (*Trichoplusia ni*)**: Engineered baculoviruses with enhanced virulence have successfully controlled cabbage loopers in vegetable crops.

» **Fall Armyworm (*Spodoptera frugiperda*)**: Recombinant baculoviruses have been effective in controlling this pest, which poses a major threat to maize and other crops worldwide.

Advantages Over Chemical Pesticides

» The use of recombinant baculoviruses offers several benefits over conventional pesticides:

» **Environmental Safety**: Baculoviruses are biodegradable and do not leave toxic residues.

» **Reduced Resistance Development**: The dual-action mechanisms of recombinant baculoviruses (viral infection plus expressed toxins) lower the risk of resistance in pest populations.

» **Compatibility with IPM**: Baculoviruses can be integrated with other pest control methods for a holistic approach to pest management.

Challenges and Future Directions

Despite their advantages, the adoption of recombinant baculoviruses faces challenges, including:

» **Regulatory Hurdles**: Approval processes for genetically modified organisms (GMOs) can be lengthy and complex.

» **Production Costs**: While scalable, recombinant baculoviruses are still more expensive than some chemical alternatives.

» **Public Perception**: Concerns about GMOs may limit acceptance in certain regions.

3.4 *Bacillus thuringiensis* (*Bt*) and Genetic Applications

***Bacillus thuringiensis* (*Bt*)** has revolutionized pest management through its ability to produce insecticidal Cry (crystalline) proteins that specifically target insect pests. This bacterium has been developed into commercial biopesticides and genetically engineered crops that offer sustainable, environmentally friendly pest control solutions. *Bt*-based products address challenges in agriculture, vector control, and environmental sustainability, minimizing reliance on chemical pesticides.

3.4.1 *Bt* as a Biopesticide

Bt biopesticides represent one of the earliest examples of microbial pest control agents. Unlike chemical insecticides, *Bt* formulations combine precision targeting with environmental safety. They are used globally to control pests in crops, forests, and even vector populations, such as mosquitoes, that transmit human diseases.

Mechanism of Action

The success of *Bt* biopesticides lies in their sophisticated and highly specific mode of action. Once ingested by a susceptible insect, *Bt* spores and Cry proteins activate in the insect's midgut. This activation follows several steps:

1. **Solubilization of Cry Proteins**: The alkaline conditions in the insect's midgut dissolve the crystalline inclusions, releasing protoxins. These protoxins are inactive precursors of the toxic proteins.
2. **Activation by Proteolytic Cleavage**: Proteolytic enzymes in the insect's gut process the protoxins into their active forms.
3. **Binding to Midgut Receptors**: The activated Cry proteins bind to specific receptors on the insect's midgut epithelial cells. The specificity of this binding determines the toxin's target pest range.
4. **Pore Formation and Cell Lysis**: Cry proteins aggregate to form pores in the cell membranes, leading to ion imbalance and cell death. This disrupts the midgut lining, causing paralysis, starvation, and eventual death of the insect.
5. **Septicemia**: The destruction of the gut lining allows gut bacteria to invade the insect's hemocoel (body cavity), accelerating death through septicemia.

Effectiveness Against Key Pest Groups

Bt biopesticides are effective against a wide range of insect pests, including species in the orders *Lepidoptera* (caterpillars), *Coleoptera* (beetles), and *Diptera* (flies and mosquitoes). Each Cry protein targets specific pest species:

- **Cry1 and Cry2 Proteins**: Effective against lepidopteran pests like cotton bollworms, armyworms, and diamondback moths.
- **Cry3 Proteins**: Target coleopteran pests such as Colorado potato beetles and rootworms.
- **Cry4 and Cry11 Proteins**: Effective against dipteran larvae, including mosquito larvae, contributing to public health efforts in vector control.

Advantages of *Bt* Biopesticides

Bt biopesticides offer multiple advantages, making them superior to many chemical pesticides:

- **Specificity**: They are non-toxic to humans, animals, and beneficial insects like bees and predatory beetles.
- **Environmental Safety**: *Bt* degrades rapidly in sunlight and soil, leaving no harmful residues.

» **Low Resistance Development**: When rotated with other pest control measures, *Bt* biopesticides help delay resistance evolution.

Limitations of *Bt* Biopesticides

Despite their benefits, *Bt* biopesticides face challenges that limit their widespread adoption:

» **Short Field Persistence**: UV light and environmental conditions can degrade *Bt* products rapidly, reducing their efficacy.

» **Resistance Risk**: Continuous and excessive use of the same *Bt* formulations can lead to resistance in target pest populations.

» **Cost of Production**: Although effective, the production and formulation of *Bt* biopesticides can be more expensive than some chemical alternatives.

Examples of *Bt* Biopesticides

Commercial *Bt* formulations include:

» **Dipel**: Targets lepidopteran larvae in crops such as cabbage and cotton.

» **Vectobac**: Controls mosquito larvae in aquatic habitats, reducing vector populations.

» **Trident**: A broad-spectrum *Bt* product used in forestry and agriculture.

Innovations to Address Limitations

To overcome limitations, research has focused on enhancing the field persistence of *Bt* products by incorporating UV protectants and formulation technologies that stabilize the Cry proteins. Additionally, combining *Bt* with other microbial agents, such as fungal or viral biopesticides, has shown synergistic effects in pest management.

3.4.2 Genetic Engineering of *Bt*

Genetic engineering has greatly expanded the applications of *Bt*, enabling the development of transgenic crops, improved Cry proteins, and resistance management strategies. These innovations have enhanced the versatility and efficacy of *Bt* in pest control.

Novel Cry Protein Development

Traditional *Bt* strains produce a limited number of Cry proteins, each with specific target ranges. Advances in genetic engineering have allowed scientists to design novel Cry proteins with enhanced properties:

» **Mutagenesis**: Introducing targeted mutations in Cry genes alters the structure and activity of the toxins, broadening their target range or increasing their toxicity.

» **Domain Swapping and Gene Fusion**: Combining functional domains from different Cry proteins has resulted in chimeric toxins with unique pest-targeting capabilities.

For example, engineered Cry proteins have been developed to target resistant populations of *Plutellaxylostella* (diamondback moth) and *Spodoptera frugiperda* (fall armyworm), pests that have evolved resistance to native Cry proteins.

Transgenic *Bt* Crops

Bt crops are among the most significant advancements in agricultural biotechnology. By incorporating Cry protein-encoding genes directly into plant genomes, scientists have created crops that produce *Bt* toxins in all tissues, providing continuous protection against pests throughout the growing season.

Examples of *Bt* Crops:

1. ***Bt* Cotton**:
   - Targets bollworms and pink bollworms, major pests in cotton cultivation.
   - Widely adopted in countries like India, the United States, and China.
2. ***Bt* Maize**:
   - Protects against stem borers and rootworms, reducing yield losses and pesticide dependence.
   - Common in regions such as North America and Sub-Saharan Africa.
3. ***Bt* Soybean**:
   - Effective against defoliating pests like caterpillars, preserving crop yields in Brazil and Argentina.

Advantages of *Bt* Crops

Bt crops have transformed modern agriculture by offering the following benefits:

- **Reduced Pesticide Use**: *Bt* crops significantly lower the need for chemical insecticides, reducing environmental contamination and input costs for farmers.
- **Enhanced Yield Stability**: By protecting crops from pest damage, *Bt* crops ensure more consistent yields.
- **Lower Labor Costs**: Farmers spend less time applying chemical sprays, improving efficiency and profitability.

Resistance Management in *Bt* Crops

Continuous exposure to the same *Bt* toxins can lead to the evolution of resistance in pest populations. To address this, scientists have developed strategies to delay resistance:

1. **Pyramid Stacking**: Incorporating multiple Cry proteins with different modes of action into a single crop. This approach reduces the likelihood of pests evolving resistance to all toxins simultaneously.
2. **Refuge Strategy**: Farmers plant non-*Bt* crop refuges alongside *Bt* fields. These refuges provide a breeding ground for susceptible pests, diluting resistance alleles in the population.
3. **Gene Editing**: Emerging techniques like CRISPR-Cas9 allow precise modifications to Cry genes, creating new variants that overcome resistance mechanisms.

Examples of Resistance-Resistant *Bt* Crops

» *Bt* maize varieties expressing both Cry1Ab and Cry2Ab proteins are designed to manage resistance in stem borers.

» *Bt* cotton with stacked Cry1Ac and Cry2Ab genes effectively controls resistant bollworm populations.

3.4.3 Environmental and Economic Benefits

Bt-based technologies provide significant environmental and economic advantages, making them a cornerstone of sustainable agriculture. Their widespread adoption has positively impacted ecosystems, farming practices, and food security.

Environmental Benefits

1. **Preservation of Beneficial Organisms:** *Bt* products and crops are highly specific to target pests, leaving pollinators, natural predators, and soil organisms unharmed. This specificity helps maintain biodiversity and ecological balance.
2. **Reduced Chemical Insecticide Use:** *Bt* adoption has led to a marked decrease in chemical pesticide applications, mitigating the negative impacts of pesticide runoff on aquatic ecosystems and groundwater contamination.
3. **Lower Carbon Footprint**: By reducing the need for chemical pesticide production, transportation, and application, *Bt* technologies contribute to lower greenhouse gas emissions in agriculture.
4. **Soil Health**: The reduction in chemical pesticide use protects beneficial soil microbes, promoting healthier soils and more sustainable farming practices.

Economic Benefits

1. **Increased Farmer Profits**: *Bt* crops improve yields by protecting against pest damage, resulting in higher profits for farmers. The reduction in pesticide costs further enhances economic benefits.
2. **Reduced Input Costs**: *Bt* technologies lower the need for repeated pesticide applications, saving on labor, equipment, and chemical expenses.
3. **Improved Global Food Security**: By reducing crop losses to pests, *Bt* crops contribute to higher agricultural productivity, addressing food security challenges in developing regions.

Socio-economic Impact

Bt technologies have empowered smallholder farmers, particularly in developing countries. For example:

» *Bt* cotton adoption in India has increased farmer incomes, reduced pesticide-related health issues, and improved rural livelihoods.

» In Africa, *Bt* maize has helped combat pest-induced yield losses, supporting food security initiatives.

Challenges and Future Outlook

While *Bt* technologies offer substantial benefits, challenges remain:

- » Resistance in pest populations requires ongoing innovation in Cry protein development.
- » Public perception and regulatory hurdles around genetically modified organisms (GMOs) continue to limit adoption in some regions.

3.5 Entomopathogenic Fungi

Entomopathogenic fungi, such as *Beauveria bassiana* and *Metarhizium anisopliae*, are naturally occurring microbial pathogens that infect and kill insects. These fungi have emerged as valuable biological control agents due to their ability to target a wide range of insect pests while preserving beneficial organisms and maintaining ecological balance. Advances in genetic engineering have further enhanced their utility, improving their virulence, adaptability to environmental stress, and host specificity.

3.5.1 Mode of Action

Entomopathogenic fungi exhibit a unique infection process that sets them apart from other microbial control agents like bacteria and viruses. Unlike these pathogens, fungi do not require ingestion by the host to initiate infection. Instead, they attack insects externally, making them versatile agents against pests with varying feeding habits.

Stages of Fungal Infection

1. **Adherence and Germination**:
   - The infection begins when fungal spores (conidia) land on the insect's cuticle. The hydrophobic nature of fungal spores facilitates adhesion to the waxy cuticle layer.
   - Once adhered, the spores germinate, forming germ tubes that grow and explore the cuticle surface.
2. **Cuticle Penetration:**
   - The fungus produces enzymes such as chitinases, proteases, and lipases that degrade the insect's cuticle, allowing the germ tube to penetrate.
   - Mechanical pressure from the growing fungal hyphae further aids in breaching the insect's protective barriers.
3. **Invasion of Hemocoel**:
   - After penetrating the cuticle, the fungal hyphae enter the hemocoel (the insect's body cavity), where they proliferate, spreading throughout the host's body.
4. **Toxin Production**:
   - During colonization, the fungus produces secondary metabolites, including toxins like destruxins (produced by *Metarhizium anisopliae*) and beauvericin (produced by *Beauveria bassiana*).

- These compounds disrupt cellular functions, weaken the immune system, and eventually kill the host.

5. **Death and Sporulation**:
   - The infected insect succumbs to systemic fungal growth and toxin effects.
   - Under favorable conditions, the fungus emerges from the cadaver, producing new conidia that can infect other insects.

Advantages of Entomopathogenic Fungi

» **Broad Host Range**: These fungi can infect numerous pest species, including aphids, whiteflies, grasshoppers, and beetles.

» **Non-Ingestive Action**: Unlike *Bt* or viruses, fungi can infect pests with piercing-sucking mouthparts, such as aphids, making them versatile agents.

» **Environmental Compatibility**: Fungi are biodegradable, leaving no harmful residues in the environment.

» **Persistence**: Fungi can persist in the soil or on plant surfaces, providing long-term pest suppression.

Challenges in Field Applications

» **Environmental Sensitivity**: Fungal spores are susceptible to UV radiation, desiccation, and temperature extremes, limiting their efficacy under harsh conditions.

» **Slow Speed of Kill**: The infection process may take several days, allowing pests to cause damage before succumbing.

3.5.2 Genetic Improvement of Fungi

Genetic engineering has opened new avenues for enhancing the performance of entomopathogenic fungi. These modifications aim to address their natural limitations, such as sensitivity to environmental stressors, low virulence, and non-specificity, while maximizing their potential as effective biocontrol agents.

Increased Virulence

To enhance the pathogenicity of entomopathogenic fungi, researchers have introduced genes encoding potent insecticidal proteins. Examples include:

» ***Bt* Cry Proteins**: Incorporating genes for Cry toxins from *Bacillus thuringiensis* into fungal genomes increases their ability to kill pests quickly. These toxins act synergistically with the fungus, accelerating pest mortality.

» **Neurotoxins**: Genes encoding scorpion or spider neurotoxins have been expressed in fungi like *Metarhizium anisopliae*. These neurotoxins disrupt the nervous system of pests, leading to rapid paralysis and death.

» **Enzyme Overexpression**: Enhancing the production of cuticle-degrading enzymes such as chitinase and protease improves the fungus's ability to penetrate the insect cuticle.

Thermal and UV Tolerance

One of the primary challenges in deploying entomopathogenic fungi is their sensitivity to environmental conditions. Genetic modifications have been employed to improve fungal survival in field environments:

» **UV Resistance**: By introducing genes that encode UV-protective compounds, such as melanin, scientists have developed fungal strains that can withstand prolonged exposure to sunlight.

» **Thermal Adaptation**: Engineering fungi to express heat-shock proteins enhances their ability to survive at higher temperatures, making them suitable for deployment in tropical and arid regions.

Host-Specificity Engineering

Genetic modifications can tailor fungi to target specific pest species, reducing risks to non-target organisms. Strategies include:

» **Specific Promoters**: Using pest-specific promoters ensures that fungal virulence factors are expressed only in target species.

» **Targeted Toxin Expression**: Engineering fungi to produce toxins that are active only against particular pests minimizes unintended impacts on beneficial insects like pollinators and natural predators.

Examples of Genetically Improved Fungi

» **Metarhizium anisopliae expressing *Bt* Cry1Ac**: Demonstrated increased lethality against lepidopteran pests.

» **Beauveria bassiana with enhanced melanin production**: Exhibited improved UV resistance and longer field persistence.

3.5.3 Formulation and Deployment

Effective field application of entomopathogenic fungi requires careful formulation and delivery strategies. Advances in formulation technology have improved the stability, efficacy, and shelf life of fungal products, enabling their large-scale deployment in agriculture.

Formulation Types

Entomopathogenic fungi can be formulated into various products tailored for specific pest control scenarios:

1. **Wettable Powders and Aqueous Suspensions**:
   - Fungal spores are mixed with wetting agents to allow even dispersion in water.
   - These formulations are applied as sprays, targeting pests on crop surfaces.
2. **Granules and Pellets:**
   - Granular formulations are designed for soil application, targeting root pests or soil-dwelling stages of insects.
   - Pellets often contain UV protectants and nutrients to enhance fungal spore viability.

3. **Oil-Based Formulations:**
    - Oil carriers improve spore adhesion to insect cuticles and protect spores from desiccation and UV degradation.
    - Ideal for targeting pests in arid environments.
4. **Baits:**
    - Fungal spores are incorporated into bait formulations, attracting and infecting specific pests.
    - Used for controlling locusts, grasshoppers, and other mobile pests.

Field Deployment

Genetically improved fungi have shown promising results in managing various pests in agricultural and forestry systems. Examples include:

» **Aphids and Whiteflies**: Sprays containing *Beauveria bassiana* effectively suppress sap-feeding pests on crops like tomatoes and cucumbers.

» **Grasshoppers and Locusts**: Granular formulations of *Metarhizium anisopliae* have been deployed successfully in locust-infested regions, reducing populations over large areas.

» **Weevils and Beetles**: Soil-applied *Metarhizium anisopliae* granules control root-feeding pests in crops like sugarcane and maize.

Advantages of Genetically Improved Fungi

1. **Enhanced Efficacy**: Genetic modifications ensure faster pest mortality and greater field persistence.
2. **Eco-Friendly**: Fungal biopesticides degrade naturally, leaving no harmful residues in the environment.
3. **Reduced Chemical Dependency**: Using fungal biocontrol agents reduces the need for synthetic insecticides, promoting sustainable agriculture.

Challenges and Considerations

» **Regulatory Approval**: Genetically modified fungal strains must undergo rigorous safety evaluations before commercial release.

» **Public Perception**: Concerns about genetically engineered organisms may limit acceptance in some regions.

» **Cost of Production**: Scaling up production of genetically improved fungi remains a challenge for commercialization.

3.6 Challenges and Future Directions

The genetic improvement of natural enemies, such as entomopathogenic fungi, baculoviruses, and genetically engineered organisms like *Bt* crops, has revolutionized pest management and biological control. These advancements have introduced eco-friendly alternatives to chemical pesticides, reducing environmental damage and enhancing agricultural sustainability. However, despite these successes, several challenges persist in ensuring the safe, effective, and widespread application of these technologies. Addressing these challenges requires concerted efforts in research, regulation, education, and innovation.

3.6.1 Challenges

While genetically improved biological agents offer immense promise, their implementation faces significant hurdles. These include regulatory barriers, public skepticism, and the need for sustainable resistance management strategies.

Regulatory Hurdles

The deployment of genetically modified organisms (GMOs) in pest control is subject to stringent regulatory scrutiny worldwide. These regulations aim to ensure that GMOs do not pose risks to human health, non-target organisms, or ecosystems. However, these processes can be time-consuming, costly, and inconsistent across regions.

1. **Biosafety Assessments**: Before approval, GMOs must undergo rigorous biosafety evaluations. This includes testing for allergenicity, toxicity, and potential gene flow to non-target species. These assessments often require years of laboratory and field studies, delaying the commercialization of promising technologies.
2. **Regulatory Variability**: Different countries have varying regulations regarding GMOs. For instance, while the United States focuses on product-based evaluations, the European Union employs a process-based approach, often leading to stricter controls. This inconsistency complicates international collaboration and technology transfer.
3. **Costand Time Barriers**: The regulatory process can be prohibitively expensive for smaller companies or academic researchers. Lengthy approval timelines discourage innovation and limit the availability of improved biocontrol agents to farmers.

Public Perception

Public skepticism about genetically modified organisms poses another significant challenge. Negative perceptions, often fueled by misinformation, can hinder the acceptance of genetically improved natural enemies.

1. **Safety Concerns**: Many stakeholders, including farmers and consumers, worry about the potential risks of GMOs to health and the environment. Concerns about gene flow, unintended ecological effects, and unknown long-term impacts persist.
2. **Ethical Considerations**: The genetic modification of living organisms raises ethical questions about human intervention in nature. Critics argue that manipulating natural enemies could disrupt ecosystems or lead to unforeseen consequences.
3. **Media Mis-information**: Media portrayal of GMOs often amplifies fears and misconceptions. Misleading headlines or incomplete information can exacerbate public resistance to these technologies.

Resistance Management

The evolution of resistance in target pest populations is a major challenge to the long-term efficacy of genetically improved natural enemies.

1. **Selection Pressure**: Continuous exposure to genetically improved agents, such as *Bt* crops or engineered fungi, exerts strong selection pressure on pests, accelerating the development of resistance.
2. **Limited Genetic Diversity**: Reliance on a narrow range of genetic modifications increases the likelihood of resistance. Pests adapt more quickly when exposed to uniform control measures.
3. **Lack of Monitoring**: Inadequate monitoring and surveillance systems fail to detect early signs of resistance, allowing it to spread unchecked.
4. **Management Costs**: Implementing resistance management strategies, such as crop rotation or refuge planting, can be labor-intensive and costly, discouraging farmers from adopting them.

3.6.2 Future Directions

To address these challenges and maximize the potential of genetic technologies in pest management, future efforts should focus on improving regulatory frameworks, fostering public understanding, and developing innovative solutions for resistance management.

Improving Regulatory Frameworks

1. **Harmonization of Regulations**: Establishing global standards for the evaluation and approval of GMOs can streamline the regulatory process. Harmonization would facilitate international trade and technology transfer while ensuring consistent safety measures.
2. **Risk-Based Approaches**: Shifting from a precautionary approach to a risk-based framework would enable faster approval of low-risk technologies. This involves focusing resources on evaluating the most significant potential risks while expediting the approval of safer products.
3. **Support for Small Innovators**: Governments and funding agencies should provide financial and technical support to small businesses and academic institutions developing innovative biocontrol agents. Simplifying the regulatory process for low-risk innovations can encourage participation from diverse stakeholders.
4. **Post-Release Monitoring**: Enhancing post-release monitoring ensures that any unforeseen effects of GMOs are promptly identified and mitigated. Transparent reporting mechanisms can build trust among regulators, scientists, and the public.

Fostering Public Understanding

1. **Education Campaigns**: Public awareness campaigns should communicate the benefits and safety of genetically improved natural enemies. These campaigns must target key stakeholders, including farmers, policymakers, and consumers, using accessible and relatable messaging.
2. **Engaging Communities**: Involving local communities in the development and deployment of GMOs can build trust and acceptance. Farmer participatory programs, for example, allow stakeholders to witness firsthand the efficacy and safety of genetically improved agents.

3. **Countering Mis-information**: Leveraging social media and digital platforms to counter misinformation can help shape public opinion. Collaborations with credible influencers, scientists, and organizations can amplify accurate information.
4. **Highlighting Success Stories**: Showcasing successful case studies of genetically improved natural enemies can demonstrate their tangible benefits. Examples include the reduction of pesticide use through *Bt* crops or the effective control of locusts using engineered fungi.

Innovative Resistance Management

1. **Pyramiding Traits**: Incorporating multiple genetic traits into a single agent, such as expressing different Cry proteins in *Bt* crops, reduces the likelihood of pests developing resistance to all traits simultaneously.
2. **Rotational Deployment**: Alternating between different bio-control agents or genetic traits can minimize selection pressure on pests. This approach mirrors crop rotation, which disrupts pest life cycles and slows resistance evolution.
3. **Gene Editing for Dynamic Adaptation**: Advanced gene-editing technologies, such as CRISPR-Cas9, allow for rapid modification of natural enemies in response to emerging pest resistance. This dynamic adaptation ensures that biocontrol agents remain effective over time.
4. **Integration with Other Control Measures**: Combining genetically improved agents with cultural practices, chemical controls, and biological diversity creates a multi-pronged approach to pest management. This integrated strategy reduces reliance on any single method, mitigating resistance risks.

Embracing Emerging Technologies

1. **Synthetic Biology**: Synthetic biology enables the design of entirely new genetic constructs for natural enemies, enhancing their specificity, efficacy, and environmental compatibility. For example, engineering fungi to produce custom toxins could target previously uncontrollable pests.
2. **Artificial Intelligence and Big Data**: AI-powered tools can analyze pest population dynamics, predict resistance development, and optimize deployment strategies for genetically improved agents. These insights ensure precision and effectiveness in pest control programs.
3. **Climate-Resilient Strains**: Developing bio-control agents that thrive under changing climate conditions, such as higher temperatures or prolonged droughts, ensures their efficacy in diverse environments.
4. **Gene Drives**: Gene drive technology could be employed to spread desirable traits, such as sterility or vulnerability to biocontrol agents, throughout pest populations. This approach offers long-term pest suppression but requires careful ecological and ethical considerations.

Chapter 4

Genes of Interest in Entomology

1. Introduction

The field of entomology has witnessed remarkable advancements with the advent of molecular biology and genetic research, enabling scientists to delve deeper into the genetic mechanisms underlying insect biology. Among these developments, identifying and studying genes of interest have become pivotal for understanding insect physiology, behavior, and interactions with their environment. These genes provide vital insights into the functional dynamics of insects and offer novel avenues for addressing critical challenges, including pest management, disease vector control, and the conservation of beneficial insect species. By targeting specific genes, researchers have not only enhanced the understanding of fundamental biological processes but also developed innovative tools and strategies to manipulate insect populations, improving agricultural productivity and public health outcomes.

Genes of interest in entomology encompass a wide range of genetic elements that play crucial roles in diverse physiological and developmental processes. Marker genes for sex identification, for instance, are integral to distinguishing between male and female individuals at early developmental stages, particularly in pest control programs like the Sterile Insect Technique (SIT). This technique, which involves mass-rearing and releasing sterile males to suppress pest populations, requires precise sex identification to eliminate females from the process, thereby reducing costs and increasing efficiency. Marker genes such as *doublesex* (*dsx*), which regulates sex-specific traits, have been extensively studied and utilized in this context. Advances in genetic tools like CRISPR-Cas9 have further enhanced the precision and efficiency of modifying such genes, enabling large-scale implementation of sex-specific genetic interventions.

Another significant area of focus in entomological research is the study of neuropeptides, small signaling molecules that regulate a wide range of physiological and behavioral functions in insects. Neuropeptides and their receptors are central to processes such as feeding, reproduction, diapause, and locomotion. By targeting the genes encoding neuropeptides or their receptors, researchers have developed novel pest control strategies. For example, allatostatins, a class of neuropeptides, inhibit juvenile hormone synthesis, which is essential for insect development and reproduction. Manipulating these genes can disrupt critical life processes in pest species, leading to reduced populations. Additionally,

DOI: 10.1201/9781003757405-4

synthetic analogs of neuropeptides have been developed to interfere with natural signaling pathways, offering environmentally friendly alternatives to traditional chemical pesticides. The use of RNA interference (RNAi) technology to silence neuropeptide genes has further demonstrated the potential for precision-targeted pest control, with minimal impact on non-target species.

Juvenile hormone (JH) and its regulatory enzyme, juvenile hormone esterase (JH esterase), have also been focal points in entomological genetics. JH plays a pivotal role in regulating insect development, metamorphosis, and reproduction, while JH esterase ensures the proper balance of hormone levels by hydrolyzing excess JH. Genes encoding JH esterase have been targeted to manipulate developmental processes and disrupt reproductive cycles in pest insects. For instance, transgenic approaches that enhance the expression of JH esterase can effectively reduce JH levels, leading to developmental arrest or sterility in pest populations. Conversely, the use of small-molecule inhibitors of JH esterase can elevate JH levels artificially, disrupting normal developmental transitions and preventing pests from reaching maturity. These genetic interventions not only offer innovative pest control solutions but also reduce reliance on chemical insecticides, contributing to sustainable agriculture and integrated pest management practices.

Scorpion toxins and venoms represent another intriguing category of genes of interest in entomology, with significant potential for biopesticide development. Scorpion toxins, which are naturally evolved molecules designed to target specific ion channels in insect nervous systems, provide highly specific and potent mechanisms for pest control. Genes encoding these toxins have been introduced into microbial or fungal biopesticide formulations, enhancing their efficacy and specificity against target pests. Additionally, genetic engineering has enabled the expression of scorpion toxin genes in plants or other delivery systems, providing innovative solutions for protecting crops from pest damage. These approaches are environmentally friendly, as they target pests with minimal effects on beneficial organisms, including pollinators and natural enemies.

Another critical focus in entomological genetics is the study of chitinase genes. Chitin, a structural polysaccharide, is a major component of the insect exoskeleton and gut lining. Chitinase enzymes, which degrade chitin, have been explored for their potential in disrupting insect molting and exoskeletal integrity. Genes encoding chitinase have been manipulated to enhance the efficacy of microbial pathogens, such as fungi and bacteria, used in biological pest control. For instance, genetically engineered fungi expressing chitinase genes exhibit enhanced virulence and faster kill rates, making them more effective as biopesticides. These innovations not only improve pest management outcomes but also reduce environmental contamination associated with conventional chemical pesticides.

The genes encoding crystal protein toxins, also known as Cry proteins, from *Bacillus thuringiensis* (*Bt*) are among the most well-known and widely utilized in pest management. These toxins specifically target the midgut cells of susceptible insects, leading to paralysis and death. Cry protein genes have been incorporated into transgenic crops such as *Bt* cotton, maize, and soybean, providing season-long protection against key pests. These *Bt* crops have significantly reduced the need for chemical insecticides, preserving beneficial organisms and promoting sustainable agricultural practices. However, the widespread use of *Bt* crops has raised concerns about the potential for pests to develop resistance. To address this issue, researchers have developed strategies such as pyramiding multiple Cry protein genes within a single crop and implementing refuge areas to delay resistance development.

Lectins and proteases are additional gene categories that have garnered attention in entomological research. Lectins are carbohydrate-binding proteins involved in various physiological and defense mechanisms, including immune responses and recognition processes. In pest control, lectins have been explored for their ability to disrupt insect digestion and nutrient absorption. Genes encoding lectins have been expressed in transgenic plants to confer resistance against sap-sucking and chewing pests. Proteases, on the other hand, play essential roles in digestion and cellular processes. Protease inhibitors, derived from plants or synthetic sources, have been used to interfere with insect gut proteases, reducing the nutritional uptake and growth of pest species. The manipulation of lectin and protease genes has opened new avenues for developing resistant crop varieties and biological control agents.

While the study and application of these genes of interest have yielded promising results, there are challenges to be addressed. Regulatory frameworks governing genetically modified organisms (GMOs) require rigorous testing and assessment to ensure safety and environmental sustainability. Public perception and acceptance of genetic technologies in agriculture and pest management also play a crucial role in their widespread adoption. Additionally, the potential for pests to develop resistance to genetic interventions necessitates the development of integrated and synergistic strategies, combining multiple approaches for long-term efficacy.

Marker Genes for Sex Identification: Importance in Pest Control and Breeding Programs

Sex identification through marker genes has become a vital tool in genetics, pest control, and breeding programs. The ability to distinguish between male and female organisms at an early stage has wide-ranging applications, from controlling pest populations to enhancing efficiency in livestock, aquaculture, and silkworm production. This process involves specific genes or molecular markers that show sex-specific expression patterns or structures. By leveraging these markers, scientists can target or manipulate sexes for desired outcomes.

The Basis of Sex Determination and the Role of Marker Genes

Sex determination systems across organisms vary widely. These mechanisms include chromosomal sex determination (e.g., XX/XY in mammals, ZW/ZZ in birds), haplodiploidy in bees and wasps, and environmental factors like temperature-dependent sex determination in reptiles.In genetic studies, **marker genes** are specific genes or sequences that differ in expression or structure between sexes, allowing researchers to distinguish between males and females. These markers are not only instrumental in understanding the fundamental biology of sex determination but also enable precise manipulation of populations for pest control or breeding purposes.

Key Marker Genes and Their Mechanisms

1. **Doublesex (Dsx):**
    - Origin: Found across many insect species, including *Drosophila melanogaster*, mosquitoes, and agricultural pests.
    - Mechanism: Produces sex-specific isoforms through alternative splicing, determining male or female somatic traits.
    - Applications: Widely used for genetic interventions in pest management to create sex-biased populations.

2. **Transformer (Tra):**
    - Origin: Identified in *Drosophila* and various pest species like *Aedes aegypti* mosquitoes.
    - Mechanism: Controls female development by regulating the splicing of downstream genes like *Dsx*. Functional in females and inactive in males.
    - Applications: Used for creating female-lethal strains in pest control programs.
3. **Sex-lethal (Sxl):**
    - Origin: Found in *Drosophila melanogaster*.
    - Mechanism: Determines sex by initiating a female-specific splicing pathway. Active only in females during early embryogenesis.
    - Applications: Serves as an early marker for sex differentiation in genetic research.
4. **Complementary Sex Determiner (Csd):**
    - Origin: Unique to honeybees (*Apis mellifera*).
    - Mechanism: Controls sex determination based on heterozygosity. Heterozygous individuals develop as females (workers or queens), while homozygous or hemizygous individuals become males or sterile diploid males.
    - Applications: Used in bee breeding programs to optimize colony productivity.
5. **Male-specific Lethal (Msl):**
    - Origin: Found in various insect species.
    - Mechanism: Regulates dosage compensation in males by hypertranscribing X-linked genes.
    - Applications: Marker for identifying and targeting male insects in genetic studies.

Marker Genes in Pest Control

Marker genes play a pivotal role in pest control by enabling sex-specific interventions. This is particularly important for pest species where only one sex contributes to the damage or problem. Marker-based technologies have revolutionized pest management by providing targeted, eco-friendly alternatives to chemical pesticides.

Sterile Insect Technique (SIT)

The Sterile Insect Technique is a widely adopted strategy for controlling pest populations. It involves rearing and releasing large numbers of sterilized males into the environment. These sterile males mate with wild females, resulting in no offspring and a gradual decline in the pest population.

Role of Marker Genes in SIT:

1. **Early Sex Identification:** Marker genes like *Dsx* and *Tra* allow for the early identification of males and females in laboratory-reared populations. This enables the selective removal of females, which simplifies rearing and reduces resource wastage.

2. **Creating Male-Only Populations:** Male-only populations are essential for effective SIT programs since sterile males compete with wild males for mating opportunities. Marker genes ensure the efficient sorting of males from females during production.

Examples in Pest Control:

» **Mosquitoes (*Aedes aegypti*):**
   - Marker genes like *Tra* have been manipulated to produce male-biased populations for SIT programs targeting disease vectors like dengue and Zika virus.

» **Mediterranean Fruit Fly (*Ceratitis capitata*):**
   - Sex-specific splicing of *Dsx* has been used to generate sterile male populations, significantly reducing fruit fly infestations.

Gene Drive Systems

Gene drives are genetic engineering tools designed to spread specific traits through pest populations. Marker genes play a crucial role in driving sex-specific traits, enabling population suppression or replacement.

Population Suppression:

» Marker genes like *Dsx* are used in gene drives to create male-only offspring, reducing the number of reproducing females and eventually collapsing the population.

Population Replacement:

» Marker-linked gene drives can introduce traits such as disease resistance in mosquitoes, reducing their capacity to transmit pathogens like malaria.

Examples of Marker Gene Applications in Gene Drives:

» **Anopheles gambiae (Malaria Vector):**
   - Female-specific expression of modified *Dsx* genes results in sterile females, effectively controlling vector populations.

Marker Genes in Breeding Programs

Marker genes are indispensable in breeding programs, particularly in species where one sex is more economically or biologically valuable. These genes facilitate early sex identification, selective breeding, and population management, leading to increased efficiency and productivity.

Silkworm Breeding

The silkworm (*Bombyx mori*) is a prime example of marker gene applications in breeding programs. In silkworm production, males are preferred for silk yield due to their superior cocoon quality, while females are valuable for egg-laying.

Marker Genes in Silkworms:

» **Bm-Sxl (Sex-lethal in Silkworms):**

- Determines female-specific pathways, allowing breeders to distinguish between sexes at an early stage.

Applications:

1. **Selective Breeding:**
    - Males are selected for silk production, while females are retained for breeding, maximizing overall productivity.
2. **Male Bias for Industrial Use:**
    - Marker-based techniques ensure male-biased populations, optimizing cocoon yield.

Honeybee Breeding

Honeybee colonies rely heavily on sex dynamics, with female workers performing most tasks and males (drones) playing a limited role in reproduction.

Csd (Complementary Sex Determiner):

» Governs the development of females (workers and queens) in honeybee colonies.

» Heterozygous *Csd* leads to females, while homozygous results in males or sterile diploids.

Applications in Apiculture:

1. **Optimizing Worker Populations:**
    - Marker genes are used to increase the proportion of female workers, improving honey production and pollination efficiency.
2. **Queen Rearing:**
    - Ensures the development of healthy queens for colony establishment and maintenance.

Aquaculture and Livestock Breeding

Sex identification is critical in aquaculture and livestock, where one sex often exhibits faster growth, higher yield, or better quality.

Examples:

1. **Tilapia Farming:**
    - Male tilapia grow faster and larger than females. Marker genes enable the production of male-only populations, increasing productivity.
2. **Cattle Breeding:**
    - Sex-specific markers allow for the selective breeding of females for dairy production and males for meat production.

Advantages of Marker Genes

1. **Efficiency:**
    - Marker genes enable early and accurate sex identification, reducing time and resource investments in pest control and breeding programs.
2. **Cost-Effectiveness:**
    - By allowing the selective rearing of desired sexes, marker genes minimize waste and operational costs.
3. **Environmental Benefits:**
    - Marker-based pest control reduces the reliance on chemical pesticides, preserving biodiversity and minimizing ecological damage.
4. **Precision:**
    - Genetic tools based on marker genes provide unparalleled precision in sex identification and manipulation.

Challenges and Future Directions

Challenges:

1. **Technical Complexity:**
    - Developing and deploying marker-based systems require advanced genetic tools and expertise, which may not be accessible in all regions.
2. **Regulatory Hurdles:**
    - The use of genetically modified organisms (GMOs) for marker-based applications is subject to strict regulations.
3. **Public Perception:**
    - Negative perceptions of GMOs can hinder the adoption of marker-based technologies.

Future Directions:

1. **Advanced Gene Editing:**
    - Tools like CRISPR-Cas9 will enable the creation of more efficient and precise marker-based systems.
2. **Integration with AI and Big Data:**
    - Artificial intelligence can optimize breeding and pest control programs using data from marker gene research.
3. **Sustainable Practices:**
    - Marker-based methods will increasingly align with ecological conservation and sustainability goals.

Table: Marker Genes for Sex Identification:

S. No.	Marker Gene	Origin	Application	Key Features/Notes
1	Dsx (Doublesex)	*Drosophila melanogaster*	» Sex differentiation marker in insects » Used for genetic control programs, including sterile insect techniques (SIT)	» Expressed differently in males and females, producing sex-specific isoforms. » Conserved across many insect species.
2	Tra (Transformer)	*Drosophila melanogaster*	» Determines female development. » Used in genetic sexing in insect pest management.	» Males lack functional *tra*, while females require it for differentiation.
3	M (Maleness gene)	*Musca domestica* (housefly)	» Identifies male flies for SIT programs.	» Located on the Y chromosome, governing male differentiation.
4	Sxl (Sex-lethal)	*Drosophila melanogaster*	» Essential for female-specific splicing of downstream genes. » Marker for early sex determination.	» Expressed only in females during early embryogenesis.
5	Fru (Fruiting Body)	*Drosophila melanogaster*	» Used in behavioral studies and sex identification. » Regulates courtship behaviors in males.	» Male-specific splicing regulates male sexual behaviors.
6	Md-tra	*Musca domestica* (housefly)	» Important for genetic sexing strategies in flies. Female-specific splicing factor.	» Manipulation allows for male-only populations in pest management.
7	Bm-Sxl	*Bombyx mori* (silkworm)	» Marker for female silkworms in breeding programs.	» Homolog of *Sxl* in *Drosophila*.
8	Csd (Complementary Sex Determiner)	*Apis mellifera* (honeybee)	» Determines female development in bees. » Used in bee breeding to ensure colony health.	» Heterozygous state leads to female development; homozygous or hemizygous leads to males or sterile males.
9	F (Female Determination Factor)	*Anastrepha* spp. (fruit flies)	» Used in sterile insect techniques. » Determines female-specific development.	» Critical for female-specific splicing pathways.
10	Piwi	*Drosophila melanogaster* and others	» Germline maintenance and sex identification in genetic studies.	» Expression levels vary between males and female

Double sex (dsx) Gene: Role and Applications in Sterile Insect Technique (SIT)

The **Doublesex (dsx) gene** plays a central role in the genetic sex determination pathways of many insect species. It is a master regulator of somatic sexual differentiation, governing the development of male and female characteristics through sex-specific isoforms. The **dsx gene** has become a cornerstone in pest control strategies, particularly in the **Sterile Insect Technique (SIT)**, a biological control method designed to suppress pest populations by releasing sterilized insects into the environment.

The Doublesex (dsx) Gene:

The *dsx* gene is a highly conserved transcription factor that directs the development of sexually dimorphic traits in insects. It is part of the *sex-determination hierarchy*, downstream of genes like **Sex-lethal (Sxl)** in *Drosophila* and **Transformer (Tra)** in other species. The alternative splicing of *dsx* pre-mRNA produces sex-specific mRNA transcripts, resulting in the expression of maleor female-specific isoforms.

Structure and Function of the *dsx* Gene

1. **Sex-Specific Isoforms:**
   - Male-specific isoform (*DsxM*): Promotes male differentiation by activating male-specific genes and repressing female-specific genes.
   - Female-specific isoform (*DsxF*): Facilitates female differentiation by activating female-specific genes and suppressing male-specific genes.
2. **Conserved Role Across Insects:**
   - Found in diverse species, including *Drosophila melanogaster*, mosquitoes (*Anopheles gambiae*), and agricultural pests like *Helicoverpaarmigera*.
   - Regulates external morphology (e.g., genitalia, antennae) and internal reproductive systems in a sex-specific manner.
3. **Critical for Reproductive Fitness:**
   - *Dsx* ensures proper sexual development, mating behaviors, and fertility, making it a pivotal gene for population control strategies.

Examples of the Doublesex Gene in Insects

The *dsx* gene has been extensively studied across a wide range of insect species, with key findings highlighting its importance in sex determination and pest control:

Insect Species	Role of *dsx*	Significance in Pest Control
Drosophila melanogaster	Regulates sexually dimorphic traits like genitalia.	Model organism for studying *dsx* gene function and splicing.
Aedes aegypti (Mosquito)	Controls male genital and female reproductive traits.	Target for developing male-biased populations in vector control.
Anopheles gambiae	Governs reproductive traits in malaria vectors.	Used to drive female sterility through gene editing.

Ceratitis capitata (Medfly)	Directs male and female morphology.	Engineered to create male-only sterile populations for SIT.
Helicoverpaarmigera	Affects moth sexual differentiation.	Studied for potential applications in agricultural pest control.

Applications of *dsx* in Sterile Insect Technique (SIT)

The Sterile Insect Technique (SIT) is a biological pest control strategy aimed at suppressing or eradicating pest populations by releasing large numbers of sterile males into the environment. These males compete with wild males to mate with females, but no viable offspring are produced, leading to a gradual population decline. The *dsx* gene plays a vital role in enhancing the effectiveness of SIT programs by enabling the production of male-only populations.

Importance of Male-Only Populations in SIT

1. **Male Efficiency:**
    - Male insects are the active participants in mating, while sterile females add unnecessary costs to production and logistics.
    - Male-only releases ensure that all resources are focused on effective population suppression.
2. **Minimizing Female Impact:**
    - In many pest species, females are responsible for direct damage or disease transmission. For example:
        - ◊ Female mosquitoes transmit malaria, dengue, and Zika viruses.
        - ◊ Female agricultural pests like moths lay eggs that cause crop damage.
    - Removing females from release populations enhances SIT efficiency and safety.

How *dsx* Supports SIT

1. **Genetic Manipulation:**
    - Gene-editing tools like CRISPR-Cas9 are used to modify the *dsx* gene to produce sex-specific sterility or lethality.
    - By targeting the female-specific isoform (*DsxF*), researchers can induce female sterility or prevent female development, resulting in male-only populations.
2. **Sex Sorting:**
    - *Dsx* markers enable early sex identification, allowing for the separation of males from females during mass rearing.

Examples of *dsx*-Based SIT Programs

1. **Aedes aegypti (Dengue Mosquito):**
    - *Dsx* gene modifications create male-biased populations by disrupting female development.

- Male-only sterile mosquito releases have successfully reduced mosquito populations in field trials.

2. **Anopheles gambiae (Malaria Mosquito):**
   - CRISPR-based gene drives linked to *dsx* have been designed to spread female sterility traits, significantly reducing vector populations.
3. **Ceratitis capitata (Mediterranean Fruit Fly):**
   - *Dsx*-based technologies are used to produce sterile male medflies for SIT programs, reducing fruit crop losses.
4. **Helicoverpa armigera (Cotton Bollworm):**
   - Targeting *dsx* in agricultural pests can potentially create male-only sterile moth populations, improving pest control in cotton fields.

Advantages of Using *dsx* in SIT

1. **Precision:**
   - Targeting the *dsx* gene ensures highly specific effects on sex determination, minimizing unintended impacts on non-target species.
2. **Cost-Effectiveness:**
   - Male-only rearing reduces production costs by eliminating the need to rear or sterilize females.
3. **Eco-Friendly:**
   - Unlike chemical pesticides, *dsx*-based SIT is environmentally sustainable, leaving no toxic residues.
4. **Improved Efficacy:**
   - Male-only releases increase the likelihood of mating with wild females, enhancing the suppressive impact on pest populations.

Challenges and Future Directions

Challenges:

1. **Resistance Development:**
   - Continuous reliance on *dsx*-based systems could lead to resistance in target populations.
2. **Regulatory Hurdles:**
   - Approval processes for genetically modified insects are complex and vary by region.
3. **Public Acceptance:**
   - Misinformation and skepticism about genetic technologies can hinder adoption.

Future Directions:

1. **Advanced Gene Drives:**
   - Integrating *dsx* with self-limiting gene drives to enhance safety and control.
2. **Automation in Sex Sorting:**
   - Developing automated systems to sort sexes based on *dsx* markers for large-scale rearing.
3. **Expanding Target Species:**
   - Extending *dsx*-based SIT applications to more pest and vector species for global impact.

Advances in CRISPR-Cas9 for Sex Identification

The advent of CRISPR-Cas9 technology has ushered in a new era of precision in genetic engineering. It allows for targeted manipulation of sex-determination pathways, offering a transformative tool for applications in pest control, aquaculture, animal breeding, and conservation. By enabling specific edits in sex-determining genes, CRISPR-Cas9 technology has expanded our ability to manage populations efficiently, optimize production, and protect biodiversity.

Mechanisms of CRISPR-Cas9 in Sex Identification

The CRISPR-Cas9 system enables researchers to manipulate the genetic pathways involved in sex determination by targeting critical genes responsible for male or female development. The process involves introducing double-strand breaks in specific DNA sequences, followed by repair mechanisms that allow gene disruption, insertion, or replacement.

Key Steps in Using CRISPR-Cas9 for Sex Identification

1. **Target Identification:**

   Scientists identify sex-determining genes that regulate key pathways in target organisms. These genes often vary between species but include highly conserved regulators such as *Dsx* (Doublesex), *Tra* (Transformer), and *Amh* (Anti-Müllerian Hormone).

2. **Guide RNA Design:**

   A single-guide RNA (sgRNA) is designed to match the DNA sequence of the target gene. The specificity of the sgRNA ensures precision, minimizing off-target effects.

3. **Cas9 Nuclease Activation:**

   The Cas9 protein binds to the sgRNA and forms a complex that scans the genome for matching sequences. Once bound, Cas9 creates a double-strand break in the DNA at the target site.

4. **Genome Editing Outcomes:**
   - **Gene Knockout:** Non-homologous end joining (NHEJ) repairs the break but often introduces errors, disrupting the gene's function.
   - **Gene Insertion or Replacement:** Homology-directed repair (HDR) incorporates new genetic material, allowing for the introduction of desired traits or markers.

By targeting sex-determination genes, CRISPR-Cas9 can skew sex ratios, create sterile populations, or introduce markers for early sex identification.

Applications of CRISPR-Cas9 in Sex Identification

The applications of CRISPR-Cas9 in sex identification are vast and include diverse fields such as pest control, aquaculture, livestock breeding, and conservation biology. The ability to manipulate sex ratios has far-reaching implications for population management and productivity.

1. Pest Control

One of the most transformative applications of CRISPR-Cas9 is in controlling pest populations, particularly insects that damage crops or transmit diseases. The manipulation of sex-determination genes reduces pest populations or eliminates the harmful sex (e.g., females responsible for disease transmission).

Male-Only Production for SIT (Sterile Insect Technique):

- » CRISPR-Cas9 enables the production of male-only populations by targeting female-specific genes like *Dsx* or *Tra*.
- » Male mosquitoes, for example, do not bite or transmit diseases, making male-only releases safer and more effective.

Case Studies:

- » **Aedes aegypti (Dengue Vector):** Disruption of *Tra* results in male-biased populations, reducing dengue transmission.
- » **Mediterranean Fruit Fly (*Ceratitis capitata*):** CRISPR-Cas9 editing of *Dsx* generates sterile males, effectively suppressing agricultural pests.

Gene Drives for Population Suppression:

- » CRISPR-based gene drives linked to *Dsx* propagate female-sterility traits throughout pest populations, leading to long-term population suppression.
- » Example: *Anopheles gambiae* (malaria vector) uses *Dsx*-linked gene drives to target female development, reducing disease transmission.

2. Aquaculture

In aquaculture, sex-specific traits often determine productivity and profitability. CRISPR-Cas9 allows for precise manipulation of sex-determination pathways, enabling the creation of monosex populations optimized for growth or reproduction.

Male-Biased Populations:

- » In species like tilapia, males grow faster and larger than females. CRISPR-Cas9 can target genes like *Amh* or *Dmrta2* to promote male-biased populations.
- » **Example:** CRISPR-induced knockout of *Amh* in tilapia increases male production, improving growth rates and yield.

Female-Biased Populations:

» In species such as shrimp, females are preferred for their larger size and higher reproductive output. CRISPR modifications ensure female-biased populations, optimizing production.

Improved Reproductive Management:

» CRISPR-Cas9 can regulate the timing and intensity of reproductive cycles by modifying genes involved in gonadal development, allowing for synchronized breeding and increased efficiency.

3. Livestock and Poultry Breeding

Sex identification and manipulation are critical in livestock and poultry industries, where one sex is often more economically valuable. CRISPR-Cas9 provides tools for early sex identification and population management.

Poultry Industry:

» Male chicks in egg-laying breeds are often culled due to their low economic value. CRISPR-Cas9 can:

- Introduce markers like fluorescent proteins into male embryos, allowing early, non-invasive sex identification.
- Knock out genes essential for male development, resulting in female-only populations for egg production.

Livestock Breeding:

» Female offspring are preferred in dairy farming for milk production, while male offspring are valued in meat production.

» CRISPR can selectively enhance the expression of sex-specific markers to favor desired traits.

4. Conservation Biology

CRISPR-Cas9 offers solutions for species conservation by addressing sex ratio imbalances, preventing hybridization, and enhancing reproductive success in endangered populations.

Balancing Sex Ratios:

» In species with skewed sex ratios, such as certain reptiles and amphibians, CRISPR can edit sex-determining genes to produce the underrepresented sex, stabilizing populations.

Preventing Hybridization:

» CRISPR-Cas9 can be used to engineer sterility in hybrids or modify genes involved in reproductive compatibility, protecting native species from genetic dilution.

Case Example:

» In efforts to conserve sea turtles, where temperature-dependent sex determination often produces more females due to climate change, CRISPR could be used to balance sex ratios by editing temperature-sensitivity genes.

Advances in CRISPR-Cas9 Technology for Sex Identification

CRISPR-Cas9 technology is constantly evolving, with advancements aimed at increasing precision, efficiency, and scalability.

1. Base Editing

Base editing allows for precise single-nucleotide changes without inducing double-strand breaks, reducing off-target effects. This is particularly valuable for editing critical sex-determination genes where small changes can have significant impacts.

Example:

» Base editing in *Dsx* genes can create male-specific or female-specific populations without affecting other genomic regions.

2. Prime Editing

Prime editing expands the possibilities of CRISPR by enabling insertions, deletions, and base substitutions with high precision. This is useful for complex edits in sex-determination pathways.

3. Multiplexing

CRISPR-Cas9 can now target multiple genes simultaneously, allowing for comprehensive edits in sex-determination networks.

Example:

» Targeting both *Tra* and *Dsx* in mosquitoes ensures robust male-biased populations for SIT programs.

Advantages of CRISPR-Cas9 in Sex Identification

1. **Precision:**
   - CRISPR-Cas9 provides unparalleled precision in targeting specific genes, minimizing unintended effects.
2. **Scalability:**
   - Gene-editing technologies are scalable for use in large populations, making them suitable for commercial applications.
3. **Cost-Effectiveness:**
   - By reducing resource wastage (e.g., rearing undesired sexes), CRISPR-based strategies lower operational costs.
4. **Eco-Friendly Solutions:**
   - Unlike chemical interventions, CRISPR offers species-specific and environmentally sustainable solutions.

5. **Global Applicability:**
   - CRISPR systems can be adapted for diverse species, from insects to vertebrates, addressing local and global challenges.

Challenges and Limitations

Despite its potential, CRISPR-Cas9 faces challenges that must be addressed for widespread adoption in sex identification.

1. **Off-Target Effects:**
   - Unintended edits in non-target regions can lead to unforeseen consequences, affecting organism fitness or ecosystem dynamics.
2. **Resistance Development:**
   - Continuous use of CRISPR-based interventions may lead to the evolution of resistance in target populations.
3. **Ethical Concerns:**
   - Manipulating sex ratios or creating sterile populations raises ethical questions about human intervention in natural ecosystems.
4. **Regulatory Barriers:**
   - Approval processes for genetically modified organisms vary widely, creating hurdles for the deployment of CRISPR-based solutions.
5. **Public Perception:**
   - Misinformation and skepticism about genetic technologies may hinder acceptance and implementation.

3. Neuropeptides: Role in Insect Physiology and Behavior

Neuropeptides are small protein-like molecules that serve as signaling molecules in the nervous systems of insects. They are crucial regulators of a wide array of physiological processes and behaviors, including reproduction, feeding, molting, circadian rhythms, and stress responses. By acting on specific receptors, neuropeptides influence the activity of target tissues, coordinating complex physiological and behavioral patterns essential for insect survival.

a. Neuropeptides in Insects

Structure and Synthesis

1. **Structure:**
   - Neuropeptides are typically short chains of amino acids (peptides) encoded by specific neuropeptide precursor genes.
   - These precursors are processed into active peptides by post-translational modifications, including cleavage, amidation, and glycosylation.

2. **Synthesis and Release:**
   - Synthesized in neurons, neuropeptides are stored in vesicles at synaptic terminals and released in response to specific stimuli.
   - Upon release, they bind to G-protein-coupled receptors (GPCRs) on target cells, initiating intracellular signaling cascades.
3. **Mode of Action:**
   - Neuropeptides act over short and long distances, either as neurotransmitters, neuromodulators, or hormones.
   - They exert effects on diverse physiological systems, often working in concert with other neurohormones or biogenic amines.

Key Classes of Insect Neuropeptides

Insects produce a wide variety of neuropeptides, each with distinct roles in physiological regulation and behavioral modulation. Prominent classes include:

Neuropeptide Class	Role in Physiology and Behavior	Examples
Allatotropins	Stimulate juvenile hormone (JH) synthesis; regulate growth and reproduction.	Allatotropin (AT)
Allatostatins	Inhibit juvenile hormone production; control molting and development.	Allatostatin-A, B, and C
Prothoracicotropic Hormones	Regulate ecdysteroid biosynthesis, initiating molting.	PTTH
Insulin-like Peptides (ILPs)	Control metabolism, growth, and reproduction.	ILP
Corazonin	Coordinate stress responses and mating behaviors.	Corazonin
FMRFamide-related Peptides	Regulate muscle activity, feeding, and gut motility.	FMRFamide
Tachykinin-related Peptides	Influence feeding, gut function, and nociception (pain responses).	Tachykinins (TKs)
Diuretic Hormones (DHs)	Regulate water balance and excretion.	DH31, DH44

Role of Neuropeptides in Insect Physiology

Neuropeptides are pivotal in maintaining the homeostasis and coordination of physiological systems in insects.

1. Reproduction

Neuropeptides regulate key aspects of insect reproduction, including sexual maturation, pheromone production, mating behaviors, and oviposition.

» **Juvenile Hormone Regulation:**
   - Allatotropins and allatostatins control juvenile hormone levels, which are critical for ovarian development and egg production in females.
   - Example: Allatotropin promotes vitellogenesis (yolk formation) in *Bombyx mori.*

» **Mating Behavior:**

- Neuropeptides like corazonin and neuropeptide F (NPF) modulate courtship and mating.
- Example: Corazonin drives reproductive behaviors in *Drosophila melanogaster* males.

2. Feeding and Metabolism

Neuropeptides govern feeding behavior, energy balance, and nutrient utilization.

» **Regulation of Appetite:**

- Neuropeptide F (NPF) and short neuropeptide F (sNPF) act as appetite modulators. High NPF activity correlates with increased feeding, while low activity suppresses it.
- Example: In *Drosophila melanogaster*, NPF mediates responses to food availability.

» **Energy Metabolism:**

- Insulin-like peptides (ILPs) regulate glucose and lipid metabolism, ensuring energy homeostasis.
- Example: ILP mutants in *Drosophila* exhibit impaired growth and altered energy storage.

3. Growth and Development

Neuropeptides orchestrate growth and developmental transitions such as molting and metamorphosis.

» **Ecdysteroid Production:**

- Prothoracicotropic hormone (PTTH) stimulates the release of ecdysone, triggering molting and pupation.
  - ◊ Example: PTTH levels peak before each molt in *Manduca sexta* larvae.

» **Inhibition of Development:**

- Allatostatins inhibit juvenile hormone synthesis, delaying developmental transitions when conditions are unfavorable.

4. Water Balance and Excretion

Neuropeptides like diuretic hormones (DHs) regulate ion and water homeostasis, critical for survival in variable environments.

» **Diuresis:**

- DH31 and DH44 promote water excretion by stimulating Malpighian tubules.
- Example: DHs are activated during feeding to handle osmotic stress in *Aedes aegypti*.

» **Anti-Diuresis:**

- CAPA peptides counteract diuretic hormones, ensuring water retention during dehydration.

5. Stress Responses

Neuropeptides enable insects to respond to environmental stressors such as predation, temperature extremes, and starvation.

» **Stress Hormones:**

- Corazonin and tachykinin-related peptides mediate physiological and behavioral responses to stress.
- Example: Corazonin triggers escape behaviors in *Locusta migratoria* during predator encounters.

Role of Neuropeptides in Insect Behavior

Neuropeptides are equally important in shaping insect behaviors, from foraging and social interaction to circadian rhythms and migration.

1. Foraging and Feeding Behavior

Neuropeptides modulate decisions about when, where, and how insects forage for food.

» **Food-Seeking Behavior:**

- NPF promotes foraging under starvation, while sNPF enhances feeding efficiency.
- Example: In honeybees, NPF levels influence foraging intensity and preference for nectar or pollen.

» **Hunger and Satiety:**

- FMRFamide-related peptides (FaRPs) regulate gut motility and satiety signals, balancing food intake.

2. Social Behavior

In social insects like ants, bees, and termites, neuropeptides influence caste differentiation, communication, and colony coordination.

» **Caste Development:**

- Juvenile hormone-regulating neuropeptides determine whether larvae develop into workers, queens, or soldiers.
- Example: Allatostatins influence soldier caste development in termites.

» **Pheromone Communication:**

- Neuropeptides regulate pheromone production, enabling coordinated behaviors such as foraging and defense.
- Example: Corazonin enhances alarm pheromone release in ants during predator threats.

3. Circadian Rhythms

Neuropeptides like pigment-dispersing factor (PDF) synchronize circadian rhythms, governing sleep-wake cycles and behavioral patterns.

- » **Day-Night Cycles:**
    - PDF coordinates locomotor activity and feeding during specific times of day.
    - Example: In *Drosophila melanogaster*, PDF mutants show disrupted circadian behaviors.

4. Migratory and Locomotor Behavior

Neuropeptides drive migration and long-distance movements in insects like locusts and butterflies.

- » **Locust Swarming:**
    - Corazonin triggers swarming behavior in *Locusta migratoria*, facilitating group migration.
- » **Butterfly Migration:**
    - Neuropeptides modulate energy reserves and directional movement in migrating monarch butterflies.

Applications in Research and Pest Management

1. Pest Control

Neuropeptides are potential targets for developing new pest control strategies.

- » **Disrupting Growth and Reproduction:**
    - Synthetic allatostatins can inhibit juvenile hormone, preventing reproduction and development in pests.
- » **Modulating Feeding Behavior:**
    - NPF antagonists reduce appetite, leading to starvation in crop pests.

2. Biotechnology

Neuropeptides are being harnessed for innovations in biotechnology and agriculture.

- » **Transgenic Crops:**
    - Crops engineered to produce neuropeptide inhibitors can deter pest feeding.
- » **Biopesticides:**
    - Synthetic neuropeptide analogs can selectively target pest species without harming beneficial insects.

b. Allatostatins and Juvenile Hormone Regulation

Allatostatins (ASTs) are a group of neuropeptides that serve as negative regulators of juvenile hormone (JH) synthesis in insects. Juvenile hormone plays an essential role in the regulation of growth, development, reproduction, and behavior in insects. By suppressing JH biosynthesis, allatostatins act as key modulators of physiological transitions, ensuring insects adapt to environmental cues and developmental requirements. The regulation of JH by allatostatins also highlights their potential as targets for pest control strategies aimed at disrupting growth or reproduction in pest populations.

Allatostatins

Structure and Classification

Allatostatins belong to a diverse family of small neuropeptides that exhibit conserved functions across insect species. These peptides are defined by their ability to inhibit juvenile hormone production in the corpora allata (CA), a gland responsible for JH biosynthesis.

Classification of Allatostatins: Allatostatins are categorized into three distinct families based on their structural motifs:

1. **Allatostatin-A (AST-A):**
    - The largest and most widely studied allatostatin family, characterized by the conserved -YXFGLamide C-terminal sequence.
    - Found in insects like *Drosophila melanogaster* and *Bombyx mori*.
    - Plays roles in inhibiting JH production and regulating feeding behavior.
2. **Allatostatin-B (AST-B):**
    - Structurally related to vertebrate somatostatin, with different motifs compared to AST-A.
    - Found in a more limited range of species, including cockroaches and crickets.
    - Implicated in inhibiting juvenile hormone and modulating stress responses.
3. **Allatostatin-C (AST-C):**
    - Identified by a distinct -PISCF C-terminal motif, primarily in dipteran species (e.g., flies).
    - Functions as a neuropeptide hormone, regulating both juvenile hormone and neural activity.

Synthesis and Distribution:

» Allatostatins are synthesized in neurosecretory cells located in the brain and retrocerebral complex.

» They are distributed to target tissues, such as the corpora allata, through hemolymph or neural connections.

Mechanisms of Action

Allatostatins regulate juvenile hormone synthesis by targeting specific receptors located on the corpora allata.

1. **GPCR Binding:**
    - Allatostatins bind to G-protein-coupled receptors (GPCRs) on the surface of corpora allata cells, initiating intracellular signaling pathways.
    - This interaction results in the inhibition of key enzymes, such as juvenile hormone acid O-methyltransferase (JHAMT), required for JH biosynthesis.

2. **Suppression of Enzymatic Activity:**
    - Allatostatins reduce the flux through the JH biosynthetic pathway by downregulating the activity of upstream precursors, including farnesyl pyrophosphate and methyl farnesoate.
3. **Dynamic Regulation:**
    - The allatostatin system integrates environmental signals, developmental cues, and hormonal feedback to fine-tune JH levels during various life stages.

Role of Allatostatins in Juvenile Hormone Regulation

1. Growth and Development

The regulation of juvenile hormone by allatostatins is pivotal for coordinating the growth and developmental transitions in insects. Juvenile hormone maintains larval states, delays metamorphosis, and governs molting cycles.

Molting and Larval Transitions:

» Juvenile hormone levels fluctuate during larval stages to ensure synchronized molting. Allatostatins inhibit JH during pre-molting phases to allow the rise of ecdysteroid hormones, which drive molting.

» **Example:** In *Manduca sexta*, allatostatin-A reduces JH levels prior to larval molts, facilitating the transition to the next instar.

Metamorphosis:

» During the final larval stage, allatostatins play a critical role in terminating JH production, triggering the transition from larval to pupal stages. This ensures proper differentiation of tissues during metamorphosis.

» **Example:** In *Bombyx mori*, allatostatin suppression of JH at the late larval stage signals the onset of pupation and adult formation.

Diapause:

» Allatostatins regulate diapause by inhibiting JH production during unfavorable environmental conditions, such as extreme temperatures or resource scarcity. This adaptive mechanism conserves energy and promotes survival.

» **Example:** In *Culex pipiens* mosquitoes, allatostatins mediate diapause in response to shorter photoperiods.

2. Reproduction

Reproductive processes in insects, including ovarian development, mating behavior, and vitellogenesis, are tightly regulated by juvenile hormone. Allatostatins modulate these processes by influencing JH levels.

Oogenesis and Vitellogenesis:

- » Juvenile hormone stimulates yolk protein production (vitellogenesis) in female insects. Allatostatins delay or inhibit these processes by reducing JH levels, ensuring reproduction occurs only under favorable conditions.
- » **Example:** In locusts like *Locusta migratoria*, allatostatins prevent premature ovarian maturation by suppressing JH synthesis.

Mating and Post-Mating Regulation:

- » In mated females, allatostatin inhibition is reduced, allowing JH synthesis to resume and stimulate egg production. This ensures that egg-laying is synchronized with mating events.
- » **Example:** In *Drosophila melanogaster*, allatostatin-C levels decline post-mating to promote reproductive behaviors.

Behavioral Impacts of Allatostatins

Allatostatins influence behavior indirectly by modulating juvenile hormone, which affects feeding, foraging, mating, and diapause behaviors.

Foraging and Feeding Behavior

Juvenile hormone is known to regulate foraging in social insects. Allatostatins, by suppressing JH, delay foraging behaviors and maintain younger individuals in in-nest roles.

- » **Social Insects:**
  - In honeybees (*Apis mellifera*), young worker bees remain inside the hive performing nursing tasks due to allatostatin suppression of JH. As JH levels increase, foraging behaviors are triggered.
  - **Example:** The transition from nurse bees to foragers is regulated by allatostatin-mediated JH inhibition.
- » **Feeding Behavior:**
  - Allatostatins influence satiety signals and feeding efficiency in solitary insects.
  - **Example:** In cockroaches, allatostatin analogs reduce feeding activity by lowering JH-dependent hunger signals.

Mating and Courtship

Allatostatins delay courtship behaviors in immature insects by maintaining low JH levels, ensuring reproductive readiness only after complete maturation.

- » **Example:** In *Drosophila melanogaster*, allatostatin-C regulates the timing of courtship pheromone production by controlling JH.

Seasonal and Diapause Behavior

Allatostatins synchronize seasonal behaviors by inducing diapause through juvenile hormone suppression.

» **Example:** In temperate regions, mosquitoes like *Culex pipiens* rely on allatostatins to halt reproduction and enter diapause during winter.

Applications in Pest Management

The regulatory role of allatostatins in juvenile hormone synthesis makes them attractive targets for pest control strategies. Synthetic allatostatinanalogs or genetic manipulations can be used to disrupt growth, reproduction, and behavior in pest species.

1. Growth Inhibition

» Allatostatin mimics can be designed to inhibit JH synthesis, preventing pests from molting or transitioning to reproductive stages.

» **Example:** In *Helicoverpaarmigera* (cotton bollworm), allatostatin-based treatments could halt larval growth, reducing crop damage.

2. Reproductive Suppression

» Allatostatins could suppress vitellogenesis and ovarian development, curbing pest populations.

» **Example:** Locust swarms could be controlled by allatostatin analogs that prevent egg production in females.

3. Biopesticides

» Synthetic allatostatin analogs offer species-specific pest control, reducing environmental harm and protecting beneficial insects.

» **Example:** Crops engineered to express allatostatin peptides could resist herbivorous pests without affecting pollinators.

4. Genetic Manipulation

» Gene-editing tools like CRISPR-Cas9 could target allatostatin pathways to produce sterile pest populations or disrupt JH-dependent processes.

» **Example:** Editing allatostatin genes in mosquitoes could reduce their reproductive capacity and limit disease transmission.

Future Research Directions

1. **Receptor Identification**:
    - Understanding allatostatin receptor diversity across pest species can guide the development of selective inhibitors.
2. **Synthetic Development:**
    - Designing stable and bioavailable allatostatin analogs for large-scale agricultural use.
3. **Integrated Pest Management (IPM):**
    - Combining allatostatin-based treatments with other pest control methods for sustainable solutions.

4. **Cross-Species Studies:**

   - Investigating allatostatin roles in non-model organisms could reveal new pest control targets.

c. Tachykinins and Their Functional Impacts

Tachykinins are a highly conserved family of neuropeptides found in both vertebrates and invertebrates, including insects. In insects, tachykinins play crucial roles in regulating physiological processes and behavioral responses. These small peptides are characterized by their conserved C-terminal motif – FXGLM amide and exert their effects by binding to G-protein-coupled receptors (GPCRs), triggering downstream signaling cascades. Tachykinins influence various functions such as feeding, gut motility, stress response, nociception (pain perception), and reproduction, making them key players in the neural and hormonal control systems of insects.

Molecular Characteristics of Tachykinins

1. Structure and Conservation

Tachykinins are short-chain peptides composed of 8–15 amino acids. Despite some species-specific variations, they are defined by their highly conserved C-terminal pentapeptide sequence -FXGLMamide (where X is variable).

» **Precursor Proteins:** Tachykinins are synthesized as part of larger precursor proteins called preprotachykinins (PPTs). These precursors undergo enzymatic cleavage to release active peptides.

» **Amidation:** Post-translational modification, such as amidation at the C-terminus, is critical for their bioactivity and receptor binding.

2. Tachykinin Receptors

Tachykinins exert their effects through specific GPCRs located on the surface of target cells. Binding to these receptors initiates intracellular signaling cascades involving secondary messengers like cyclic AMP (cAMP) or calcium ions, which modulate the activity of target tissues.

» **Diversity of Receptors:** Insects often possess multiple tachykinin receptors, each with varying affinities for different tachykinin isoforms. This receptor diversity allows tachykinins to regulate a wide range of physiological and behavioral processes.

Functional Roles of Tachykinins

Tachykinins have a broad spectrum of effects on insect physiology and behavior. Their functional impacts can be classified into several key domains:

1. Regulation of Feeding and Digestion

Tachykinins are central to controlling feeding behavior and gut motility, ensuring efficient nutrient acquisition and digestion.

» **Modulation of Feeding Behavior:**

   - Tachykinins suppress feeding in response to satiety signals or stress.

- Example: In *Drosophila melanogaster*, tachykinins regulate food intake by acting on neurons that process hunger and satiety signals.

» **Gut Motility and Digestion:**

- Tachykinins stimulate the contraction of visceral muscles, promoting peristalsis and efficient food passage through the gut.
- Example: In locusts (*Locusta migratoria*), tachykinins enhance gut motility during active feeding periods.

» **Regulation of Energy Balance**:

- By influencing feeding behavior and gut function, tachykinins indirectly regulate energy storage and utilization.

2. Stress Response and Nociception

Tachykinins play a critical role in mediating stress responses and nociception, helping insects adapt to adverse environmental conditions.

» **Stress-Induced Behavioral Changes:**

- Tachykinins are involved in the neural pathways that mediate escape behaviors and stress-induced feeding suppression.
- Example: In locusts, tachykinins are upregulated during predator encounters, enhancing the insect's ability to evade threats.

» **Nociception and Pain Response:**

- Tachykinins are key mediators of nociceptive signaling, enabling insects to detect and respond to harmful stimuli.
- Example: In *Drosophila*, tachykinins are released in response to noxious heat or mechanical injury, triggering withdrawal reflexes.

» **Hormonal Regulation of Stress:**

- Tachykinins interact with other neurohormones, such as corazonin and adipokinetic hormone (AKH), to orchestrate comprehensive stress responses.

3. Reproductive Regulation

Tachykinins influence reproductive processes by modulating behaviors and physiological changes associated with mating and egg-laying.

» **Mating Behavior:**

- Tachykinins are involved in the neural circuits that regulate courtship and copulatory behaviors.
- Example: In crickets (*Gryllus bimaculatus*), tachykinins modulate male courtship song production.

» **Oviposition:**

- Tachykinins stimulate oviposition by inducing contractions in the reproductive tract and promoting egg-laying behaviors.
- Example: In *Helicoverpaarmigera* (cotton bollworm), tachykinins facilitate egg deposition after mating.

4. Neuromodulation and Locomotor Activity

As neuromodulators, tachykinins influence locomotor activity and other movement-related behaviors.

» **Coordination of Locomotion:**

- Tachykinins act on motor neurons to regulate muscle contractions involved in walking, flying, and swimming.
- Example: In cockroaches (*Periplaneta americana*), tachykinins enhance the rhythmic activity of leg muscles during locomotion.

» **Swarming Behavior in Locusts**:

- Tachykinins are implicated in the neural mechanisms underlying swarming behavior, facilitating synchronized movements in group migrations.

5. Regulation of Water and Ion Balance

Tachykinins contribute to osmoregulation by influencing water retention and ion transport in the Malpighian tubules and gut.

» **Diuretic Effects**:

- Tachykinins stimulate water and ion excretion, maintaining homeostasis during feeding or stress.
- Example: In *Rhodniusprolixus* (kissing bug), tachykinins enhance diuresis after a blood meal.

» **Anti-Diuretic Coordination:**

- Tachykinins interact with antidiuretic hormones to balance water loss during dehydration.

6. Circadian Rhythms and Sleep Regulation

Tachykinins are involved in the regulation of circadian rhythms and sleep-wake cycles.

» **Circadian Modulation:**

- Tachykinins influence the timing of locomotor and feeding activities based on day-night cycles.
- Example: In *Drosophila*, tachykinin mutants exhibit altered sleep patterns and disrupted circadian rhythms.

» **Sleep Induction:**

- Tachykinins promote rest periods by suppressing arousal signals in the central nervous system.

Applications of Tachykinin Research

The functional versatility of tachykinins makes them attractive targets for applications in pest management, biotechnology, and medical research.

1. Pest Control

Tachykinins and their receptors are promising targets for developing novel pest control strategies.

» **Disrupting Feeding Behavior:**
- Synthetic tachykinin agonists or antagonists can modulate feeding, reducing crop damage by agricultural pests.
- Example: Tachykinin-based treatments could suppress the appetite of pests like *Spodoptera frugiperda* (fall armyworm).

» **Inhibiting Reproduction:**
- Tachykinin inhibitors could block oviposition in pest species, limiting population growth.
- Example: In mosquitoes (*Aedes aegypti*), targeting tachykinins may reduce egg-laying capacity.

» **Neuromuscular Disruption:**
- By interfering with locomotor activity, tachykinin analogs can impair the mobility of pests, making them more vulnerable to predators or environmental hazards.

2. Biotechnology

Tachykinins have potential applications in biotechnology, particularly in designing bioactive peptides for industrial and pharmaceutical use.

» **Peptide-Based Insecticides:**
- Synthetic tachykinin analogs could serve as eco-friendly insecticides, selectively targeting pest species while minimizing harm to beneficial insects.

» **Model Systems for GPCR Research:**
- Tachykinin receptors in insects provide valuable models for studying GPCR function, aiding drug discovery efforts in humans.

3. Insights into Human Health

The similarities between insect and vertebrate tachykinins make insect models valuable for understanding human health.

» **Pain Management:**
- Insect tachykinins offer insights into nociception and could inform the development of novel analgesics.

» **Stress Biology:**
- Research on tachykinins in insects enhances our understanding of stress responses and their regulation across species.

Future Directions in Tachykinin Research

Continued research on tachykinins promises to uncover new functions and applications. Key areas of focus include:

1. **Receptor Specificity:**
   - Identifying and characterizing tachykinin receptors in diverse insect species will guide the development of selective pest control agents.
2. **Synthetic Tachykinin Development:**
   - Designing stable and bioavailable tachykinin analogs for agricultural and pharmaceutical applications.
3. **Integration with Pest Management Programs:**
   - Combining tachykinin-based strategies with other biocontrol methods for sustainable pest management.
4. **Comparative Studies:**
   - Investigating tachykinin functions across taxa to reveal evolutionary patterns and conserved mechanisms.

d. Applications in Pest Control via RNAi and Synthetic Analogs

The use of neuropeptides, including tachykinins, in pest control has gained significant attention as an innovative, eco-friendly, and species-specific strategy to mitigate agricultural losses and vector-borne diseases. Neuropeptide-based approaches, particularly those targeting tachykinins, utilize RNA interference (RNAi) and synthetic peptide analogs to disrupt critical physiological and behavioral processes in pest species. These methods provide an alternative to traditional chemical pesticides, minimizing environmental harm and reducing non-target effects.

Tachykinin Applications in Pest Control

Tachykinins are neuropeptides that regulate vital processes in insects, such as feeding, reproduction, locomotion, and stress responses. Disrupting these processes through RNAi or synthetic analogs can effectively reduce pest populations.

1. **RNA Interference (RNAi):**
   - RNAi is a gene-silencing technique that leverages double-stranded RNA (dsRNA) to inhibit the expression of specific genes.
   - Targeting tachykinin genes via RNAi suppresses their synthesis, disrupting associated physiological and behavioral functions in pest species.
2. **Synthetic Analogs:**
   - Synthetic tachykinin analogs are chemically designed molecules that mimic or inhibit the action of natural tachykinins.
   - These analogs can overstimulate or block tachykinin receptors, leading to physiological disruptions in pests.

RNAi-Based Applications in Pest Control

RNAi is a highly specific and effective tool for pest control. It relies on delivering dsRNA corresponding to the tachykinin gene, which is processed by the insect's RNAi machinery into small interfering RNAs (siRNAs). These siRNAs guide the degradation of the target mRNA, silencing the gene.

Mechanism of RNAi in Tachykinin Silencing

1. **Delivery of dsRNA:**
   - dsRNA targeting the tachykinin gene is introduced into the pest via ingestion, injection, or topical application.
   - Delivery systems include dsRNA-coated plant tissues, nanoparticles, or microcapsules for enhanced stability and uptake.
2. **Gene Silencing Process:**
   - The dsRNA is processed by the insect's Dicer enzyme into siRNAs.
   - The siRNAs bind to the RNA-induced silencing complex (RISC), guiding it to the complementary mRNA of the tachykinin gene.
   - The RISC complex cleaves the target mRNA, effectively silencing the gene.
3. **Disruption of Physiological Functions:**
   - Silencing tachykinin genes disrupts vital processes like feeding, reproduction, and stress responses, leading to reduced survival and reproduction rates in pests.

Examples of RNAi Applications Targeting Tachykinins

1. **Feeding Disruption:**
   - RNAi targeting tachykinins involved in appetite regulation suppresses feeding behavior in pests, leading to starvation and mortality.
   - Example: In *Helicoverpaarmigera* (cotton bollworm), RNAi against tachykinin genes reduced feeding rates and larval weight.
2. **Reproductive Suppression:**
   - Silencing tachykinins involved in oviposition behavior prevents egg-laying, significantly reducing pest populations.
   - Example: In mosquitoes (*Aedes aegypti*), RNAi targeting tachykinins inhibited egg production.
3. **Locomotion and Stress Response:**
   - RNAi disrupting tachykinins involved in locomotion impairs the pest's ability to forage or escape predators.
   - Example: In locusts (*Locusta migratoria*), tachykinin knockdown reduced swarming behavior.

Advantages of RNAi in Pest Control

1. **Species Specificity:**
   - RNAi is highly specific to the target pest species, minimizing impacts on non-target organisms.
2. **Eco-Friendly Approach:**
   - RNAi-based methods reduce the use of chemical pesticides, lowering environmental contamination.
3. **Versatility:**
   - RNAi can be used against a broad range of pests, from agricultural pests to disease vectors.

Synthetic Tachykinin Analogs in Pest Control

Synthetic tachykinin analogs are designed to mimic or antagonize the action of natural tachykinins, either overstimulating or blocking their receptors. These compounds disrupt essential physiological functions in pests.

Mechanism of Synthetic Analog Action

1. **Receptor Agonists:**
   - Agonists overstimulate tachykinin receptors, causing physiological dysregulation. For example, excessive gut motility can lead to nutrient malabsorption and reduced fitness.
2. **Receptor Antagonists:**
   - Antagonists block tachykinin receptors, preventing natural tachykinins from exerting their effects. This inhibition disrupts processes like feeding, reproduction, or stress response.

Examples of Synthetic Analog Applications

1. **Appetite Suppression:**
   - Synthetic tachykinin receptor antagonists reduce feeding behavior in pests, leading to starvation.
   - Example: In *Spodoptera frugiperda* (fall armyworm), tachykinin receptor antagonists decreased feeding activity and larval growth.
2. **Reproductive Disruption:**
   - Synthetic analogs targeting tachykinins involved in reproductive behaviors reduce mating success and egg-laying capacity.
   - Example: In *Helicoverpazea*, tachykinin receptor blockers impaired oviposition.
3. **Neuromuscular Impairment:**
   - Overstimulation of tachykinin receptors causes spasms and loss of motor coordination, incapacitating pests.

- Example: In cockroaches (*Periplaneta americana*), tachykinin agonists induced muscle paralysis.

Advantages of Synthetic Tachykinin Analogs

1. **High Potency:**
    - Synthetic analogs are highly effective, even at low doses, making them economically viable.
2. **Rapid Action:**
    - Synthetic compounds produce immediate effects, such as feeding inhibition or motor impairment.
3. **Selective Targeting:**
    - Tachykinin analogs can be tailored to target specific pest species, reducing non-target effects.

Challenges and Limitations

While RNAi and synthetic tachykinin analogs offer promising pest control solutions, several challenges need to be addressed:

1. RNAi Challenges

1. **Delivery Efficiency:**
    - Delivering dsRNA effectively to target tissues remains a major hurdle, especially in field conditions.
    - Solutions: Development of nanoparticle-based delivery systems to enhance stability and uptake.
2. **RNAi Resistance**:
    - Prolonged exposure to RNAi treatments may lead to resistance in pest populations due to genetic mutations or enhanced RNAi degradation mechanisms.
    - Solutions: Rotating RNAi targets or combining RNAi with other pest control methods.
3. **Off-Target Effects**:
    - Although RNAi is species-specific, improper design of dsRNA can lead to unintended silencing of non-target genes.
    - Solutions: Improved bioinformatics tools for dsRNA design.

2. Synthetic Analog Challenges

1. **Stability:**
    - Synthetic tachykinin analogs may degrade rapidly in the environment, reducing their effectiveness.
    - Solutions: Formulating analogs with stabilizing agents or controlled-release systems.

2. **Cost of Production:**
   - The synthesis of tachykinin analogs can be expensive, limiting large-scale application.
   - Solutions: Streamlining production processes to reduce costs.
3. **Environmental Persistence:**
   - Overuse of synthetic analogs may lead to bioaccumulation or unintended ecological effects.
   - Solutions: Developing biodegradable analogs to ensure environmental safety.

Future Directions in Tachykinin-Based Pest Control

1. **Improved Delivery Systems:**
   - Advancing delivery technologies for RNAi, such as dsRNA nanoparticles or viral vectors, will enhance the efficiency and applicability of tachykinin-targeted pest control.
2. **Synthetic Analog Optimization**:
   - Designing analogs with higher stability, selectivity, and bioavailability for cost-effective and sustainable pest management.
3. **Integration into IPM Programs:**
   - Combining RNAi and synthetic tachykinin analogs with other biological control methods, such as predators or parasitoids, for integrated pest management.
4. **Gene Editing for Pest Suppression:**
   - CRISPR-Cas9 technology can be used to disrupt tachykinin genes permanently in pest populations, providing a long-term control strategy.
5. **Cross-Species Comparisons:**
   - Studying tachykinin systems across diverse insect species will uncover conserved mechanisms and novel targets for pest control.

4. Juvenile Hormone Esterase (JH Esterase)

Juvenile Hormone Esterase (JH Esterase or JHE) is a highly specialized enzyme involved in the regulation of juvenile hormone (JH) levels in insects. JH is an essential hormone that governs critical life processes, including growth, molting, metamorphosis, reproduction, and diapause. By hydrolyzing JH into inactive forms, JH Esterase ensures the fine-tuned hormonal balance required for precise physiological transitions. This regulation enables insects to synchronize developmental and reproductive processes with environmental cues.

a. Role in Developmental and Reproductive Processes

Juvenile Hormone Esterase is indispensable for the proper regulation of insect life cycles. By precisely controlling the concentration of JH in hemolymph, JH Esterase orchestrates key developmental milestones and reproductive events. Below, we explore its roles in these two critical areas.

1. Role in Developmental Processes

In insects, developmental stages are controlled by the interplay between juvenile hormone (JH) and ecdysteroids. JH maintains the larval stage, while ecdysteroids drive molting and metamorphosis. JH Esterase plays a critical role in modulating this hormonal balance, enabling transitions between developmental stages.

i. Molting Regulation

Molting is an essential process in the insect life cycle, allowing growth by shedding and regenerating the exoskeleton. The precise timing of JH degradation by JH Esterase is crucial for initiating and completing this process.

» **Mechanism:**

- Molting involves a delicate balance between juvenile hormone and ecdysteroids. During early larval stages, JH levels remain high to ensure continued larval development.
- As the insect approaches the end of a larval instar, JH Esterase activity increases, hydrolyzing JH and reducing its concentration. This reduction allows ecdysteroids to dominate, triggering molting.

» **Physiological Importance:**

- Ensures sequential progression through larval instars.
- Prevents premature development of pupal structures during early larval stages.

» **Examples:**

- In *Manduca sexta* (tobacco hornworm), peak JH Esterase activity occurs just before the larval cuticle is shed, signaling the transition to the next instar.
- In *Tenebrio molitor* (mealworm beetle), JH Esterase levels closely correlate with molting cycles, ensuring proper development.

ii. Metamorphosis

The transition from larva to pupa and subsequently to the adult stage is a defining feature of holometabolous insects. JH Esterase facilitates this transformation by reducing JH levels, which are essential for maintaining the larval state.

» **Role of JH Esterase:**

- During the final larval instar, JH Esterase activity rises dramatically, effectively reducing JH levels to near-zero. This hormonal shift permits the action of ecdysteroids, which drive metamorphic processes such as pupation and adult tissue differentiation.

» **Physiological Effects:**

- Promotes the apoptosis (programmed cell death) of larval tissues and the differentiation of adult structures.
- Ensures the proper timing of metamorphic events to avoid developmental abnormalities.

» **Examples:**

- In *Bombyx mori* (silkworm), high JH Esterase activity at the end of the larval stage facilitates the metamorphic transition by removing the inhibitory effects of JH.
- In *Drosophila melanogaster*, mutations that impair JH Esterase function lead to developmental delays and malformed adult structures.

iii. Environmental Adaptation

Environmental factors such as temperature, photoperiod, and resource availability influence the timing of insect development. JH Esterase acts as a mediator, adjusting JH levels to align developmental processes with environmental conditions.

» **Diapause Regulation:**

- Diapause is a state of arrested development that helps insects survive adverse conditions. JH Esterase plays a pivotal role in diapause initiation by degrading JH, which inhibits development.
- **Examples:**
  ◊ In *Culex pipiens* (mosquito), JH Esterase activity rises in response to shorter photoperiods, preparing larvae for overwintering diapause.
  ◊ In *Chilo suppressalis* (rice stem borer), JH Esterase ensures the proper entry and maintenance of diapause during unfavorable seasons.

» **Seasonal Synchronization:**

- By modulating JH levels, JH Esterase ensures that developmental transitions, such as pupation or emergence, occur in synchronization with seasonal changes.

2. Role in Reproductive Processes

In adult insects, juvenile hormone is a key regulator of reproduction, controlling gonadal development, vitellogenesis (yolk formation), and mating behaviors. JH Esterase modulates reproductive events by fine-tuning JH levels to optimize energy allocation and reproductive success.

i. Oogenesis and Vitellogenesis

Oogenesis and vitellogenesis are highly energy-intensive processes regulated by juvenile hormone. JH Esterase ensures these processes occur only when environmental and physiological conditions are favorable.

» **Mechanism:**

- Juvenile hormone stimulates the production of vitellogenin, a yolk protein precursor, in the fat body. Vitellogenin is then transported to the ovaries for incorporation into developing eggs.
- JH Esterase reduces JH levels after the completion of oogenesis, ensuring that resources are conserved for future reproductive cycles.

- **Examples:**
  - In *Locusta migratoria* (migratory locust), JH Esterase activity increases after oviposition, resetting the reproductive system for subsequent cycles.
  - In *Aedes aegypti* (dengue mosquito), JH Esterase regulates the post-mating decline in JH levels, allowing females to focus on egg maturation.

ii. Mating and Post-Mating Regulation

Juvenile hormone influences mating behaviors and post-mating physiological changes in many insect species. JH Esterase modulates these processes to optimize reproductive outcomes.

- **Mating Readiness:**
  - In virgin females, low JH levels maintained by JH Esterase delay reproductive maturation, ensuring that mating occurs only when conditions are ideal.
- **Post-Mating Effects:**
  - In mated females, JH Esterase activity decreases, allowing a temporary surge in JH levels to stimulate egg-laying and reproductive behaviors.
  - **Examples:**
    - ◊ In *Drosophila melanogaster*, mating triggers a decline in JH Esterase activity, leading to increased JH levels that promote vitellogenesis and oviposition.
    - ◊ In honeybees (*Apis mellifera*), JH Esterase activity is modulated to balance colony-level reproductive needs with individual energy conservation.

iii. Resource Allocation:

Reproduction demands significant energy and nutrient resources. JH Esterase plays a role in optimizing energy allocation by modulating JH levels.

- **Trade-Offs:**
  - High JH levels favor reproduction, while low JH levels prioritize survival and somatic maintenance. JH Esterase ensures this balance is maintained.
- **Examples:**
  - In *Tenebrio molitor* (mealworm beetle), JH Esterase activity rises under nutrient-limited conditions, suppressing reproduction to conserve resources.

b. Gene Manipulation for Pest Control

Gene manipulation has become a transformative tool in pest control, offering innovative, sustainable, and highly specific solutions to reduce the impact of pests on agriculture, public health, and ecosystems. By directly targeting genes responsible for survival, reproduction, or disease transmission, genetic technologies bypass the limitations of conventional chemical pesticides, such as environmental toxicity, pest resistance, and non-target effects. The advent of cutting-edge techniques, including CRISPR-Cas9, RNA interference (RNAi), gene drives, and transgenics, has enabled unparalleled precision and efficacy in pest management.

1. Methodologies of Gene Manipulation

Gene manipulation employs advanced tools to edit, silence, or replace specific genes in pest organisms. Each methodology has distinct mechanisms and applications:

i. CRISPR-Cas9 Genome Editing

Overview: CRISPR-Cas9 is a revolutionary gene-editing technology that enables precise alterations to an organism's DNA. The system consists of a guide RNA (gRNA) that directs the Cas9 enzyme to a specific DNA sequence, where it induces a double-strand break. This break is repaired through mechanisms that allow gene deletion, insertion, or modification.

Applications in Pest Control:

1. **Sterility Induction:**
   - CRISPR-Cas9 can disrupt genes critical for reproduction, creating sterile populations.
   - Example: In *Aedes aegypti*, CRISPR has been used to disrupt the *Doublesex (Dsx)* gene, producing sterile females and male-biased populations for vector control.
2. **Gene Knockouts:**
   - Targeting essential genes can cause mortality or reduced fitness in pests.
   - Example: In *Helicoverpaarmigera* (cotton bollworm), knockout of metabolic genes leads to starvation.
3. **Disease Blocking:**
   - Editing genes in vectors like mosquitoes can block pathogen transmission.
   - Example: Transgenic *Anopheles gambiae* mosquitoes with altered genes block malaria parasite development.

ii. RNA Interference (RNAi)

Overview: RNA interference is a gene-silencing technique that uses double-stranded RNA (dsRNA) to target and degrade specific messenger RNA (mRNA). This inhibits protein synthesis, disrupting key physiological processes.

Applications in Pest Control:

1. **Feeding Disruption:**
   - RNAi can silence genes involved in appetite regulation, causing starvation.
   - Example: In *Spodoptera frugiperda* (fall armyworm), RNAi targeting midgut proteases reduces feeding activity.
2. **Reproductive Inhibition:**
   - Silencing vitellogenesis genes prevents egg production in female pests.
   - Example: In *Aedes aegypti*, RNAi has been used to silence yolk protein precursors.

3. **Molting Disruption:**
   - RNAi targeting chitinase or cuticle genes interferes with molting, leading to death.
   - Example: In *Locusta migratoria* (migratory locust), RNAi silences chitin synthesis genes.

iii. Gene Drives

Overview: Gene drives are genetic constructs designed to spread specific traits through a population by biasing inheritance patterns. This enables engineered genes to propagate rapidly, even when they reduce individual fitness.

Applications in Pest Control:

1. **Population Suppression**:
   - Gene drives carrying sterility genes or lethal alleles collapse pest populations.
   - Example: In *Anopheles gambiae*, gene drives linked to female-sterility traits reduce vector populations.
2. **Population Replacement**:
   - Gene drives can introduce traits that block pathogen transmission.
   - Example: Mosquitoes engineered to resist the dengue virus spread resistance traits throughout wild populations.

iv. Transgenics

Overview: Transgenic techniques involve introducing foreign genes into pests to alter their biology, reduce their fitness, or block disease transmission.

Applications in Pest Control:

1. **Pathogen Resistance:**
   - Transgenic mosquitoes have been engineered to block malaria parasites or dengue viruses.
   - Example: *Aedes aegypti* expressing anti-dengue genes significantly reduces disease spread.
2. **Behavioral Modification**:
   - Introducing genes that alter sensory pathways disrupts pest-host interactions.
   - Example: Transgenic pests engineered to avoid crops reduce agricultural damage.

2. Mechanisms of Gene Manipulation

Gene manipulation targets critical physiological and behavioral traits in pests. Key mechanisms include:

i. Reproductive Suppression

Sterility: Gene editing can disable reproductive genes, preventing gamete production or fertilization.

Sex Ratio Manipulation:

CRISPR and gene drives can bias populations toward males, reducing reproductive capacity. Male-only populations are especially valuable for Sterile Insect Technique (SIT) programs.

Examples:

- In *Ceratitis capitata* (Mediterranean fruit fly), CRISPR disrupts spermatogenesis genes to produce sterile males.
- Gene drives in *Anopheles gambiae* propagate male-biased traits, reducing malaria transmission.

ii. Disease Transmission Blocking

Pathogen Resistance:

Engineering vectors to resist pathogens interrupts the transmission cycle of diseases like malaria, dengue, and Zika.

Behavioral Modification:

Editing sensory genes reduces host-seeking behavior in disease vectors.

Examples:

- Transgenic *Anopheles gambiae* incapable of hosting *Plasmodium* parasites prevent malaria spread.
- Gene editing in *Aedes aegypti* reduces feeding on humans, limiting dengue transmission.

iii. Survival Disruption

Essential Gene Silencing:

Targeting genes critical for metabolic or structural functions, such as chitin synthesis, causes death.

Feeding Inhibition:

Gene editing can reduce feeding efficiency by silencing digestive enzymes or appetite-regulating neuropeptides.

Examples:

- RNAi targeting chitinase genes in *Helicoverpazea* disrupts molting, leading to mortality.
- RNAi in *Bemisiatabaci* (whitefly) reduces salivary enzyme production, weakening feeding ability.

3. Applications of Gene Manipulation

Gene manipulation is transforming pest control across diverse fields:

i. Agricultural Pest Management

Gene-editing technologies are being applied to manage pests like moths, beetles, and aphids.

Examples:

- RNAi targeting midgut genes in *Spodoptera frugiperda* protects maize from larval damage.

- » CRISPR editing of detoxification genes in *Bemisiatabaci* enhances susceptibility to pesticides.

ii. Vector-Borne Disease Control

Genetic approaches are being used to suppress or replace populations of disease vectors like mosquitoes and tsetse flies.

Examples:

- » Gene drives in *Aedes aegypti* propagate anti-dengue genes, reducing vector capacity.
- » Transgenic *Glossina morsitans* (tsetse fly) impairs reproduction, reducing sleeping sickness vectors.

iii. Invasive Species Management

Gene manipulation helps manage invasive species that threaten native ecosystems.

Examples:

- » Gene drives in invasive rodents have been proposed for island ecosystems.
- » CRISPR editing in *Culex quinquefasciatus* reduces mosquito populations in non-native regions.

4. Advantages of Gene Manipulation

Gene manipulation offers several advantages:

i. Specificity and Precision

Gene-editing targets only the intended species, minimizing non-target effects.

ii. Sustainability

Genetic strategies provide long-term solutions without repeated applications

iii. Environmental Safety

Gene manipulation reduces the need for broad-spectrum pesticides, lowering ecological harm.

iv. Cost-Effectiveness

Self-propagating technologies like gene drives lower long-term costs.

5. Challenges and Ethical Considerations

i. Resistance Development

Pests may develop resistance to gene-editing approaches.

Solution: Rotate gene targets and integrate with traditional methods.

ii. Off-Target Effects

Unintended genetic changes may disrupt ecosystems.

Solution: Use advanced bioinformatics for precise gene-editing designs.

iii. Ethical Concerns

Gene manipulation raises concerns about human intervention in ecosystems.

Solution: Engage stakeholders in transparent decision-making processes.

iv. Regulatory Barriers

Complex regulations can delay implementation.

Solution: Harmonize global regulatory standards for GMOs.

c. Use of Juvenile Hormone Esterase (JH Esterase) Inhibitors in Developmental Disruption

Juvenile Hormone Esterase (JH Esterase) plays a crucial role in regulating juvenile hormone (JH) levels in insects, ensuring proper timing of developmental and reproductive processes. By breaking down JH, this enzyme governs transitions between life stages such as molting, metamorphosis, and reproduction. However, inhibiting JH Esterase disrupts these processes, leading to developmental abnormalities or mortality. The use of **JH Esterase inhibitors** as a pest control strategy has garnered significant attention due to its precision in targeting hormonal regulation and their potential to reduce pest populations in an eco-friendly manner.

1. Mechanisms of JH Esterase Inhibitors

JH Esterase inhibitors block the enzymatic activity of JH Esterase, resulting in elevated levels of juvenile hormone. This disrupts the tightly regulated hormonal balance required for critical life processes, causing developmental delays, abnormal molting, or reproductive failure.

i. How JH Esterase Inhibitors Work

- » **Binding to JH Esterase:** JH Esterase inhibitors are designed to bind to the active site of the enzyme, preventing it from hydrolyzing JH into inactive forms.
- » **Prolonged JH Activity:** With JH levels remaining high, developmental transitions such as molting or metamorphosis are delayed or completely disrupted.
- » **Cascading Effects:** Elevated JH levels lead to physiological imbalances, including disrupted cuticle formation, impaired energy metabolism, and failure to progress through life stages.

ii. Targeting Life Stages

- » **Larval Stages:** Inhibitors are most effective during larval stages, where JH governs molting and growth. Prolonged JH activity prevents larvae from transitioning to the pupal stage.
- » **Pupal and Adult Stages:** Excessive JH during these stages disrupts reproductive maturation, leading to sterility or reduced fecundity.

2. Applications in Pest Control

JH Esterase inhibitors are being explored as innovative tools for controlling pest populations by disrupting their growth and reproduction.

i. Disruption of Molting and Development

- » **Mechanism:** By inhibiting JH Esterase, larval insects fail to undergo normal molting cycles, leading to developmental arrest or death.

» **Examples:**

- In *Helicoverpaarmigera* (cotton bollworm), JH Esterase inhibitors delay molting, causing accumulation of energy deficits and mortality.
- In *Aedes aegypti* (mosquitoes), inhibitors prevent larvae from transitioning to pupae, reducing vector populations.

ii. Interruption of Metamorphosis

» **Mechanism:** Metamorphic transitions require precise JH degradation to permit ecdysteroid dominance. Inhibitors maintain high JH levels, resulting in failed pupation or malformed adults.

» **Examples:**

- In *Bombyx mori* (silkworm), JH Esterase inhibitors disrupt metamorphosis, leading to the production of non-viable adults.
- In *Tenebrio molitor* (mealworm beetle), treated larvae exhibit developmental delays and increased mortality.

iii. Reproductive Failure

» **Mechanism:** JH regulates vitellogenesis and egg-laying in adults. Prolonged JH activity due to JH Esterase inhibitors disrupts these processes, leading to reduced reproductive output.

» **Examples:**

- In *Locusta migratoria* (migratory locust), inhibitors reduce egg production by interfering with ovarian maturation.
- In *Anopheles gambiae* (malaria vector), reproductive cycles are disrupted, indirectly reducing disease transmission.

iv. Diapause Prevention

» **Mechanism:** Elevated JH levels prevent the entry into diapause, making pests vulnerable to environmental stressors.

» **Examples:**

- In *Culex pipiens* (mosquito), JH Esterase inhibitors prevent overwintering diapause, exposing larvae to fatal conditions.

3. Advantages of Using JH Esterase Inhibitors

JH Esterase inhibitors offer several advantages over traditional pest control methods:

i. Specificity and Targeted Action

» **Species-Specific Applications:** These inhibitors target hormonal pathways unique to insects, reducing risks to non-target organisms, including pollinators and predators.

» **Developmental Stage Targeting:** By focusing on larval or pupal stages, inhibitors ensure precise disruption of pest life cycles.

ii. Reduced Environmental Impact

- **Eco-Friendly Approach:** Unlike chemical pesticides, JH Esterase inhibitors degrade rapidly and do not accumulate in the environment.
- **Minimized Resistance Development:** Hormonal pathways are less prone to rapid resistance development compared to traditional pesticide targets like sodium channels or acetylcholinesterase.

iii. Compatibility with Integrated Pest Management (IPM)

- **Synergistic Use:** JH Esterase inhibitors can be combined with biological controls, such as parasitoids or predators, for enhanced efficacy.
- **Sustainability:** Their use aligns with sustainable agriculture practices, reducing chemical inputs.

4. Challenges in Using JH Esterase Inhibitors

While promising, the use of JH Esterase inhibitors faces technical and practical challenges:

i. Delivery and Stability

- **DeliveryChallenges:** Effective delivery methods are needed to ensure inhibitors reach their target tissues in pest larvae or adults.
- **Stability Issues:** JH Esterase inhibitors must be stable under field conditions to ensure efficacy.

ii. Non-Target Effects

- **Risk to Beneficial Insects:** While specific, JH Esterase inhibitors may affect beneficial insects if misapplied.
- **Environmental Interactions:** Factors such as UV light and microbial degradation may reduce inhibitor potency.

iii. Cost of Development

- **High Research Costs:** Developing and scaling up production of inhibitors requires significant investment.
- **Economic Feasibility:** Ensuring affordability for widespread use in agricultural and public health settings is a key challenge.

5. Scorpion Toxins and Venoms: An In-Depth Exploration

Scorpions are among the most ancient arthropods on Earth, with a history stretching back over 400 million years. Their survival across such a vast evolutionary timescale is partly attributable to the sophisticated venom systems they employ to subdue prey and defend against predators. Scorpion venoms are complex cocktails of bioactive molecules, including peptides, proteins, and small organic compounds, many of which are potent neurotoxins. These molecules, produced and delivered via specialized venom glands, have evolved to target specific physiological pathways in insects, vertebrates, and other arthropods. This specificity has made scorpion toxins a subject of increasing interest for a variety of applications, especially in pest management and the development of environmentally friendly biopesticides.

1. Role of Toxin Genes in Pest Management
2. Engineering Toxin-Producing Biopesticides
3. Environmental Benefits and Target Specificity

1. Role of Toxin Genes in Pest Management

A. Overview of Scorpion Toxin Genes

Scorpion venoms are rich in peptides that selectively target ion channels in the nervous systems of insects and other pests. These toxins typically function by modulating sodium (Na^+), potassium (K^+), calcium (Ca^{2+}), and chloride (Cl^-) ion channels. Some toxins also interfere with neurotransmitter release or receptor signaling pathways. The genes encoding these toxins have been identified and characterized in various scorpion species, revealing a high degree of specificity for particular ion channel subtypes.

Advances in genomics and transcriptomics have enabled researchers to sequence scorpion venom gland cDNA libraries, leading to the identification of numerous toxin genes. For instance, the venom of the scorpion *Mesobuthus martensii* contains peptides like BmK IT-1 and BmK IT-2, which are specific to insects and have minimal effects on vertebrates. These genes hold significant promise for pest management due to their precision in targeting agricultural pests.

B. Applications in Integrated Pest Management (IPM)

Pest management strategies increasingly rely on sustainable and environmentally safe solutions, making scorpion toxin genes valuable tools. These genes can be used in two primary ways:

1. **Genetically Modified Crops:** Scientists have successfully incorporated scorpion toxin genes into crops to create transgenic plants with enhanced pest resistance. For example, a scorpion toxin gene like *AaIT* (from *Androctonus australis*) has been expressed in plants to produce peptides toxic to lepidopteran pests such as caterpillars. These crops deter pests by interfering with their neurological functions, reducing crop damage and the need for chemical pesticides.
2. **Bio-control Agents:** By incorporating toxin genes into microbial platforms, researchers have developed biocontrol agents that produce scorpion toxins in situ. These agents, often based on bacteria, fungi, or viruses, can be deployed in agricultural fields to target specific pest populations without harming non-target organisms.

2. Engineering Toxin-Producing Biopesticides

A. Development of Biopesticides

Biopesticides derived from scorpion toxins represent an innovative approach to pest control. These biological formulations use either purified toxins or living organisms engineered to produce the toxins. The production process involves identifying genes encoding insect-specific toxins, synthesizing these genes, and inserting them into microbial systems like *Bacillus thuringiensis* (*Bt*) or baculoviruses.

1. **Recombinant Toxin Production:** Recombinant DNA technology enables the mass production of scorpion toxins in microbial hosts like *Escherichia coli* or yeast. This approach ensures scalability and cost-efficiency while maintaining high levels of purity and activity. Recombinant toxins can be formulated into sprays or coatings for direct application in agriculture.

2. **Toxin-Expressing Microbial Systems:** Advances in synthetic biology allow scientists to engineer bacteria or fungi that produce scorpion toxins when introduced into pest habitats. For example, a bacterium expressing the *AaIT* toxin can colonize the rhizosphere of crops, producing the toxin in proximity to pests.

B. Synergistic Combinations

Scorpion toxins can be used in combination with other biocontrol agents or chemical pesticides to enhance their efficacy. For instance, combining a scorpion toxin gene with a *Bt* toxin gene in a single microbial platform can broaden the spectrum of pest control while minimizing resistance development. These synergistic strategies are especially valuable for combating pests with diverse feeding habits or physiological adaptations.

C. Challenges and Solutions

While the potential of toxin-producing biopesticides is immense, several challenges must be addressed:

1. **Regulatory Approvals:** Biopesticides containing genetically engineered organisms must undergo rigorous testing to ensure safety for humans, animals, and the environment. Developing comprehensive risk assessment protocols is crucial for regulatory approval.
2. **Resistance Management:** Prolonged use of any pest control agent can lead to resistance in target populations. To mitigate this, biopesticides should be integrated into rotational or combinatorial pest management programs.
3. **Public Acceptance:** Educating the public and stakeholders about the safety and benefits of biopesticides is essential to overcome resistance to the adoption of genetically engineered products.

3. Environmental Benefits and Target Specificity

A. Reduced Environmental Impact

The use of scorpion toxin-based biopesticides offers significant environmental advantages compared to conventional chemical pesticides:

1. **Minimized Chemical Run off:** Unlike chemical pesticides, which often contaminate soil and water through runoff, scorpion toxin biopesticides degrade rapidly in the environment. This reduces their persistence and the risk of non-target exposure.
2. **Selective Targeting:** Scorpion toxins are highly specific to certain ion channels found in pest species. This specificity minimizes harm to beneficial insects like pollinators, predators, and parasitoids, as well as to vertebrates, including humans and livestock.
3. **Reduced Carbon Footprint:** Producing bio-pesticides in microbial systems requires less energy and generates fewer greenhouse gases than manufacturing synthetic pesticides. This aligns with global efforts to reduce carbon emissions in agriculture.

B. Target Specificity and Biodiversity Conservation

The precision of scorpion toxins in targeting specific pests helps preserve biodiversity by minimizing collateral damage to ecosystems. For example:

1. **Pollinator Protection:** Pollinators like bees are essential for the reproduction of many crops and wild plants. Scorpion toxins that are harmless to bees ensure that pest control measures do not disrupt pollination services.
2. **Predator Conservation:** Many natural pest predators, such as lady beetles and lacewings, are susceptible to chemical pesticides. Using scorpion toxin-based biopesticides preserves these predators, supporting natural pest regulation.
3. **Mitigating Secondary Pest Outbreaks:** The non-disruptive nature of scorpion toxin biopesticides reduces the likelihood of secondary pest outbreaks caused by the elimination of natural enemies.

C. Compatibility with Sustainable Agriculture

Scorpion toxin-based pest control aligns with the principles of sustainable agriculture by promoting soil health, water conservation, and reduced reliance on chemical inputs. These biopesticides are particularly valuable in organic farming systems, where synthetic pesticides are prohibited. Furthermore, their rapid biodegradation and low toxicity profiles support the long-term health of agroecosystems.

5. Scorpion Toxins and Venoms

Scorpion venoms are complex cocktails of bioactive molecules, including peptides, proteins, and small organic compounds, many of which act as potent neurotoxins. These molecules have evolved over millions of years to target specific physiological pathways in prey and predators, making them highly effective tools for immobilization or defense. This specificity and potency have made scorpion toxins increasingly relevant in agricultural pest management, biotechnology, and environmentally friendly pest control solutions.

a. Role of Toxin Genes in Pest Management

Scorpion venom peptides have gained attention in pest management due to their ability to target specific ion channels and disrupt the nervous systems of pests. Advances in molecular biology have facilitated the identification, characterization, and utilization of toxin genes for targeted pest control strategies.

1. **Mechanism of Toxin Action**

   Scorpion toxins primarily target ion channels, such as voltage-gated sodium (Na^+), potassium (K^+), and chloride (Cl^-) channels, which are crucial for neural signal transmission.

   - **Alpha-Toxins**: Bind to sodium channels and prolong their activation, causing persistent depolarization and paralysis in insects.
   - **Beta-Toxins**: Lower the activation threshold of sodium channels, increasing neuron excitability.

These mechanisms lead to rapid immobilization and death in insect pests, making these toxins ideal candidates for pest management.

2. **Genetic Basis of Toxin Specificity**

   Advances in genomics and transcriptomics have revealed the genetic basis of scorpion toxins' specificity. Toxin genes, such as *AaIT* (from *Androctonus australis*), are known to selectively target insect-specific ion channels while sparing vertebrates. This precision reduces risks to non-target organisms and enhances safety in agricultural applications.

3. **Applications in Pest Management**

   - **Genetically Modified Crops**: Scorpion toxin genes have been incorporated into crops to confer resistance to pests. For instance, crops expressing *AaIT* effectively repel lepidopteran pests like caterpillars.
   - **Biocontrol Agents**: Bacteria, fungi, and viruses engineered to express scorpion toxin genes can target specific pest populations without harming beneficial species.

4. **Benefits in Integrated Pest Management (IPM)**

   The inclusion of scorpion toxins in IPM strategies provides a sustainable approach to reducing pest populations while minimizing environmental harm. Their high specificity ensures compatibility with other biocontrol measures.

b. Engineering Toxin-Producing Biopesticides

Biopesticides derived from scorpion toxins represent a cutting-edge approach to eco-friendly pest control. Advances in biotechnology have enabled the mass production and formulation of these biopesticides for agricultural use.

1. **Production of Recombinant Toxins**

   - **Microbial Systems**: Scorpion toxin genes are expressed in bacterial systems like *Escherichia coli* for scalable production. These recombinant toxins can be purified and used in pest control formulations.
   - **Synthetic Biology**: By optimizing toxin genes for expression in various microbial or plant systems, researchers have developed stable and active toxin-producing platforms.

2. **Toxin-Expressing Biocontrol Agents**

   - **Bacterial Platforms**: Genetically engineered bacteria, such as modified *Bacillus thuringiensis* (*Bt*), can produce scorpion toxins to target pests more effectively.
   - **Fungal and Viral Vectors**: Entomopathogenic fungi like *Beauveria bassiana* and baculoviruses have been engineered to deliver scorpion toxins to pests, increasing their lethality.

3. **Field Applications**

   - **Sprays and Coatings**: Toxin-based formulations can be directly applied to crops for immediate pest control.

- **Soil Treatments**: Microbial agents producing scorpion toxins can colonize the rhizosphere, providing localized pest deterrence.

4. **Combining Toxins for Synergy**

   Combining scorpion toxins with other bioactive agents, such as *Bt* toxins or chemical pesticides, enhances efficacy and delays resistance development in pest populations. This synergistic approach maximizes the utility of toxin-based biopesticides.

5. **Challenges and Innovations**

   - **Regulatory Approvals**: Biopesticides must meet stringent safety standards to gain approval.
   - **Resistance Management**: Rotational use and integration with other IPM strategies are essential to prevent resistance.
   - **Cost-Effective Production**: Advances in fermentation and synthetic biology are reducing the costs associated with toxin production.

c. Environmental Benefits and Target Specificity

The use of scorpion toxins in pest control provides numerous environmental benefits, especially when compared to conventional chemical pesticides. Their specificity and biodegradability make them ideal for sustainable agriculture.

1. **Reduced Environmental Impact**

   - **Minimized Runoff**: Unlike chemical pesticides, scorpion toxins degrade quickly, reducing contamination of soil and water systems.
   - **Lower Toxicity**: These toxins specifically target pest organisms, sparing non-target species and reducing ecological disruption.

2. **Preservation of Beneficial Species**

   - **Pollinators**: Scorpion toxins do not affect bees and other pollinators, ensuring the continued health of agricultural ecosystems.
   - **Natural Predators**: Beneficial predatory insects and parasitoids are largely unaffected, supporting natural pest control processes.

3. **Biodiversity Conservation**

   By minimizing the collateral damage caused by broad-spectrum chemical pesticides, scorpion toxin-based biopesticides contribute to preserving biodiversity and ecological balance.

4. **Target Specificity and Ecosystem Health**

   The high degree of target specificity of scorpion toxins ensures they affect only pests with chitin-containing exoskeletons or specific ion channel types. This precision reduces unintended consequences for vertebrates, plants, and soil microbiomes.

5. **Alignment with Sustainable Agriculture Goals**

   - **Organic Farming**: Scorpion toxin-based biopesticides align with organic farming principles, offering an alternative to synthetic pesticides.

- **Reduction in Chemical Inputs**: These biopesticides support the transition to lower pesticide use in conventional farming systems.
- **Carbon Footprint**: The production and application of scorpion toxin-based products have a smaller carbon footprint than manufacturing chemical pesticides.

6. Chitinase

Chitinase is a crucial enzyme in biological systems that degrades chitin, an essential polysaccharide found in the exoskeletons of arthropods, fungal cell walls, and nematode eggshells. Its ability to hydrolyze chitin into smaller, bioavailable components has made it an invaluable tool in pest management, agriculture, and biotechnology.

a. Importance in Exoskeleton Degradation

Chitinase is central to the degradation of chitin, a vital structural component in pests, fungi, and nematodes. Understanding its role provides insights into its applications in agriculture and pest control.

1. **Chitin as a Structural Polymer**
   - Chitin is a long-chain polysaccharide made of β-1,4-linked N-acetylglucosamine. It is the second most abundant natural polymer after cellulose.
   - It forms the backbone of:
     - ◊ **Insect Exoskeletons**: Providing rigidity and protection.
     - ◊ **Fungal Cell Walls**: Supporting growth and spore dispersal.
     - ◊ **Nematode Eggshells**: Safeguarding embryos against environmental stressors.

   Chitin's mechanical strength and insolubility make it a formidable barrier against natural degradation, necessitating specialized enzymes like chitinase.

2. **Mechanism of Chitinase Action**

   Chitinase breaks down chitin into smaller oligomers or monomers (e.g., N-acetylglucosamine) through two types of activity:
   - **Endochitinase Activity**: Cleaves chitin randomly at internal points, producing short-chain oligomers.
   - **Exochitinase Activity**: Cleaves chitin from the chain ends, releasing monomers or dimers.

   This dual action ensures the complete breakdown of chitin, compromising the structural integrity of pest exoskeletons and fungal cell walls.

3. **Role in Pest Mortality**
   - Chitinase disrupts critical processes in insects, such as molting and exoskeleton formation, leading to mortality.
   - In fungi, chitinase damages the cell walls, causing cell lysis and inhibiting growth.
   - For nematodes, chitinase compromises eggshells, reducing hatching rates and population viability.

4. **Natural Producers of Chitinase**

   Chitinases are produced by a variety of organisms, including:

   - **Bacteria**: Such as *Serratia marcescens* and *Bacillus thuringiensis*, which degrade chitin in soil ecosystems.
   - **Fungi**: Like *Trichoderma spp.*, which use chitinase to parasitize other fungi.
   - **Plants**: Produce chitinases as a defense mechanism against fungal infections.

These natural sources form the foundation for the biotechnological application of chitinases.

b. Genetic Engineering for Enhanced Pathogen Efficacy

Chitinase's potential can be amplified through genetic engineering to improve its activity, specificity, and stability, enabling more effective applications in pest control and agriculture.

1. **Improving Enzyme Efficiency**

   Genetic engineering focuses on optimizing the performance of chitinases to maximize their impact on pests:

   - **Directed Evolution**: Introducing random mutations to generate enzyme variants with enhanced activity and stability.
   - **Site-Directed Mutagenesis**: Modifying specific amino acid residues in the active site to improve substrate binding and catalytic efficiency.

For example, engineered chitinases with additional chitin-binding domains (CBDs) show significantly higher degradation rates.

2. **Expression in Heterologous Systems**

   Mass production of chitinases is achieved by expressing their genes in efficient host systems, such as:

   - **Bacteria**: *Escherichia coli* and *Bacillus subtilis* are commonly used for high-yield enzyme production.
   - **Yeast**: Systems like *Pichia pastoris* offer the advantage of post-translational modifications that improve enzyme stability.
   - **Plants**: Transgenic plants expressing chitinase genes exhibit enhanced resistance to fungal infections and insect damage.

3. **Engineering Multi-Functional Chitinases**

   Combining chitinase activity with other pest control traits has led to the development of multifunctional enzymes:

   - **Fusogenic Proteins**: Chitinases fused with proteins that enhance membrane disruption for quicker pest mortality.
   - **Combination with *Bt* Toxins**: Chitinase breaks down the protective exoskeleton, enhancing the access and efficacy of *Bacillus thuringiensis* toxins.

4. **CRISPR-Cas9 Applications**
    - The CRISPR-Cas9 system has been used to enhance the production of chitinase by editing regulatory regions of its genes.
    - This approach also facilitates the creation of pest-resistant crops by inserting optimized chitinase genes directly into plant genomes.
5. **Enhancing Field Stability**

    Chitinase enzymes are often vulnerable to environmental conditions like heat, pH changes, and UV radiation. Genetic modifications can address these issues:
    - Creating thermostable variants that retain activity under high temperatures.
    - Engineering pH-tolerant chitinases for diverse soil conditions.

c. Applications in Pest Control Formulations

Chitinase enzymes are extensively used in pest control due to their ability to degrade chitin-based structures. Their versatility allows for incorporation into a variety of formulations.

1. **Direct Application of Chitinases**
    - **Spray Formulations**: Chitinase enzymes can be applied directly as sprays to degrade the exoskeletons of pests on crops.
    - **Soil Treatments**: Enzymes applied to soil target nematode eggs and soil-borne fungal pathogens.
    - **Seed Coatings**: Seeds treated with chitinase enhance protection against fungal infections during germination.
2. **Integration with Biocontrol Agents**

    Chitinase can be paired with other biological agents for enhanced pest control:
    - **Entomopathogenic Fungi**: Chitinase boosts the virulence of fungi like *Metarhizium* and *Beauveria bassiana* by breaking down the host's exoskeleton.
    - **Nematode-Trapping Fungi**: Enzymes weaken nematode eggshells, facilitating fungal parasitism.
3. **Role in Transgenic Crops**
    - **Chitinase-Expressing Plants**: Genetically modified plants that produce chitinase enzymes show enhanced resistance to fungal and insect pests. For example:
        - ◊ Rice and wheat expressing chitinase genes are less susceptible to blast and sheath blight diseases.
        - ◊ Cotton plants with chitinase genes exhibit reduced bollworm infestations.
    - **Gene Stacking**: Combining chitinase genes with *Bt* toxin genes in crops provides broader pest resistance.

4. **Encapsulation for Controlled Release**
   - **Nanoparticles**: Encapsulating chitinase in nanoparticles protects the enzyme from environmental degradation and allows controlled release.
   - **Biodegradable Polymers**: These coatings ensure sustained activity of the enzyme in field conditions.
5. **Chitinase in Stored Grain Protection**
   - Chitinase formulations can protect stored grains from pests like weevils and beetles by degrading their exoskeletons.
   - These formulations reduce post-harvest losses and reliance on chemical fumigants.
6. **Sustainability in Pest Management**

   Chitinase-based pest control aligns with sustainable agricultural practices:
   - Reducing dependency on chemical pesticides.
   - Preserving beneficial insects and other non-target organisms due to specificity for chitin-containing pests.
   - Promoting eco-friendly alternatives to synthetic pesticides.

7. Crystal Protein Toxin Genes (CPTI)

Crystal protein toxin genes (CPTI) encode insecticidal proteins produced by *Bacillus thuringiensis* (*Bt*), a soil-dwelling bacterium. These proteins, commonly referred to as Cry proteins, are highly specific to certain insect pests and are widely used in agricultural biotechnology for pest control.

a. Origin and Role in Pest Control

1. Origin of Crystal Protein Toxin Genes

Crystal protein toxin genes (CPTI) were first discovered in *Bacillus thuringiensis*, a bacterium identified in 1911 for its insecticidal properties. Key aspects of their origin include:

- » **Cry Protein Structure**: CPTI genes encode crystalline inclusions made of insecticidal proteins, which are produced during the sporulation phase of *Bt*.
- » **Mode of Action**: Cry proteins are highly selective and lethal to specific insect orders, including Lepidoptera (moths and butterflies), Coleoptera (beetles), and Diptera (flies and mosquitoes).
- » **Gene Diversity**: Over 200 Cry proteins have been identified, each with specificity for distinct insect groups. For instance:
  - ◊ Cry1 and Cry2 target Lepidoptera.
  - ◊ Cry3 targets Coleoptera.
  - ◊ Cry4 is effective against Diptera.

2. Role in Pest Control

Cry proteins disrupt the midgut epithelium of susceptible insects, leading to their death. The process involves:

- **Ingestion**: Insects ingest Cry proteins from treated crops or biopesticide formulations.
- **Activation**: Cry proteins are solubilized and activated in the insect's alkaline midgut, where they are cleaved by proteases into active toxins.
- **Binding and Pore Formation**: The active toxins bind to specific receptors on gut epithelial cells, forming pores that disrupt ion balance, causing paralysis and death.

3. Advantages in Pest Control

- **Specificity**: Cry proteins target specific pests, minimizing harm to non-target organisms like pollinators and natural predators.
- **Safety**: As *Bt* toxins are non-toxic to vertebrates, including humans, they are considered environmentally safe.
- **Eco-Friendliness**: The use of Cry proteins reduces reliance on chemical pesticides, supporting sustainable agricultural practices.

b. *Bt* Crops and Integrated Pest Management

1. Development of *Bt* Crops

Bt crops are genetically modified (GM) plants that express crystal protein toxin genes, providing inherent resistance to specific insect pests. Milestones in *Bt* crop development include:

- **First *Bt* Crop**: *Bt* corn was the first genetically engineered crop approved in 1996, targeting the European corn borer.
- **Expansion**: *Bt* cotton, *Bt* rice, *Bt* eggplant, and other crops followed, addressing pests like bollworms, stem borers, and caterpillars.

2. Mechanism of *Bt* Crop Protection

Bt crops continuously produce Cry proteins in their tissues, offering season-long pest protection. When pests feed on these plants, the Cry proteins act in the same way as in biopesticide applications, leading to pest mortality.

3. Benefits of *Bt* Crops in IPM

Bt crops play a crucial role in integrated pest management by:

- **Reducing Chemical Pesticide Use**: Farmers can significantly reduce the application of synthetic pesticides, cutting costs and environmental damage.
- **Targeting Major Pests**: *Bt* crops effectively control key pest species, minimizing crop losses and boosting yields.
- **Compatibility with IPM Practices**: *Bt* crops integrate well with other pest management strategies, such as biological control agents and cultural practices.

4. Examples of *Bt* Crops

- ***Bt* Cotton**: Protects against bollworms (*Helicoverpa* spp.) and pink bollworms (*Pectinophora gossypiella*).
- ***Bt* Corn**: Provides resistance to the European corn borer (*Ostrinia nubilalis*) and corn rootworms (*Diabrotica* spp.).
- ***Bt* Brinjal (Eggplant)**: Targets fruit and shoot borers (*Leucinodes orbonalis*).

5. Environmental and Economic Impacts

- **Higher Yields**: *Bt* crops reduce pest-related crop losses, leading to increased agricultural productivity.
- **Environmental Benefits**: Reduced pesticide use protects water quality and preserves biodiversity.
- **Cost Savings**: Lower pesticide expenses and labor costs benefit farmers economically.

c. Resistance Management Strategies

1. The Problem of Resistance

Repeated exposure to Cry proteins in *Bt* crops and biopesticides has led to the evolution of resistance in some pest populations. Resistance arises when:

- Insects develop mutations in Cry protein receptors, reducing toxin binding.
- Overuse of *Bt* technology exerts strong selective pressure, favoring resistant individuals.

2. Strategies to Delay Resistance

To sustain the efficacy of Cry proteins, various resistance management strategies have been implemented:

a. High-Dose/Refuge Strategy

- **High-Dose Expression**: *Bt* crops are engineered to produce high levels of Cry proteins, ensuring lethal doses for nearly all pests, including heterozygous resistant individuals.
- **Refuge Areas**: Non-*Bt* crops are planted nearby to maintain a population of susceptible pests, diluting the frequency of resistance genes when resistant and susceptible pests interbreed.

b. Gene Pyramiding

- **Stacking Cry Genes**: *Bt* crops are engineered to express multiple Cry proteins with different modes of action. This reduces the likelihood of pests developing resistance to all Cry proteins simultaneously.
- Example: *Bt* corn expressing Cry1A.105, Cry2Ab2, and Cry3Bb1 for broad-spectrum pest control.

c. Crop Rotation and Diversification

- **Rotating *Bt* and Non-*Bt* Crops**: Alternating crop types disrupts the lifecycle of pests and reduces selective pressure for resistance.

- **Diversified Cropping Systems:** Planting a mix of crops ensures that pests encounter varied environments, limiting their adaptation to any single control method.

d. Monitoring and Early Detection

- **Resistance Monitoring Programs**: Regular field surveys detect early signs of resistance in pest populations.
- **Adaptive Management**: Adjusting pest control practices based on monitoring data helps maintain the effectiveness of *Bt* crops.

e. Combining *Bt* with Other IPM Tools

- **Biological Control**: Using natural predators, parasitoids, or entomopathogenic fungi in combination with *Bt* crops can reduce pest pressure.
- **Cultural Practices**: Techniques like crop sanitation and intercropping minimize pest populations and enhance the efficacy of *Bt* technology.

3. Advances in Resistance Management

- **CRISPR-Cas9 in Pest Management**: Gene editing can create susceptibility alleles in pest populations, counteracting resistance development.
- **RNAi-Based Synergy**: RNA interference (RNAi) can target pest resistance genes, restoring sensitivity to Cry proteins.
- **Next-Generation Cry Proteins**: Scientists are designing synthetic Cry proteins with novel binding sites to overcome existing resistance.

8. Lectins and Proteases

Lectins and proteases are bioactive proteins that play essential roles in various biological processes, including digestion, defense mechanisms, pest control, and host-pathogen interactions. These molecules have garnered attention for their potential applications in agriculture and biotechnology, particularly in pest management and disease resistance.

a. Role in Digestion and Defense Mechanisms

1. Lectins: Structure and Function

Lectins are carbohydrate-binding proteins that recognize and bind to specific sugar moieties on cell surfaces or within glycoconjugates. Their biological functions include:

- **Defense Against Pathogens**: Lectins bind to glycoproteins and glycolipids on the surface of pathogens, inhibiting their attachment and invasion.
- **Anti-Insect Activity**: Lectins interfere with the digestion and nutrient absorption of insect pests by binding to glycans in their gut lining.
- **Signaling and Communication**: In plants, lectins participate in cell signaling, recognition of symbiotic microorganisms, and defense responses.

2. Proteases: Enzymatic Role in Digestion

Proteases, also known as peptidases, are enzymes that hydrolyze peptide bonds in proteins, breaking them down into smaller peptides or amino acids. Their roles include:

- **Nutrient Acquisition**: Proteases facilitate protein digestion in organisms by degrading dietary proteins into absorbable units.
- **Defense Mechanisms**: In plants, proteases are key components of defense against herbivores and pathogens by:
  - Activating defense-related proteins.
  - Producing anti-nutritional peptides that deter herbivores.
  - Inhibiting microbial growth by degrading essential microbial proteins.

3. Combined Role in Defense

Lectins and proteases often act synergistically in defense mechanisms. For instance:

- Lectins target pest gut cells, while proteases degrade essential gut proteins, collectively impairing pest digestion and survival.
- In host-pathogen interactions, lectins block pathogen attachment, and proteases degrade virulence factors.

b. Potential for Genetic and Biochemical Pest Control

1. Lectins in Pest Control

The carbohydrate-binding properties of lectins make them effective in targeting pest and pathogen physiology:

- **Anti-Insect Properties**: Lectins bind to glycoconjugates in the insect gut, disrupting epithelial integrity and interfering with nutrient absorption, leading to mortality or reduced fecundity.
- **Plant Defense Enhancement**: Transgenic plants expressing lectin genes exhibit enhanced resistance to pests like aphids and whiteflies.
- **Specificity**: Lectins can be engineered to target specific pest species, minimizing non-target effects.

2. Proteases in Pest Control

Proteases disrupt the digestion and physiological processes of pests, making them valuable for pest control:

- **Gut Protein Degradation**: Proteases degrade key proteins in pest midguts, impairing digestion and causing starvation.
- **Anti-Fungal Activity**: Proteases degrade fungal cell wall proteins and enzymes critical for fungal growth.
- **Transgenic Plants**: Plants engineered to produce proteases exhibit resistance to chewing and sucking pests.

3. Biochemical Formulations

Both lectins and proteases can be formulated into biopesticides for direct application in agricultural fields:

- **Spray Formulations:** These proteins can be applied as foliar sprays to target pests directly.
- **Seed Treatments**: Coating seeds with lectins or proteases protects them from soil-borne pathogens and pests during germination.

4. Gene Stacking in Crops

Combining lectin and protease genes in genetically modified crops provides broad-spectrum pest resistance:

- **Lectin for Gut Binding**: Disrupts pest feeding behavior.
- **Protease for Digestion Interference**: Causes mortality by impairing essential physiological processes.

5. Synthetic Biology and Protein Engineering

- **Optimized Lectins and Proteases**: Synthetic biology tools can be used to design proteins with enhanced stability, specificity, and efficacy.
- **Fusion Proteins**: Creating hybrid proteins that combine lectin and protease functions increases their pest control effectiveness.

c. Implications in Host-Pathogen Interactions

1. Role in Host Defense

Lectins and proteases are integral to the immune defense of plants and other organisms:

- **Lectin-Mediated Pathogen Recognition**: Lectins bind to pathogen-associated molecular patterns (PAMPs) on microbial surfaces, activating plant immune responses.
- **Protease Activation in Defense**: Proteases activate defense-related proteins, such as pathogenesis-related (PR) proteins, which inhibit pathogen growth.

2. Inhibition of Pathogen Virulence Factors

- **Lectins:** Inhibit pathogen attachment to host tissues by binding to their surface carbohydrates.
- **Proteases**: Degrade microbial proteins involved in invasion, adhesion, and toxin production.

3. Pathogen Countermeasures

Pathogens have evolved strategies to overcome lectinand protease-mediated defenses:

- **Lectin Inhibition**: Some pathogens produce proteins that inhibit host lectins, neutralizing their effects.
- **Protease Inhibitors**: Pathogens secrete inhibitors that block host protease activity, enabling their survival and proliferation.

4. Engineering Enhanced Host Defenses

Advances in genetic engineering allow for the development of crops with improved resistance to pathogens:

» **Overexpression of Lectins**: Enhances pathogen recognition and inhibition.

» **Introduction of Protease Genes**: Produces enzymes that degrade pathogen virulence factors.

» **Gene Editing**: CRISPR-Cas9 technology enables precise modifications to enhance the activity of lectins and proteases in host defense.

5. Symbiotic Interactions

» **Mutualistic Microbes**: Lectins mediate beneficial symbioses between plants and microorganisms, such as nitrogen-fixing rhizobia.

» **Microbial Modulation of Protease Activity**: Proteases regulate symbiotic interactions by controlling microbial population dynamics.

Chapter 5

Transgenic Plants for Pest Resistance

1. Introduction to Transgenic Plants for Pest Resistance and Diseases

In the context of modern agriculture, the advent of biotechnology has revolutionized the way crops are cultivated and protected against pests and diseases. Traditional agricultural methods, while significant, are often insufficient to meet the demands of a growing global population and the challenges posed by climate change. This inadequacy has driven the search for innovative solutions to improve crop yields, reduce reliance on chemical pesticides, and ensure food security. Among these solutions, the development of transgenic plants represents a monumental achievement, offering precise and sustainable strategies to combat pests and diseases that threaten agricultural productivity.

Transgenic plants are genetically modified organisms (GMOs) engineered to possess specific traits beneficial for agriculture. These traits are achieved by introducing genes from other organisms, which confer characteristics such as pest resistance, herbicide tolerance, or improved nutritional content. The process involves identifying and isolating genes with desired properties, transferring them into the target plant's genome, and cultivating plants that exhibit these traits. Pest resistance is one of the most critical applications of transgenic technology, as pests are a major source of crop loss worldwide. Diseases caused by bacteria, fungi, and viruses further exacerbate the challenges faced by farmers, necessitating the development of disease-resistant plants.

The significance of pest resistance in transgenic plants cannot be overstated. Pests, including insects, nematodes, and weeds, cause yearly economic losses. For instance, the cotton bollworm and corn borer are notorious for decimating cotton and maize crops. Chemical pesticides have traditionally been used to mitigate these losses, but their overuse has led to environmental degradation, pest resistance, and negative impacts on non-target organisms, including beneficial insects. Transgenic plants offer an environmentally friendly alternative, reducing the need for chemical interventions while providing durable and effective protection against pests.

One of the most well-known examples of transgenic pest-resistant plants is *Bt* crops, which have been engineered to express Bacillus thuringiensis (*Bt*) toxins. These toxins are lethal to specific insect pests but are harmless to humans and most non-target organisms. *Bt* crops, including cotton, maize, and soybean, have been widely adopted in several countries, significantly reducing crop losses

DOI: 10.1201/9781003757405-5

and pesticide use. The success of *Bt* crops underscores the potential of transgenic technology to address global agricultural challenges.In addition to pest resistance, transgenic plants play a vital role in combating plant diseases caused by pathogens such as fungi, bacteria, and viruses. Diseases like late blight in potatoes, caused by *Phytophthora infestans*, and bacterial wilt in tomatoes result in severe yield losses and pose a threat to food security. Traditional methods of disease management, such as crop rotation and chemical fungicides, have limitations in terms of effectiveness and sustainability. Transgenic plants equipped with disease-resistance genes provide a promising solution to these challenges. For example, plants engineered with resistance (R) genes can recognize and counteract specific pathogens, preventing the onset of disease.

The development of transgenic plants for disease resistance relies on advanced molecular techniques, including CRISPR-Cas9, RNA interference (RNAi), and marker-assisted selection. These tools enable precise genetic modifications, allowing researchers to target specific genes associated with susceptibility or resistance to diseases. RNAi technology, for instance, has been used to silence genes essential for viral replication in plants, effectively rendering them resistant to viral infections. Similarly, CRISPR-Cas9 has revolutionized the field by providing a versatile and efficient method for editing plant genomes, paving the way for the creation of disease-resistant crops with minimal off-target effects.

Despite the undeniable benefits of transgenic plants, their adoption is not without challenges. One of the primary concerns is the potential impact on biodiversity and non-target organisms. For example, the widespread use of *Bt* crops has raised concerns about the development of resistance in target pests and the unintended effects on beneficial insects, such as pollinators. To address these issues, strategies such as refuge planting and gene stacking have been implemented. Refuge planting involves cultivating non-*Bt* crops alongside *Bt* crops to slow down the evolution of resistant pest populations, while gene stacking combines multiple resistance traits to enhance durability and reduce the likelihood of resistance development.Another challenge lies in the regulatory and ethical aspects of transgenic technology. The release of genetically modified organisms into the environment is a contentious issue, with debates surrounding their safety, environmental impact, and ethical implications. Public perception and acceptance of transgenic plants vary across regions, influenced by cultural, economic, and political factors. In some countries, stringent regulations and public opposition have hindered the adoption of transgenic crops, despite their potential benefits. Addressing these concerns requires transparent risk assessment, effective communication of scientific evidence, and the development of policies that balance innovation with precaution.

The economic implications of transgenic plants are also significant, particularly for smallholder farmers in developing countries. While transgenic crops can increase yields and reduce input costs, the high initial investment and dependence on proprietary seeds may pose challenges for resource-poor farmers. Ensuring equitable access to transgenic technology is essential to realize its full potential in addressing global agricultural challenges. Initiatives such as public-private partnerships and the development of open-access technologies can help bridge the gap and promote inclusive adoption. Looking ahead, the future of transgenic plants for pest and disease resistance is promising, with ongoing advancements in biotechnology and genomics. Researchers are exploring novel approaches, such as synthetic biology and genome editing, to create crops with enhanced resistance traits and improved adaptability to changing environmental conditions. For example, the use of synthetic promoters and regulatory elements can fine-tune gene expression, enabling plants to respond dynamically to pest and disease pressures. Additionally, integrating traditional breeding methods with modern biotechnological tools can accelerate the development of resilient crop varieties, ensuring a sustainable and secure food supply.

Transgenic plants also have the potential to contribute to broader goals of sustainable agriculture, such as reducing greenhouse gas emissions and conserving natural resources. By minimizing the reliance on chemical pesticides and fertilizers, transgenic crops can lower the carbon footprint of agricultural practices and mitigate the environmental impact of farming. Moreover, the development of drought-tolerant and nutrient-efficient transgenic plants can enhance productivity in marginal lands, addressing the challenges of climate change and resource scarcity.

a. Overview of Biotechnology in Agriculture

Biotechnology has emerged as a transformative force in agriculture, introducing innovative solutions to address challenges in food security, crop productivity, and environmental sustainability. By applying biological processes and genetic manipulation techniques, biotechnology allows us to create new plant varieties, enhance resistance to pests and diseases, and improve yields under challenging environmental conditions. This introduction serves as the foundation for understanding the broader applications and impacts of biotechnology in modern agriculture.Historically, agriculture relied heavily on selective breeding and traditional methods to improve crop performance and productivity. While these methods were effective, they were time-intensive and limited in scope. The advent of biotechnology marked a significant shift in agricultural practices, offering precision and speed that were previously unattainable. Genetic engineering, one of the most notable applications of biotechnology, allows scientists to directly modify the DNA of crops to introduce desirable traits such as drought tolerance, pest resistance, and enhanced nutritional content. This ability to tailor crops at the molecular level has revolutionized the way we approach food production.

A crucial aspect of biotechnology in agriculture is its role in addressing the pressing global challenge of feeding a growing population. The United Nations projects that the global population will reach approximately 10 billion by 2050, necessitating a significant increase in food production. Traditional agricultural practices alone cannot meet this demand without causing severe environmental degradation. Biotechnology offers a solution by enabling the development of crops that require fewer resources, produce higher yields, and are more resilient to environmental stresses.One of the earliest success stories of agricultural biotechnology is the development of genetically modified (GM) crops. Introduced in the 1990s, GM crops such as *Bt* cotton and herbicide-tolerant soybeans quickly gained popularity among farmers due to their improved productivity and cost-effectiveness. *Bt* cotton, for instance, contains a gene from the bacterium *Bacillus thuringiensis* that provides resistance to specific insect pests. This innovation reduced the need for chemical pesticides, lowering production costs and minimizing environmental harm.

Beyond genetic engineering, biotechnology encompasses a range of techniques that contribute to agricultural advancement. For example, tissue culture techniques enable the rapid propagation of disease-free plants, while marker-assisted selection accelerates the breeding process by identifying desirable traits at the genetic level. Additionally, advances in synthetic biology and gene editing technologies like CRISPR-Cas9 have opened new possibilities for precision agriculture, allowing scientists to edit genes with unprecedented accuracy.Despite its potential, the adoption of biotechnology in agriculture has faced challenges, including public skepticism and regulatory hurdles. Concerns about the safety and ethics of genetically modified organisms (GMOs) have sparked debates worldwide, leading to stringent regulations in many countries. However, extensive scientific research has consistently demonstrated that GM crops are safe for human consumption and the environment when developed and used responsibly.

Applications of Biotechnology in Agriculture

Biotechnology has found extensive applications in agriculture, ranging from improving crop traits to combating diseases and pests. This section explores the diverse ways biotechnology is transforming agricultural practices, with each subheading focusing on a specific application.

Genetic Engineering and Crop Improvement

Genetic engineering lies at the heart of agricultural biotechnology, offering precise tools to modify the genetic makeup of crops. By introducing desirable traits directly into a plant's genome, scientists can create varieties with enhanced productivity, nutritional value, and adaptability. For instance, biofortified crops such as Golden Rice, which is enriched with vitamin A, address nutrient deficiencies in vulnerable populations. Similarly, genetic modifications have been employed to develop crops that thrive in saline soils or withstand prolonged droughts, ensuring food production in regions with harsh climates.

Biotechnology for Pest and Disease Resistance

Pests and diseases have long been significant threats to agriculture, causing substantial yield losses and economic damage. Biotechnology offers a solution through the development of crops with built-in resistance to specific pests and pathogens. *Bt* crops, as mentioned earlier, are a prime example, significantly reducing the reliance on chemical pesticides. In addition, advances in RNA interference (RNAi) technology have enabled scientists to silence genes in pests or pathogens, providing a novel approach to pest control.

Enhancing Crop Yields and Stress Tolerance

As climate change intensifies, crops face increasing pressure from extreme weather events, heat stress, and water scarcity. Biotechnology addresses these challenges by developing crops with improved stress tolerance. For example, drought-tolerant maize varieties have been engineered to maintain productivity under limited water availability. Similarly, crops with enhanced photosynthetic efficiency are being developed to maximize yield potential in changing environmental conditions.

These applications underscore the versatility of biotechnology in addressing diverse agricultural challenges. By harnessing the power of genetic engineering and other biotechnological tools, we can create a more resilient and sustainable agricultural system.

Environmental Impact of Agricultural Biotechnology

Biotechnology has profound implications for the environment, offering both benefits and challenges. This section examines the environmental impact of agricultural biotechnology, focusing on its role in reducing chemical use, promoting sustainability, and addressing potential risks.

Reducing Chemical Use

One of the most significant environmental benefits of biotechnology is its ability to reduce the reliance on chemical inputs such as pesticides and fertilizers. *Bt* crops, for instance, have dramatically decreased the need for insecticides, resulting in lower chemical runoff into water bodies and reduced harm to non-target organisms. Similarly, herbicide-tolerant crops have enabled farmers to adopt conservation tillage practices, minimizing soil erosion and preserving soil health.

Potential Risks and Controversies

Despite its benefits, agricultural biotechnology has also raised environmental concerns. Critics argue that the widespread adoption of GM crops could lead to unintended consequences, such as the emergence of resistant pests and weeds. Additionally, there are concerns about gene flow from GM crops to wild relatives, potentially disrupting ecosystems. Addressing these challenges requires robust risk assessment protocols and the development of strategies to mitigate potential impacts.

Promoting Sustainable Practices

Biotechnology can play a pivotal role in promoting sustainable agricultural practices. For example, the development of nitrogen-use-efficient crops reduces the need for synthetic fertilizers, minimizing greenhouse gas emissions and nutrient runoff. Furthermore, biotechnological innovations such as biopesticides and biofertilizers offer eco-friendly alternatives to conventional chemicals, supporting sustainable farming systems.

.Definition and History of Transgenic Plants

Transgenic plants, also known as genetically modified organisms (GMOs), represent one of the most significant advancements in agricultural biotechnology. These plants are created by introducing genes from other species into their genomes to confer desirable traits, such as pest resistance, herbicide tolerance, and improved nutritional quality. The creation of transgenic plants has revolutionized agriculture, addressing global challenges like food security, climate change, and resource efficiency.

Definition of Transgenic Plants

A transgenic plant is defined as a plant that contains a gene or genes that have been artificially inserted rather than inherited through traditional breeding. This process involves the introduction of foreign DNA, often from entirely different species, into the plant's genome using advanced genetic engineering techniques. The inserted gene called a **transgene**, can originate from bacteria, viruses, animals, or even other plants.

Transgenic plants are developed to express traits that are either not present or poorly expressed in their natural form. For instance, the introduction of a bacterial gene into a plant can enable it to produce a protein that is toxic to specific pests, effectively providing built-in pest resistance. Similarly, genes encoding for drought tolerance or enhanced nutritional content can be introduced, making transgenic plants an essential tool for modern agriculture.

The development of transgenic plants involves several steps, including gene identification, isolation, cloning, and transformation. Techniques such as Agrobacterium-mediated transformation and biolistic particle delivery (gene gun) are commonly used to transfer the desired genes into plant cells. Once the gene is integrated into the plant's genome, the transformed cells are cultured to regenerate a whole plant, which can then be tested and propagated. Transgenic plants differ from traditionally bred plants in their precision and scope. While traditional breeding relies on crossing plants and selecting offspring with desirable traits, genetic engineering allows for the direct introduction of specific genes, even from unrelated species. This capability expands the range of traits that can be incorporated into crops, enabling solutions to complex agricultural challenges.

Table: History of transgenic plants

S. No.	Year	Event/Development	Details
1	1973	Recombinant DNA Technology Introduced	Stanley Cohen and Herbert Boyer developed techniques to cut and paste DNA, laying the foundation for genetic engineering.
2	1983	First Genetically Modified Plant Created	Scientists successfully inserted an antibiotic-resistant gene into tobacco plants using Agrobacterium tumefaciens.
3	1987	Field Trials of Transgenic Crops	The first field trials of genetically modified crops, including herbicide-resistant tobacco and tomatoes, were conducted.
4	1994	Commercialization of Transgenic Crops	The Flavr Savr tomato became the first genetically modified crop approved for commercial sale in the U.S., designed to resist rotting.
5	1996	*Bt* Crops Introduced	Bacillus thuringiensis (*Bt*) crops, such as *Bt* cotton and *Bt* maize, were commercially introduced to combat insect pests.
6	2000	Completion of Arabidopsis Genome	The genome of *Arabidopsis thaliana*, a model plant, was sequenced, advancing genetic research and transgenic crop development.
7	2003	Golden Rice Development	The first nutritionally enhanced transgenic crop, Golden Rice, was developed to combat vitamin A deficiency in developing countries.
8	2010	Drought-Tolerant Transgenic Crops	Monsanto introduced the first genetically engineered drought-tolerant maize, aimed at improving productivity in water-scarce regions.
9	2012	CRISPR-Cas9 Revolution	CRISPR-Cas9 emerged as a groundbreaking genome-editing tool, simplifying the development of transgenic and gene-edited crops.
10	2018	Approval of CRISPR-Edited Crops	Several countries began approving CRISPR-edited crops, such as high-fiber wheat, marking a new era in crop genetic engineering.
11	2020	Gene Stacking in Transgenic Crops	Advanced gene-stacking technologies allowed the integration of multiple traits, such as pest resistance and herbicide tolerance, into single crops.

12	Ongoing	Synthetic Biology Applications	The integration of synthetic biology with transgenic technologies is enabling the creation of plants with novel traits and enhanced resistance.

History of Transgenic Plants

The history of transgenic plants is a story of scientific discovery, technological innovation, and societal transformation. The journey began in the mid-20th century with the advent of molecular biology and the discovery of DNA as the genetic material. These foundational breakthroughs paved the way for the development of genetic engineering techniques that would later be applied to plants.

The Early Foundations(1950s–1970s)

The concept of genetic engineering traces back to the discovery of the structure of DNA by James Watson and Francis Crick in 1953. This discovery laid the groundwork for understanding how genetic information is stored, replicated, and expressed. In the following decades, scientists developed techniques to manipulate DNA, such as restriction enzymes for cutting DNA and ligases for joining DNA fragments.

In the 1970s, the first recombinant DNA experiments were conducted, marking the birth of modern genetic engineering. Researchers demonstrated that genes could be transferred between organisms, opening up the possibility of creating transgenic organisms. The development of plasmid vectors, particularly in bacteria like *Escherichia coli*, provided the tools for gene cloning and transfer.

The Birth of Transgenic Plants(1980s)

The 1980s witnessed the first successful creation of transgenic plants. Researchers discovered that the bacterium *Agrobacterium tumefaciens*, which naturally transfers DNA into plant cells, could be harnessed as a tool for genetic engineering. By modifying the bacterium's Ti plasmid, scientists were able to introduce foreign genes into plants.

In 1983, the first transgenic plants were created using *Agrobacterium*-mediated transformation. These plants expressed a bacterial gene for antibiotic resistance, demonstrating the feasibility of genetic engineering in plants. This milestone was soon followed by the introduction of traits such as herbicide tolerance and pest resistance into crops like tobacco and tomato.

Commercialization and Adoption(1990s-2000s)

The 1990s marked the commercialization of transgenic crops, beginning with the release of Flavr Savr tomatoes in 1994. These tomatoes were engineered to have a longer shelf life by silencing a gene involved in fruit softening. Although Flavr Savr tomatoes were not a commercial success, they paved the way for more robust transgenic crops.

In 1996, *Bt* cotton and Roundup Ready soybeans became the first widely adopted transgenic crops. *Bt* cotton, containing a gene from the bacterium *Bacillus thuringiensis*, provided resistance to bollworms, a major pest in cotton farming. Roundup Ready soybeans were engineered for tolerance to glyphosate, a widely used herbicide, enabling more efficient weed control.

The adoption of transgenic crops grew rapidly, with millions of hectares cultivated globally within a decade. By 2000, major crops like maize, canola, and rice had transgenic varieties, addressing challenges such as pest pressure, weed competition, and nutrient deficiencies.

Advances in Gene Editing (2010s–Present)

The 2010s saw the advent of new gene-editing technologies like CRISPR-Cas9, which revolutionized genetic engineering. Unlike traditional transgenic techniques, CRISPR allows for precise editing of a plant's existing genes without introducing foreign DNA. This innovation has blurred the lines between transgenic and non-transgenic plants, expanding the possibilities for crop improvement.

Today, transgenic plants are integral to agriculture, with applications ranging from improving yields and nutritional content to combating climate change. The development of Golden Rice, a biofortified rice variety enriched with vitamin A, exemplifies the potential of transgenic plants to address global health challenges. Similarly, drought-tolerant maize and salt-tolerant rice varieties demonstrate their role in adapting to changing environmental conditions.

c. Importance of Pest Resistance in Global Food Security

Global food security is one of the most critical challenges of the 21st century, exacerbated by a rapidly growing population, climate change, and dwindling agricultural resources. Among the numerous factors affecting food production, pest infestations remain a significant threat, capable of devastating entire crops and jeopardizing the livelihoods of millions of farmers. Pest resistance, a vital aspect of agricultural biotechnology, plays a pivotal role in safeguarding global food supplies and ensuring sustainable agricultural practices.Pest resistance refers to the ability of crops to withstand or repel pest attacks, either through natural genetic traits or through genetic modifications introduced via biotechnology. This resistance can protect yields, reduce dependency on chemical pesticides, and promote environmentally sustainable farming practices.

The Challenge of Pest Infestations in Agriculture

Pests, including insects, nematodes, fungi, and other organisms, are among the primary causes of crop losses worldwide. According to the Food and Agriculture Organization (FAO), pests and diseases account for up to 40% of global crop losses annually. This staggering statistic underscores the urgency of developing effective pest management strategies.

Historically, farmers relied on chemical pesticides to control pests. While effective, these chemicals come with significant drawbacks, including environmental contamination, harm to non-target species, and the development of pesticide-resistant pest populations. The excessive use of pesticides has also raised concerns about human health and food safety, prompting a shift toward more sustainable solutions.Pest resistance in crops offers an alternative to chemical controls by providing built-in protection against specific pests. For example, genetically modified *Bt* crops contain a gene from the bacterium *Bacillus thuringiensis*, which produces proteins toxic to certain insect pests. These crops have proven highly effective in reducing pest-related losses while minimizing pesticide use.

Importance of Pest Resistance in Global Food Security

1. **Protecting Crop Yields and Reducing Losses**

   Pest resistance is crucial for protecting crop yields and ensuring a stable food supply. Crops with built-in pest resistance can withstand infestations that would otherwise decimate yields. For instance, *Bt* cotton and *Bt* maize have significantly reduced yield losses caused by bollworms and corn borers, respectively. By safeguarding yields, pest-resistant crops contribute to food availability and economic stability, particularly in regions where agriculture is the primary source of livelihood.

In developing countries, where smallholder farmers often lack access to advanced pest control measures, pest-resistant crops can make a substantial difference. By reducing the reliance on chemical pesticides, these crops lower production costs and increase profitability for farmers, contributing to poverty alleviation and food security.

2. **Reducing Dependency on Chemical Pesticides**

The widespread use of chemical pesticides has long been a cornerstone of pest management. However, this approach is not without consequences. Chemical pesticides often have non-specific modes of action, affecting beneficial organisms and disrupting ecosystems. Moreover, the overuse of pesticides can lead to the emergence of resistant pest populations, rendering the chemicals ineffective over time.

Pest-resistant crops offer a more sustainable alternative by providing targeted protection against specific pests. For example, *Bt* crops are highly effective against lepidopteran pests while posing no harm to beneficial insects like pollinators. By reducing the need for chemical inputs, pest resistance promotes environmentally friendly farming practices and protects biodiversity.

3. **Enhancing Resilience to Climate Change**

Climate change is intensifying the challenges of pest management by altering pest distribution and behavior. Rising temperatures, changing precipitation patterns, and shifting growing seasons create favorable conditions for pests to thrive in new regions. For instance, the fall armyworm, a notorious crop pest, has spread to parts of Africa, Asia, and Australia in recent years, causing widespread damage to maize and other staple crops.

Pest-resistant crops can enhance the resilience of agricultural systems to these changing conditions. By equipping crops with the ability to withstand emerging pests, farmers can mitigate the risks posed by climate-induced pest outbreaks and maintain food production in vulnerable regions.

4. **Promoting Sustainable Agricultural Practices**

Sustainability is a key component of global food security, and pest resistance aligns with this goal by reducing the environmental footprint of agriculture. Pest-resistant crops enable farmers to adopt integrated pest management (IPM) strategies, combining biological, cultural, and genetic approaches to pest control. These strategies minimize the reliance on synthetic chemicals and promote the health of ecosystems.

Additionally, pest resistance supports conservation tillage practices, which involve minimal soil disturbance to preserve soil structure and fertility. By reducing the need for pesticide applications, pest-resistant crops contribute to the adoption of conservation tillage, enhancing soil health and carbon sequestration.

5. **Addressing Emerging Pest Threats**

As agriculture becomes increasingly globalized, the movement of pests across borders poses a growing threat to food security. The introduction of invasive pests, such as the Asian citrus psyllid and the coffee borer beetle, has caused significant economic and ecological damage in affected regions. Pest-resistant crops offer a proactive solution by providing protection against these emerging threats, reducing the risk of large-scale outbreaks and their associated impacts on food production.

Success Stories and Global Adoption

The impact of pest-resistant crops on global food security is best illustrated through real-world examples. *Bt* cotton, one of the earliest commercially successful transgenic crops, has transformed cotton production in countries like India and China. By reducing losses to bollworms, *Bt* cotton has increased yields and profitability for millions of farmers, contributing to rural development and poverty reduction.Similarly, *Bt* maize has been widely adopted in the United States, South Africa, and the Philippines, where it has significantly reduced losses to corn borers and other pests. These success stories demonstrate the potential of pest-resistant crops to address global challenges and improve agricultural productivity.

Challenges and Future Directions

While pest resistance has proven highly effective, its widespread adoption is not without challenges. The evolution of resistant pest populations poses a significant threat to the long-term efficacy of pest-resistant crops. To address this issue, scientists and policymakers must promote the implementation of resistance management strategies, such as crop rotation, refuge planting, and the use of multiple pest control methods. Additionally, the development of pest-resistant crops must be accompanied by robust regulatory frameworks and public engagement to address concerns about safety, ethics, and accessibility. Ensuring that smallholder farmers in developing countries have access to pest-resistant technologies is essential for maximizing their impact on global food security.

2: Mechanisms of Pest Resistance in Transgenic Plants

2.1 Genetic Modifications Used for Pest Resistance

2.1.1 *Bt* Crops: Bacillus thuringiensis Toxin Genes

2.1.1.1 Bacillus thuringiensis (*Bt*)

Bacillus thuringiensis (*Bt*) is a remarkable soil bacterium that has fundamentally transformed pest control in agriculture. Its ability to produce crystal (Cry) proteins toxic to specific insects has been extensively harnessed in modern farming practices, particularly through genetically engineered crops. *Bt*'s unique properties have not only provided a sustainable alternative to chemical pesticides but also played a pivotal role in addressing global challenges such as food security, environmental preservation, and agricultural productivity. To fully appreciate its importance, it is essential to explore the biology of this microorganism, the specific nature of Cry proteins, their mechanisms of action, and the implications of *Bt* technology.

Bacillus thuringiensis was first identified in 1901 by the Japanese biologist Shigetane Ishiwata, who isolated it while investigating the cause of a fatal disease in silkworms. Over a decade later, in 1911, Ernst Berliner, a German scientist, isolated the bacterium from the flour moth and named it after the region of Thuringia in Germany. These early discoveries hinted at *Bt*'s potential as a natural pest control agent, but it would take several decades of scientific exploration before its full capabilities were realized. Initially, *Bt*'s use was confined to microbial insecticides, where it was applied directly to crops to combat pest infestations. While effective, these early methods were limited by environmental factors such as rain and sunlight, which degraded the spores and crystals, reducing their efficacy.

Bt is a gram-positive, rod-shaped bacterium that thrives in diverse habitats, including soil, decaying organic matter, and on plant surfaces. It is a member of the *Bacillus* genus, characterized by its ability to form endospores, a feature that enables it to survive in extreme conditions. During sporulation—a process triggered by nutrient depletion—*Bt* produces parasporal crystals composed of Cry proteins. These crystals are the bacterium's defining feature and the primary reason for its widespread application in pest control.

Cry proteins are a family of insecticidal proteins with highly specific toxic effects on certain insect orders, such as Lepidoptera (moths and butterflies) and Coleoptera (beetles). These proteins are encoded by *cry* genes, which are among the most studied genetic elements in microbiology and agricultural biotechnology. The diversity of Cry proteins is astounding, with each type targeting a particular group of pests. This specificity is a critical advantage, as it minimizes the impact on non-target organisms, including beneficial insects, animals, and humans.

The mode of action of Cry proteins is a finely tuned process that ensures their high specificity and efficacy. When an insect pest ingests *Bt* spores or transgenic plant material containing Cry proteins, the alkaline conditions of its midgut solubilize the parasporal crystals, releasing the proteins in their protoxin form. Proteolytic enzymes in the gut then activate these protoxins, converting them into their toxic forms. Once activated, the Cry proteins bind to specific receptors on the epithelial cells lining the insect midgut. This binding triggers the formation of pores in the cell membranes, disrupting the ion balance and causing osmotic lysis of the cells. The result is severe damage to the gut lining, leading to starvation, septicemia, and eventual death of the insect.

The specificity of Cry proteins is one of their most valuable attributes. Unlike chemical pesticides, which often have broad-spectrum effects and can harm beneficial organisms, Cry proteins target only those insects that possess the requisite gut receptors. This precision not only enhances the environmental safety of *Bt*-based pest control but also supports biodiversity by preserving non-target species.

The advent of genetic engineering in the late 20th century marked a turning point in the application of *Bt*. Scientists successfully transferred *cry* genes into crop plants, enabling them to produce Cry proteins directly. This innovation eliminated the need for repeated applications of microbial insecticides and provided continuous protection against pests throughout the growing season. The first commercially available *Bt* crop, a variety of *Bt* cotton, was introduced in the 1990s. It was engineered to resist bollworms, a major pest of cotton plants. The success of *Bt* cotton was followed by the development of other *Bt* crops, including maize, brinjal (eggplant), and soybean, each designed to combat specific pests.

Bt crops have had a profound impact on agriculture. By reducing the reliance on chemical pesticides, they have not only lowered production costs for farmers but also mitigated the environmental and health risks associated with pesticide use. For example, the adoption of *Bt* cotton in India has significantly reduced the incidence of pesticide poisoning among farmers and farmworkers, while also decreasing pesticide-related runoff into water bodies. *Bt* crops have also contributed to yield stability, particularly in regions prone to severe pest infestations. This stability is crucial for ensuring food security, especially in developing countries where agriculture is a primary livelihood.

However, the widespread adoption of *Bt* technology is not without challenges. One of the primary concerns is the potential for pests to develop resistance to Cry proteins. This phenomenon, known as resistance evolution, can undermine the long-term efficacy of *Bt* crops. To address this issue, scientists and policymakers have implemented strategies such as refuge planting, where a portion of the crop area is planted with non-*Bt* varieties to provide a habitat for susceptible pest populations. This approach reduces the selective pressure for resistance and helps maintain the effectiveness of *Bt* technology.

Another challenge is public perception and regulatory scrutiny of genetically modified organisms (GMOs). Despite extensive scientific evidence supporting the safety of *Bt* crops for human consumption and the environment, concerns about their potential risks persist. These concerns have led to stringent regulatory processes in many countries, delaying the approval and adoption of *Bt* crops. Public education and transparent communication are essential for addressing these concerns and fostering greater acceptance of *Bt* technology.

Bt's role in pest management extends beyond transgenic crops. *Bt*-based biopesticides remain an important tool, particularly for organic farming and integrated pest management (IPM) systems. These biopesticides are formulated from *Bt* spores and crystals and can be applied to crops as sprays or dusts. While they do not provide the continuous protection of *Bt* crops, they offer flexibility and can be used in conjunction with other pest control methods.

The impact of *Bt* on global agriculture cannot be overstated. It has provided farmers with an effective, environmentally friendly solution to some of the most persistent challenges in pest management. By reducing pesticide use, enhancing yield stability, and supporting sustainable farming practices, *Bt* technology has made a significant contribution to global food security. As the agricultural sector continues to face pressures from climate change, population growth, and resource constraints, the role of *Bt* and similar biotechnological innovations will become even more critical.

2.1.1.2 Mechanism in Transgenic Plants

The mechanism through which transgenic plants utilize *Bacillus thuringiensis* (*Bt*) genes to provide resistance against pests represents a groundbreaking application of biotechnology. This approach has transformed pest management by embedding the insecticidal properties of *Bt* into crop plants, enabling them to autonomously combat specific pests. The process is intricate, involving the integration of *Bt* genes into plant genomes, the subsequent expression of *Bt* proteins, and their targeted action on pest populations. A detailed examination of this mechanism sheds light on its scientific basis, practical applications, and transformative impact on global agriculture.

Incorporating *Bt* Genes into Plant Genomes

The first step in creating a *Bt* transgenic plant is the transfer of specific *Bt* genes into the plant genome. This process requires advanced genetic engineering techniques to ensure that the plant can produce *Bt* proteins effectively and in sufficient quantities to combat pest infestations.

1. Selection of *Bt* Genes for Targeted Pests

The success of *Bt* transgenic plants begins with the careful selection of *cry* genes from *Bacillus thuringiensis*. These genes encode Cry proteins, which are the active agents responsible for pest control.

Each Cry protein has a specific mode of action, targeting particular groups of pests based on the receptors present in their gut. For example:

» Cry1Ac and Cry2Ab proteins are effective against lepidopteran pests such as cotton bollworms and armyworms.

» Cry3Bb proteins target coleopteran pests like rootworms, which are particularly damaging to maize crops.

By selecting the appropriate *cry* gene for a given crop and pest scenario, scientists ensure that the transgenic plant will deliver precise and effective pest control.

2. Gene Transfer Mechanisms

Transferring *Bt* genes into plant genomes involves the use of vectors, which act as carriers for the genetic material. Plasmids, circular DNA molecules commonly found in bacteria, are among the most widely used vectors in genetic engineering. The process typically involves the following steps:

» **Insertion into the Vector:** The selected *Bt* gene is inserted into a plasmid, along with a promoter sequence that regulates its expression in the plant.

» **Agrobacterium-Mediated Transformation:** One of the most common methods for introducing the *Bt* gene into plants is through *Agrobacterium tumefaciens*, a bacterium naturally capable of transferring DNA to plants. Scientists modify the *Agrobacterium* plasmid to carry the *Bt* gene, which is then integrated into the plant's genome when the bacterium infects plant cells.

» **Biolistic Transformation (Gene Gun):** In cases where *Agrobacterium*-mediated transformation is not feasible, a gene gun may be used to deliver *Bt* DNA into plant cells. This method involves coating microscopic particles with the DNA and propelling them into plant tissues.

3. Integration and Screening

Once the *Bt* gene is introduced into the plant genome, transformed cells are cultured in a laboratory to regenerate whole plants. During this stage, researchers use molecular and biochemical assays to confirm successful integration of the *Bt* gene and its expression in the plant. Only plants that exhibit stable expression of the *Bt* protein and desired traits are selected for further development.

Expression of *Bt* Proteins in Transgenic Plants

After the *Bt* gene is incorporated into the plant genome, the plant begins to express the gene, producing *Bt* proteins that provide protection against pests. This expression is carefully regulated to ensure that the protein is produced in the parts of the plant most vulnerable to pest attacks, such as leaves, stems, or fruits.

1. Role of Promoters in Gene Expression

The promoter sequence attached to the *Bt* gene determines where and when the gene is expressed in the plant. For instance:

» Constitutive promoters drive expression in all parts of the plant, providing broad protection.

» Tissue-specific promoters limit expression to specific organs, such as leaves or seeds, optimizing the plant's resources and minimizing unintended effects.

By tailoring the promoter sequence, scientists can fine-tune the production of *Bt* proteins to align with the pest's feeding habits and the crop's growth stages.

2. Continuous Protection Against Pests

One of the key advantages of *Bt* transgenic plants is their ability to provide continuous pest protection. Unlike chemical pesticides, which degrade over time and require repeated applications, *Bt* proteins are consistently produced throughout the plant's life cycle. This feature not only reduces labor and costs associated with pest management but also ensures that the plant remains protected even during critical growth periods.

3. Target-Specific Action

Bt proteins are highly specific in their mode of action, targeting only pests that possess the appropriate receptors in their gut lining. This specificity minimizes harm to non-target organisms, including beneficial insects such as pollinators and natural predators. For example, the Cry1Ac protein in *Bt* cotton targets bollworms without affecting honeybees or other beneficial species, making it an environmentally friendly pest control solution.

Mechanism of Pest Resistance

The mechanism by which *Bt* proteins protect transgenic plants from pests involves a precise series of biological interactions that culminate in the death of the target pest. This process can be broken down into several stages:

1. Ingestion of *Bt* Protein

Pest larvae feeding on the transgenic plant ingest tissues containing *Bt* proteins. These proteins are initially present in a crystalline form, rendering them inactive until they reach the pest's midgut.

2. Activation in the Alkaline Gut Environment

The alkaline pH of the insect midgut solubilizes the crystal proteins, releasing the protoxins. Proteolytic enzymes in the gut then cleave these protoxins, activating the Cry proteins. The activation step is crucial for the protein to exert its toxic effects.

3. Binding to Gut Receptors

Activated Cry proteins bind to specific receptors on the epithelial cells lining the insect's midgut. This binding is highly specific, depending on the match between the protein and the pest's receptor profile. The precise interaction ensures that only target pests are affected.

4. Formation of Pores in the Gut Epithelium

Once bound to the receptors, the Cry proteins form pores in the cell membranes of the gut lining. These pores disrupt the ionic balance within the cells, leading to osmotic lysis and cell death. The resulting damage to the gut epithelium allows gut bacteria to invade the hemolymph (insect blood), causing septicemia and rapid mortality.

Case Study: *Bt* Cotton and Bollworm Control

Bt cotton is a prime example of how transgenic technology has revolutionized pest management. Engineered to express Cry1Ac and Cry2Ab proteins, *Bt* cotton provides robust protection against lepidopteran pests such as cotton bollworms. These pests were historically responsible for severe yield losses and necessitated extensive pesticide use.

1. Impact on Yield and Pesticide Reduction

The introduction of *Bt* cotton has led to dramatic reductions in bollworm infestations, resulting in higher yields and improved profitability for farmers. In India, where cotton is a major cash crop, *Bt* cotton has been credited with reducing pesticide use by up to 50%, lowering production costs, and minimizing environmental contamination.

2. Economic Benefits for Farmers

Bt cotton has been particularly beneficial for smallholder farmers, who often lack access to expensive chemical pesticides. By providing built-in pest resistance, *Bt* cotton has increased farmer incomes and contributed to rural development.

3. Environmental Sustainability

The reduced reliance on chemical insecticides has had significant environmental benefits, including lower pesticide runoff into water bodies and decreased harm to non-target organisms. This aligns with broader goals of sustainable agriculture and biodiversity conservation.

Challenges and Future Directions

While *Bt* transgenic plants have been a resounding success, their widespread adoption is not without challenges. The primary concern is the potential for pests to develop resistance to *Bt* proteins, a phenomenon that could undermine the long-term efficacy of the technology.

1. Resistance Management Strategies

To address the issue of resistance evolution, farmers and scientists employ integrated pest management (IPM) strategies. These include:

- **Refuge Planting:** Planting non-*Bt* crops alongside *Bt* crops to maintain a population of susceptible pests, thereby slowing resistance development.
- **Stacked Traits:** Combining multiple *Bt* genes in a single plant to target pests with different modes of action, reducing the likelihood of resistance.

2. Advancements in Genetic Engineering

New technologies, such as CRISPR-Cas9, offer the potential to enhance *Bt* crops by creating novel Cry proteins or modifying pest-specific genes. These innovations could provide more durable and versatile solutions to pest control challenges.

3. Regulatory and Public Perception Issues

Despite the scientific consensus on the safety of *Bt* transgenic plants, public concerns about genetically modified organisms (GMOs) persist. Transparent communication and robust regulatory frameworks are essential to build trust and facilitate the broader adoption of *Bt* technology.

2.1.1.3 Global Impact and Adoption

The global impact and adoption of *Bt* crops are among the most significant achievements in agricultural biotechnology. Since their introduction in the mid-1990s, genetically engineered *Bt* crops have transformed pest management practices, agricultural productivity, and environmental sustainability. Spanning millions of hectares worldwide, these crops are a cornerstone of modern farming systems, providing substantial economic, ecological, and social benefits.

The Scale of *Bt* Crop Adoption

Bt crops, genetically engineered to produce insecticidal proteins derived from *Bacillus thuringiensis* (*Bt*), have been embraced on a global scale. Since their commercial introduction in 1996, the cultivation of *Bt* crops has expanded rapidly, with over 100 million hectares grown annually. This widespread adoption reflects the crops' ability to address persistent agricultural challenges, particularly those related to pest infestations, yield losses, and environmental degradation.

1. Geographical Distribution

Bt crops are cultivated in over 30 countries, spanning diverse agroecological zones. Major adopters include:

» **United States:** A leader in *Bt* crop adoption, particularly for *Bt* maize and *Bt* cotton, where pest pressures are significant.

» **India:** The largest producer of *Bt* cotton, which has revolutionized cotton farming by reducing losses to bollworms.

» **Brazil and Argentina:** Prominent growers of *Bt* maize and *Bt* soybean, addressing pests such as the fall armyworm and stem borers.

» **China:** A major producer of *Bt* cotton, particularly in regions where cotton pests historically caused severe yield losses.

These countries have demonstrated the scalability and adaptability of *Bt* crops, showcasing their ability to thrive in different climatic conditions and farming systems.

2. Adoption by Smallholder Farmers

A remarkable aspect of *Bt* crops is their adoption by smallholder farmers in developing countries. For instance:

» In India, over 7 million smallholder farmers grow *Bt* cotton, benefiting from increased yields and reduced reliance on chemical pesticides.

» In South Africa, smallholder maize farmers have experienced significant gains in productivity and income through *Bt* maize adoption.

The accessibility of *Bt* technology has empowered farmers in resource-constrained settings, enabling them to overcome pest-related challenges and improve their livelihoods.

3. Global Expansion Trends

The expansion of *Bt* crops continues to rise as more countries recognize their benefits. Nations in Asia, Africa, and Latin America are increasingly incorporating *Bt* crops into their agricultural strategies to address food security and pest management challenges.

Examples of *Bt* Crops

The success of *Bt* crops is illustrated by key examples that have transformed the cultivation of staple and cash crops worldwide.

1. ***Bt* Cotton**

- **Target Pests:** *Bt* cotton is engineered to combat lepidopteran pests such as the cotton bollworm, pink bollworm, and armyworm, which are among the most destructive pests of cotton crops.
- **Impact:** The introduction of *Bt* cotton has significantly reduced pest-related yield losses. For example:
  - In India, *Bt* cotton has increased cotton yields by 50% while reducing pesticide use by over 40%.
  - In the United States, *Bt* cotton has minimized the need for insecticide applications, lowering production costs and environmental risks.
- **Economic Benefits:** Farmers growing *Bt* cotton have reported higher incomes due to increased yields and reduced expenditures on chemical pest control.

2. ***Bt* Maize**

- **Target Pests:** *Bt* maize addresses pests like the European corn borer, fall armyworm, and corn rootworm, which cause substantial damage to maize crops globally.
- **Impact:** In the United States, *Bt* maize has increased yields and reduced the economic impact of pest infestations by billions of dollars annually. Similarly, in countries like Brazil, *Bt* maize has helped farmers combat the fall armyworm, a pest that thrives in tropical climates.
- **Environmental Benefits:** By reducing the need for broad-spectrum insecticides, *Bt* maize has helped protect beneficial organisms such as pollinators and natural predators.

3. ***Bt* Brinjal (Eggplant)**

- **Target Pests: *Bt* brinjal is designed to resist the fruit and shoot borer, a pest that** causes significant losses in eggplant production in South Asia and Southeast Asia.
- **Impact:** In Bangladesh, the adoption of *Bt* brinjal has reduced pest infestation levels by over 80%, increasing yields and reducing pesticide use dramatically.
- **Social Benefits:** *Bt* brinjal has improved the health of farmers and consumers by lowering exposure to toxic pesticides commonly used in conventional eggplant farming.

Environmental and Human Health Benefits

Bt crops have brought significant environmental and human health benefits, largely by reducing the reliance on chemical insecticides. This reduction addresses some of the most pressing concerns in agriculture, including pesticide pollution, ecosystem degradation, and public health risks.

1. Reduction in Chemical Pesticide Use

- *Bt* crops produce insecticidal proteins internally, eliminating the need for frequent pesticide applications.

- For example, *Bt* cotton has reduced insecticide use in India by nearly 40%, preventing millions of liters of pesticides from being sprayed annually.
- This reduction has also decreased the risk of pesticide runoff into water bodies, helping to protect aquatic ecosystems and downstream communities.

2. **Preservation of Beneficial Insects**
   - Unlike chemical pesticides, which often kill non-target organisms, *Bt* crops target only specific pests.
   - Beneficial insects, such as pollinators (bees) and natural pest predators (ladybugs), thrive in fields of *Bt* crops, promoting ecological balance and supporting integrated pest management systems.
3. **Improved Soil Health**
   - By reducing pesticide applications, *Bt* crops support conservation tillage practices, which help preserve soil structure and fertility.
   - In regions like the United States, where *Bt* maize is widely grown, farmers have reported improved soil health and reduced erosion due to less intensive chemical interventions.
4. **Human Health Benefits**
   - *Bt* crops have significantly reduced human exposure to harmful pesticides, particularly in developing countries where pesticide misuse is common.
   - Studies have shown that farmers growing *Bt* cotton in India experience fewer pesticide-related health issues, such as respiratory problems and skin irritations.
   - Consumers also benefit from reduced pesticide residues on food, contributing to safer and healthier diets.

Broader Implications for Agriculture

The global adoption of *Bt* crops extends beyond pest control, influencing broader agricultural trends and addressing critical challenges such as food security, sustainability, and climate resilience.

1. **Food Security**
   - *Bt* crops contribute to food security by increasing yields and reducing crop losses caused by pests.
   - In regions facing high pest pressures, such as sub-Saharan Africa and South Asia, *Bt* crops provide a reliable means of maintaining food production and meeting the dietary needs of growing populations.
2. **Economic Empowerment**
   - The economic benefits of *Bt* crops are particularly evident among smallholder farmers, who face disproportionate risks from pest infestations.
   - By reducing input costs and increasing income, *Bt* crops have empowered millions of farmers, improving livelihoods and fostering rural development.

3. **Sustainability and Climate Resilience**

   » *Bt* crops align with the principles of sustainable agriculture by reducing the environmental footprint of farming.

   » As climate change intensifies pest pressures and disrupts traditional farming practices, *Bt* crops offer a resilient solution, enabling farmers to adapt to evolving challenges.

4. **Scientific Advancements**

   » The success of *Bt* crops has spurred further research into genetic engineering and pest-resistant technologies.

   » Advances such as gene stacking (combining multiple *Bt* genes) and CRISPR-based precision breeding are paving the way for the next generation of pest-resistant crops.

Challenges and the Road Ahead

Despite their success, the widespread adoption of *Bt* crops is not without challenges. Resistance development among target pests, public skepticism about genetically modified organisms (GMOs), and regulatory hurdles remain significant obstacles.

1. **Resistance Management**

   » The evolution of resistance in pests poses a long-term threat to the efficacy of *Bt* crops.

   » Strategies such as planting non-*Bt* refuges, rotating crops, and developing new *Bt* genes are essential to mitigate resistance and sustain the benefits of *Bt* technology.

2. **Public Perception and Regulation**

   » Public concerns about the safety and ethics of GMOs have led to stringent regulatory frameworks in some countries, slowing the adoption of *Bt* crops.

   » Transparent communication and robust scientific evidence are critical to addressing these concerns and fostering greater acceptance of *Bt* technology.

3. **Equitable Access**

   » Ensuring that smallholder farmers in low-income regions have access to *Bt* seeds and support services is crucial for maximizing the technology's global impact.

2.1.2 RNA Interference (RNAi) Technology

2.1.2.1 Overview of RNAi Mechanism

RNA interference (RNAi), a natural gene-silencing mechanism, has become a cornerstone of modern biotechnology and pest management. This mechanism involves the regulation of gene expression through small RNA molecules, offering precise, targeted control over gene activity. RNAi's application in agriculture has revolutionized pest resistance strategies, reducing the need for chemical pesticides and advancing sustainable farming.

The Discovery and Fundamentals of RNAi

RNAi was first identified in the 1990s through groundbreaking research in nematodes (*Caenorhabditis elegans*). Andrew Fire and Craig Mello, who later received the Nobel Prize for their discovery, revealed that double-stranded RNA (dsRNA) could silence specific genes in a sequence-dependent manner. This revelation laid the groundwork for understanding RNAi as a universal biological process that regulates gene expression across diverse eukaryotic organisms, including plants.

1. **Biological Role of RNAi**

RNAi evolved as a natural defense mechanism against viruses and transposable elements, which threaten genome stability. By targeting and degrading foreign RNA, RNAi protects organisms from genetic instability and pathogenic attacks.

- In plants, RNAi is particularly vital for silencing viral genes, conferring immunity against infections.
- It also regulates endogenous gene expression, ensuring proper development and response to environmental stimuli.

2. **Molecular Components of RNAi**

The RNAi pathway involves several key molecular players that work together to silence target genes:

- **Double-stranded RNA (dsRNA):** The initiator molecule, often introduced artificially in transgenic plants, which triggers the RNAi response.
- **Dicer Enzyme:** Processes dsRNA into small interfering RNA (siRNA) fragments of 21-24 nucleotides.
- **RNA-Induced Silencing Complex (RISC):** A multi-protein complex that uses siRNA as a guide to identify and degrade complementary messenger RNA (mRNA) sequences.

Mechanism of RNAi in Transgenic Plants

The RNAi mechanism in plants harnesses the cell's natural machinery to silence pest-specific genes, effectively disrupting pest development, reproduction, or survival. This highly targeted approach reduces collateral damage to non-target organisms, making RNAi a sustainable alternative to chemical pesticides.

1. **Introduction of Double-Stranded RNA (dsRNA)**

- **Synthetic Design of dsRNA:** Scientists identify essential pest-specific genes, such as those involved in chitin synthesis (important for exoskeleton formation) or digestive enzymes. dsRNA sequences are then designed to match these genes.
- **Incorporation into Plant Genomes:** Through genetic engineering, dsRNA-encoding sequences are integrated into the plant genome. Promoter elements control where and when the dsRNA is expressed, optimizing its efficacy against pests.

2. **Processing of dsRNA into siRNA**

- The introduced dsRNA is recognized by the plant's Dicer enzyme, which cleaves it into siRNA fragments.

- » siRNA serves as the functional molecule that guides the silencing process, ensuring specificity to the target pest gene.

3. **Action of RNA-Induced Silencing Complex (RISC)**
   - » siRNA is loaded into RISC, where it directs the complex to mRNA molecules with complementary sequences.
   - » RISC cleaves the mRNA, preventing its translation into proteins essential for pest survival.

4. **Effects on Pests**
   - » When pests consume RNAi-enabled plants, the ingested dsRNA enters their gut cells and activates the pest's RNAi machinery.
   - » This results in the silencing of critical genes, leading to developmental defects, impaired digestion, or death.

Applications of RNAi in Pest-Resistant Crops

RNAi technology has been successfully applied to develop transgenic crops with built-in pest resistance. These crops provide an innovative solution to managing pests that are resistant to traditional chemical pesticides or those that cause widespread agricultural damage.

1. **RNAi Maize for Rootworm Resistance**
   - » **Target Pest:** Western corn rootworm, one of the most destructive pests of maize.
   - » **Mechanism:** RNAi maize produces dsRNA targeting rootworm-specific genes involved in gut function. When rootworms feed on the maize roots, the dsRNA silences these genes, disrupting the pest's digestive system and causing mortality.
   - » **Impact:** RNAi maize has significantly reduced rootworm infestations, decreasing yield losses and reliance on chemical insecticides.

2. **RNAi Potatoes for Beetle Control**
   - » **Target Pest:** Colorado potato beetle, a major pest of potato crops.
   - » **Mechanism:** dsRNA targets genes critical for beetle development and reproduction, effectively suppressing their population.
   - » **Impact:** RNAi potatoes have improved crop yields while reducing pesticide use, benefiting both farmers and the environment.

3. **RNAi Cotton for Bollworm Management**
   - » **Target Pest:** Cotton bollworm, a lepidopteran pest that damages cotton bolls and fibers.
   - » **Mechanism:** dsRNA silences genes necessary for larval development, protecting cotton plants from infestation.
   - » **Impact:** RNAi cotton complements existing pest-resistant technologies like *Bt* cotton, providing an additional layer of defense.

Advantages of RNAi Technology

RNAi-based pest resistance offers numerous advantages over traditional pest control methods, including enhanced precision, sustainability, and environmental safety.

1. **Precision Targeting**
   » RNAi's gene-specific action ensures that only pests with the targeted gene sequence are affected. This precision minimizes the impact on non-target organisms, such as pollinators, natural predators, and soil microbes.
2. **Environmental Benefits**
   » RNAi eliminates the need for broad-spectrum chemical insecticides, reducing pesticide runoff and contamination of water bodies.
   » The dsRNA molecules degrade rapidly in the environment, leaving no toxic residues behind.
3. **Reduced Risk of Resistance Development**
   » Unlike chemical pesticides, which often target broad biochemical pathways, RNAi can silence multiple genes simultaneously.
   » By targeting genes essential for pest survival, RNAi makes it more difficult for pests to evolve resistance.
4. **Integration with Other Technologies**
   » RNAi can be combined with other pest-resistant traits, such as *Bt* proteins, to create crops with stacked defenses against multiple pests. This approach enhances crop protection while extending the longevity of resistance mechanisms.

Challenges and Limitations

Despite its potential, RNAi technology faces several challenges that must be addressed to ensure its effectiveness and widespread adoption.

1. **Delivery Challenges**
   » While dsRNA can be expressed in transgenic plants, delivering it to pests that do not feed on plant tissues, such as sap-sucking insects, remains a technical hurdle.
   » Researchers are exploring alternative delivery methods, such as dsRNA sprays, to expand RNAi's applicability.
2. **Potential for Resistance Evolution**
   » Prolonged use of RNAi crops could lead to the emergence of resistant pest populations.
   » Implementing resistance management strategies, such as gene rotation and refuge planting, is critical to mitigate this risk.
3. **Cost and Accessibility**
   » The development and deployment of RNAi crops can be expensive, potentially limiting their adoption by smallholder farmers.

» Efforts to reduce costs and improve accessibility are necessary to maximize the benefits of RNAi technology.

4. **Regulatory and Public Perception**

» Regulatory frameworks for RNAi-based products are still evolving, creating uncertainty for developers and farmers.

» Public concerns about genetic modification may hinder acceptance, requiring transparent communication and public education initiatives.

Broader Implications for Agriculture

The adoption of RNAi in pest management has far-reaching implications for sustainable agriculture and global food security.

1. **Enhanced Food Security**

» By reducing crop losses caused by pests, RNAi technology contributes to stable food supplies and improved farmer incomes.

» It is particularly valuable in regions where pest pressures are severe and traditional control methods are ineffective.

2. **Environmental Sustainability**

» RNAi aligns with the principles of sustainable agriculture by minimizing the environmental footprint of pest control.

» Its compatibility with integrated pest management (IPM) strategies supports biodiversity and ecosystem health.

3. **Scientific Innovation**

» RNAi has inspired new avenues of research in gene silencing and pest management.

» Advances such as CRISPR-Cas9 are building on RNAi's foundations, offering even more precise and versatile tools for crop protection.

2.1.2.2 Application in Pest Resistance

The application of RNA interference (RNAi) in pest resistance represents a transformative advancement in agricultural biotechnology. By leveraging the gene-silencing mechanism of RNAi, transgenic plants can be engineered to produce double-stranded RNA (dsRNA) molecules that target specific pest species, disrupting their critical metabolic processes. This precision-targeted approach offers several advantages, including effective pest control, reduced environmental impact, and minimal effects on non-target organisms.

RNAi-Based Transgenic Plants and Pest Resistance

1. **Mechanism of Action in Transgenic Plants**

» RNAi-based pest-resistant transgenic plants are designed to express dsRNA that silences essential genes in pest organisms. This involves:

- **Gene Selection:** Scientists identify key genes in the target pest's genome that are critical for survival, such as those involved in digestion, reproduction, or exoskeleton formation.
- **dsRNA Production in Plants:** The selected gene sequence is synthesized as dsRNA and introduced into the plant genome through genetic engineering. The plant then produces dsRNA in specific tissues, such as leaves or roots, which pests consume during feeding.
- **Ingestion by Pests:** Once the dsRNA enters the pest's digestive system, it activates the RNAi pathway, leading to the degradation of the pest's corresponding mRNA and silencing of the targeted gene. This disruption of essential functions results in reduced fitness, developmental defects, or death.

2. **Species-Specific Pest Control**

   » RNAi technology's ability to target specific genes ensures that only the intended pest species are affected, leaving non-target organisms unharmed. This precision is achieved by designing dsRNA sequences that are unique to the pest species, minimizing the risk of off-target effects.

3. **Broad Application Potential**

   » RNAi-based transgenic plants can target a wide range of pests, including insects, nematodes, and even plant pathogens.

   » This versatility makes RNAi a valuable tool for managing diverse agricultural pests and addressing emerging pest threats caused by climate change and global trade.

Example: RNAi Corn Targeting Western Corn Rootworm

The Western corn rootworm (*Diabrotica virgifera virgifera*) is one of the most destructive pests affecting maize production in North America and parts of Europe. Traditional pest control methods, such as chemical insecticides, have been only partially effective and have raised concerns about environmental sustainability and pest resistance. RNAi technology offers a novel solution.

1. **Mechanism of Action in RNAi Corn**

   » RNAi corn is engineered to produce dsRNA molecules that target specific genes essential for the rootworm's survival, such as those involved in gut cell integrity or metabolic pathways.

   » When rootworms feed on corn roots, they ingest the dsRNA, which is processed into small interfering RNA (siRNA) by the pest's RNAi machinery.

   » The siRNA guides the silencing of the target gene, disrupting critical biological functions and leading to the rootworm's death.

2. **Impact on Rootworm Management**

   » RNAi corn has significantly reduced rootworm populations, decreasing yield losses and the need for chemical pesticides.

   » This approach addresses the rootworm's resistance to traditional pest control methods, providing a sustainable and effective alternative.

3. **Environmental and Economic Benefits**

» Reduced reliance on chemical insecticides has lowered the environmental impact of maize farming, including pesticide runoff and harm to non-target species.

» Farmers have experienced economic gains through improved yields and lower input costs, particularly in regions where rootworm infestations are severe.

Advantages of RNAi in Pest Resistance

1. **Target-Specific Pest Control**

» RNAi offers unparalleled precision in pest management by targeting genes that are unique to the pest species. This specificity minimizes unintended effects on beneficial organisms, such as pollinators and natural predators.

» Unlike broad-spectrum chemical pesticides, RNAi ensures that ecosystems remain balanced, supporting biodiversity and integrated pest management strategies.

2. **Reduced Environmental Impact**

» By eliminating or reducing the need for chemical insecticides, RNAi-based pest resistance contributes to sustainable farming practices.

» The dsRNA molecules produced by transgenic plants degrade rapidly in the environment, leaving no toxic residues behind.

3. **Mitigation of Resistance Development**

» Pests often develop resistance to chemical pesticides through genetic mutations. RNAi, by targeting essential and often highly conserved genes, makes resistance development more challenging.

» Additionally, RNAi can be combined with other pest control strategies, such as *Bt* technology, to create multi-layered defenses against pests.

4. **Compatibility with Diverse Crops and Pests**

» RNAi technology can be adapted to various crops and pest species, making it a versatile tool for addressing global agricultural challenges.

» Emerging pests, including those resistant to chemical pesticides, can be targeted using RNAi, ensuring resilience in agricultural systems.

Challenges and Considerations in RNAi Applications

Despite its advantages, the widespread implementation of RNAi in pest resistance faces several challenges that must be addressed to maximize its potential.

1. **Delivery of dsRNA to Pests**

» While transgenic plants producing dsRNA are effective against chewing pests, delivering dsRNA to sap-sucking pests or other pests with limited plant tissue consumption remains a challenge.

» Innovative delivery methods, such as topical dsRNA sprays or nanoformulations, are being developed to expand RNAi's applicability.

2. **Resistance Management**

» Prolonged exposure to RNAi-based pest control could lead to the evolution of resistance in pests.

» Strategies such as rotating target genes, stacking multiple RNAi traits, and combining RNAi with conventional methods are essential to delay resistance development.

3. **Regulatory and Public Perception Issues**

» The regulatory landscape for RNAi-based products is still evolving, creating uncertainty for developers and farmers.

» Public concerns about genetic modification and the perceived risks of RNAi technology must be addressed through transparent communication and public education.

4. **Cost and Accessibility**

» The development and deployment of RNAi crops require significant investment, which could limit accessibility for smallholder farmers in developing regions.

» Efforts to reduce production costs and provide support for farmers are critical for ensuring equitable access to RNAi technology.

Broader Implications of RNAi in Pest Resistance

The application of RNAi in pest-resistant plants has far-reaching implications for global agriculture, offering solutions to some of the most pressing challenges in food production and environmental conservation.

1. **Enhanced Food Security**

» By reducing crop losses caused by pests, RNAi-based pest resistance contributes to stable food supplies and improved farmer livelihoods.

» This is particularly critical in regions where pest pressures threaten food security and economic stability.

2. **Support for Sustainable Agriculture**

» RNAi aligns with the principles of sustainable agriculture by minimizing the environmental footprint of pest control.

» It complements other sustainable practices, such as conservation tillage and integrated pest management, fostering resilience in farming systems.

3. **Catalyst for Scientific Innovation**

» The success of RNAi has spurred further research into gene silencing technologies, leading to advancements in precision agriculture and crop improvement.

» Emerging tools, such as CRISPR-Cas9, build on RNAi's foundation, offering even more precise and versatile solutions for agricultural challenges.

2.1.2.3 Benefits and Challenges

RNA interference (RNAi) has emerged as one of the most promising biotechnological tools in agriculture, particularly for pest resistance. Its precision in targeting specific genes and its minimal impact on non-target organisms make it a preferred method for sustainable pest management. However, alongside its numerous benefits, RNAi faces several challenges that must be addressed for its widespread adoption.

Benefits of RNAi in Pest Resistance

RNAi offers a range of benefits that position it as a transformative technology in agriculture. These advantages extend across environmental, economic, and agricultural dimensions, offering a sustainable alternative to traditional pest control methods.

1. **High Precision in Targeting Specific Pests**
   » RNAi operates by silencing genes with sequences complementary to the introduced double-stranded RNA (dsRNA). This specificity ensures that only target pests are affected, minimizing harm to non-target organisms.
   » For example, RNAi designed to combat the Western corn rootworm affects only this pest, leaving beneficial insects like pollinators and natural predators unharmed. This precision contrasts sharply with broad-spectrum chemical pesticides, which often kill both pests and beneficial organisms.
2. **Reduced Ecological Risks**
   » RNAi is an environmentally friendly pest management tool. Unlike chemical pesticides that persist in the environment and can contaminate soil and water, the dsRNA molecules degrade quickly, leaving no toxic residues.
   » This rapid degradation reduces the risk of bioaccumulation and long-term environmental harm, supporting biodiversity and ecosystem health.
3. **Decreased Dependence on Chemical Pesticides**
   » The use of RNAi technology reduces the need for chemical insecticides, which have been associated with adverse effects on human health, non-target species, and the environment.
   » For instance, RNAi-based pest-resistant crops like RNAi maize have enabled farmers to control rootworm infestations without resorting to repeated chemical applications, reducing pesticide use and associated costs.
4. **Lower Risk of Pest Resistance Development**
   » RNAi targets essential genes in pests, often those that are conserved and less prone to mutation. This makes it harder for pests to develop resistance compared to traditional pesticides, which typically target biochemical pathways that pests can adapt to over time.
   » By stacking multiple RNAi traits or combining RNAi with other pest-resistant technologies (e.g., *Bt* proteins), resistance development can be further delayed, ensuring the long-term efficacy of pest control strategies.

5. **Flexibility and Adaptability**

   » RNAi can be tailored to target a wide range of pests, including insects, nematodes, and even plant pathogens.

   » Its flexibility allows for the rapid development of solutions to emerging pest threats, particularly those exacerbated by climate change or global trade.

6. **Economic Benefits for Farmers**

   » RNAi technology reduces the costs associated with pest management by minimizing the need for chemical pesticides and their associated labor.

   » Higher crop yields and quality, coupled with reduced input costs, contribute to improved profitability for farmers, especially in regions where pests cause significant damage.

Challenges Facing RNAi in Agriculture

While the benefits of RNAi are substantial, the technology also faces significant challenges that must be addressed to achieve widespread adoption. These challenges span technical, regulatory, and societal domains, highlighting the need for continued innovation and stakeholder engagement.

1. **Technical Challenges in Delivery**

   » Delivering dsRNA to pests effectively is a major hurdle, particularly for pests with feeding habits that limit direct exposure to plant tissues (e.g., sap-sucking insects like aphids).

   » While transgenic plants producing dsRNA are effective against chewing pests, alternative delivery methods, such as topical sprays or nanoparticles, are needed to expand RNAi's applicability. Developing cost-effective and scalable delivery systems remains a critical area of research.

2. **Potential for Resistance Development**

   » Although RNAi is less prone to resistance development compared to traditional pesticides, the possibility still exists. Prolonged use of RNAi targeting the same genes could lead to the selection of resistant pest populations.

   » Implementing resistance management strategies, such as rotating target genes, combining RNAi with other control methods, and planting refuges, is essential to mitigate this risk.

3. **Regulatory Barriers**

   » The regulatory landscape for RNAi-based products is still evolving, with many countries lacking clear guidelines for their approval and commercialization.

   » The complexity of RNAi's mode of action and its perceived novelty may result in lengthy regulatory processes, increasing costs for developers and delaying market entry.

   » Harmonizing international regulations and establishing transparent approval pathways are critical for enabling global adoption of RNAi technology.

4. **Public Skepticism and Consumer Acceptance**

   » As with many genetically engineered products, RNAi-based technologies face public skepticism, particularly regarding their safety and environmental impact.

» Misinformation and negative perceptions of genetic modification can hinder acceptance among consumers and policymakers.

» Effective science communication and public engagement are vital to addressing these concerns and building trust in RNAi technology.

5. **Economic Barriers for Smallholder Farmers**

   » The high initial costs of developing and deploying RNAi-based crops may limit their accessibility for smallholder farmers, particularly in developing regions.

   » Ensuring affordability and providing technical support are essential for enabling equitable access to RNAi technology and maximizing its benefits.

6. **Knowledge Gaps in Long-Term Effects**

   » While RNAi is generally considered safe and environmentally friendly, its long-term ecological effects are not yet fully understood.

   » Comprehensive studies are needed to assess potential impacts on soil health, microbial communities, and non-target organisms over extended periods.

Balancing Benefits and Challenges

Addressing the challenges associated with RNAi technology is critical to unlocking its full potential and ensuring its responsible deployment. By leveraging its benefits and overcoming its limitations, RNAi can play a pivotal role in transforming agriculture.

1. **Research and Innovation**

   » Continued investment in research is essential to address technical challenges, such as delivery mechanisms and resistance management.

   » Advances in nanotechnology and gene editing could enhance the efficiency and scalability of RNAi-based pest control systems.

2. **Policy and Regulation**

   » Establishing clear, science-based regulatory frameworks will facilitate the approval and adoption of RNAi products.

   » Policymakers must engage with scientists, industry stakeholders, and the public to ensure that regulations balance safety with innovation.

3. **Education and Outreach**

   » Building public awareness about RNAi's safety, benefits, and environmental advantages is key to overcoming skepticism.

   » Farmers must be provided with training and support to adopt RNAi-based technologies effectively and sustainably.

4. **Collaboration and Accessibility**

   » Partnerships between governments, research institutions, and private companies can help reduce the costs of RNAi technology and make it accessible to smallholder farmers.

» International collaboration is vital for addressing global pest challenges and ensuring equitable distribution of RNAi innovations.

2.1.3 Protease Inhibitors

2.1.3.1 Function and Role

Protease inhibitors play a critical role in pest management by disrupting the digestive processes of herbivorous insects and other agricultural pests. These naturally occurring proteins or peptides function by blocking the activity of proteolytic enzymes-key molecules that pests rely on to break down proteins in their diet. Without functional proteolytic enzymes, pests experience starvation and eventual death, making protease inhibitors an essential tool in integrated pest management strategies.

Protease Inhibitors: Mechanism of Action

Protease inhibitors are molecules that bind to proteases, rendering them inactive. Proteases are enzymes responsible for catalyzing the breakdown of proteins into smaller peptides and amino acids, a critical process for digestion and nutrition in all living organisms. In pests, proteases play a vital role in nutrient absorption, particularly during the larval and juvenile stages when growth and development demand high protein intake.

1. **Inhibition of Digestive Enzymes**

    » Protease inhibitors target and deactivate the proteolytic enzymes in the pest's digestive tract. This results in the inability of the pest to process ingested proteins, leading to nutrient deficiency.

    » The inhibition of proteases triggers a cascade of physiological disruptions in pests, including reduced growth rates, reproductive failure, and eventual starvation.

2. **Binding Specificity and Types of Proteases**

    » Protease inhibitors are highly specific, targeting distinct classes of proteases. The two most common types of proteases affected by inhibitors in pests are:

    - **Serine Proteases:** These are the most prevalent digestive enzymes in many insect pests, including lepidopteran species such as caterpillars. They play a critical role in breaking down dietary proteins.
    - **Cysteine Proteases:** Found in pests like beetles, these enzymes contribute significantly to protein digestion, particularly in acidic or specialized gut environments.

3. **Physiological Impact on Pests**

    » By disrupting digestion, protease inhibitors induce malnutrition in pests, severely affecting their ability to grow, reproduce, and survive.

    » Over time, this mechanism reduces pest populations, protecting crops from damage and yield losses.

Common Protease Inhibitors in Pest Resistance

Several classes of protease inhibitors have been identified and studied for their role in pest control. These inhibitors are either naturally produced by plants as part of their defense mechanisms or artificially introduced into crops through genetic engineering.

1. **Serine Protease Inhibitors**
   - » **Mechanism of Action:** Serine protease inhibitors block serine proteases by binding to their active sites, preventing the hydrolysis of protein substrates.
   - » **Examples:**
     - **Bowman-Birk Inhibitors:** Found in legumes such as soybeans, these inhibitors are highly effective against serine proteases.
     - **Kunitz-Type Inhibitors:** These are abundant in plants like potatoes and tomatoes and target specific serine proteases in pests.
   - » **Target Pests:** Lepidopteran pests, such as caterpillars, rely heavily on serine proteases and are particularly vulnerable to these inhibitors.
2. **Cysteine Protease Inhibitors**
   - » **Mechanism of Action:** Cysteine protease inhibitors deactivate cysteine proteases by forming covalent bonds with their active sites. This inhibition disrupts the pest's ability to digest proteins in acidic gut environments.
   - » **Examples:**
     - **Cystatins:** Found in plants such as rice and maize, these inhibitors are highly effective against beetle species that depend on cysteine proteases for digestion.
   - » **Target Pests:** Coleopteran pests, including the Colorado potato beetle and rootworms, are particularly susceptible to cysteine protease inhibitors.
3. **Other Protease Inhibitors**
   - » **Aspartic Protease Inhibitors:** These inhibitors target aspartic proteases, enzymes involved in protein digestion under acidic conditions.
   - » **Metalloprotease Inhibitors:** These are less common but have potential applications in pest resistance, particularly for pests that rely on metalloproteases for protein hydrolysis.

Role of Protease Inhibitors in Plant Defense

Protease inhibitors are a fundamental component of the natural defense mechanisms of many plants. By producing these inhibitors, plants deter herbivorous pests and reduce the damage caused by feeding.

1. **Induced Defense Response**
   - » Many plants produce protease inhibitors in response to pest attack. When herbivores feed on plant tissues, signaling pathways such as the jasmonic acid pathway are activated, triggering the production of protease inhibitors.
   - » This induced response provides plants with a dynamic mechanism to counter pest infestations.

2. **Protection of Vital Plant Tissues**
   - » Protease inhibitors are often concentrated in specific tissues, such as seeds, leaves, and roots, where they are most likely to encounter feeding pests.
   - » This localized defense minimizes the metabolic cost to the plant while maximizing the impact on pests.
3. **Synergistic Effects with Other Defense Compounds**
   - » Protease inhibitors often work in conjunction with other plant defense compounds, such as secondary metabolites (e.g., alkaloids and tannins) and structural defenses (e.g., trichomes and spines).
   - » This multi-layered defense strategy enhances the overall effectiveness of pest resistance.

Genetic Engineering of Protease Inhibitors in Crops

Advances in biotechnology have enabled the introduction of protease inhibitor genes into crops, providing them with enhanced pest resistance. Transgenic plants expressing protease inhibitors have been developed to target specific pests, reducing the need for chemical pesticides and improving agricultural sustainability.

1. **Development of Transgenic Plants**
   - » Scientists identify and isolate protease inhibitor genes from naturally resistant plants. These genes are then inserted into the genomes of target crops using genetic engineering techniques.
   - » The resulting transgenic plants express protease inhibitors in their tissues, providing built-in protection against pests.
2. **Examples of Transgenic Crops**
   - » **Potatoes Expressing Cysteine Protease Inhibitors:** These plants are resistant to the Colorado potato beetle, a major pest that relies on cysteine proteases for digestion.
   - » **Rice Expressing Serine Protease Inhibitors:** These plants demonstrate increased resistance to lepidopteran pests, reducing the need for insecticides.
3. **Advantages of Transgenic Crops with Protease Inhibitors**
   - » Reduced pesticide use lowers environmental contamination and production costs for farmers.
   - » Enhanced pest resistance improves yield stability, particularly in regions prone to severe pest infestations.
   - » The specificity of protease inhibitors minimizes unintended effects on non-target organisms.

Challenges and Considerations

While protease inhibitors offer significant benefits in pest resistance, their application in agriculture is not without challenges. These issues must be addressed to ensure the sustainable and effective use of this technology.

1. **Potential for Resistance Development**
   - » Prolonged exposure to protease inhibitors can lead to the evolution of resistant pest populations.

Pests may adapt by producing alternative proteases or inhibitors that neutralize the plant-derived protease inhibitors.

» Strategies such as combining protease inhibitors with other pest-resistant traits (e.g., *Bt* proteins) and rotating crops are essential to delay resistance development.

2. **Non-Specific Effects on Beneficial Organisms**

» While protease inhibitors are generally specific to target pests, there is a risk of unintended effects on beneficial organisms, such as predatory insects or soil microbes.

» Thorough ecological assessments are necessary to minimize these risks and ensure that protease inhibitors contribute to sustainable farming systems.

3. **Regulatory and Public Acceptance**

» The use of genetically modified crops expressing protease inhibitors may face regulatory and public acceptance challenges, particularly in regions with strict GMO regulations.

» Transparent communication and evidence-based assessments are critical for fostering trust and acceptance of this technology.

4. **Integration with Sustainable Farming Practices**

» Protease inhibitors should be integrated into broader pest management strategies, such as integrated pest management (IPM), to maximize their effectiveness and minimize potential drawbacks.

Future Prospects for Protease Inhibitors in Agriculture

Protease inhibitors hold immense potential for addressing the challenges of pest resistance in agriculture. Ongoing research and innovation are likely to expand their applications and improve their efficacy.

1. **Advances in Biotechnology**

» Gene editing technologies, such as CRISPR-Cas9, offer new opportunities for optimizing protease inhibitor expression in crops.

» These tools enable precise modifications to enhance specificity and potency while reducing off-target effects.

2. **Development of Multi-Trait Crops**

» Combining protease inhibitors with other pest-resistant traits, such as RNAi or *Bt* proteins, can provide crops with multi-layered defenses against diverse pest populations.

3. **Application in Emerging Pests**

» Protease inhibitors can be tailored to target emerging pest threats, particularly those that have developed resistance to traditional control methods.

4. **Sustainable Pest Management**

» As part of integrated pest management strategies, protease inhibitors can contribute to reducing pesticide use, preserving biodiversity, and promoting sustainable agricultural practices.

2.1.3.2 Genetic Engineering for Resistance

The application of genetic engineering to enhance pest resistance in crops has revolutionized agriculture, offering innovative ways to protect plants from pest damage while reducing reliance on chemical pesticides. Among these advancements, the introduction of genes encoding protease inhibitors into crops stands out as a ground breaking approach. By disrupting the digestive processes of pests, transgenic plants expressing protease inhibitors provide a sustainable and precise method of pest control.

Mechanism of Protease Inhibitor-Based Resistance in Transgenic Plants

Protease inhibitors function by targeting and disabling proteolytic enzymes in pests, which are essential for digesting proteins in their diet. Genetic engineering allows scientists to introduce genes encoding these inhibitors into crops, enabling the plants to produce protease inhibitors in their tissues.

1. **Gene Selection and Cloning**
   - » **Source of Genes:** Protease inhibitor genes are typically isolated from plants that naturally produce these compounds as part of their defense mechanisms. For example:
     - **Bowman-Birk and Kunitz inhibitors** are sourced from legumes like soybeans and beans.
     - **Cystatins** are derived from crops like rice, maize, and barley.
   - » **Gene Cloning:** Once identified, these genes are cloned and modified to optimize their expression in target crops. This involves adding regulatory elements such as promoters and terminators to ensure efficient transcription and translation in the host plant.
2. **Gene Transfer into Crop Plant**s
   - » **Agrobacterium-Mediated Transformation:** A common method for introducing genes into plants, this technique uses the bacterium *Agrobacterium tumefaciens* to transfer the desired DNA into plant cells.
   - » **Biolistic Transformation (Gene Gun):** In cases where *Agrobacterium* is ineffective, a gene gun can be used to shoot DNA-coated particles into plant cells.
3. **Expression of Protease Inhibitors**
   - » Once integrated into the plant genome, the protease inhibitor gene is expressed, leading to the production of protease inhibitors in specific plant tissues.
   - » The inhibitors are strategically produced in the parts of the plant most vulnerable to pest attacks, such as leaves, stems, and roots.
4. **Impact on Pest Digestive Systems**
   - » When pests feed on the transgenic plant, they ingest tissues containing protease inhibitors.
   - » These inhibitors bind to the active sites of proteolytic enzymes in the pest's digestive tract, rendering them inactive.
   - » Without functional enzymes, the pest cannot effectively digest proteins, leading to nutrient deficiency, growth retardation, and death.

Applications of Transgenic Plants Expressing Protease Inhibitors

The introduction of protease inhibitor genes into crops has been applied across various agricultural systems, providing targeted pest resistance and enhancing yield stability.

1. **Crops Targeting Lepidopteran Pests**

   » **Cotton:** Transgenic cotton expressing serine protease inhibitors effectively controls caterpillars, which are major pests that feed on cotton bolls and leaves.

   » **Rice:** By producing cysteine protease inhibitors, transgenic rice resists lepidopteran pests such as rice stem borers, reducing crop losses in Asia.

2. **Crops Targeting Coleopteran Pests**

   » **Potatoes:** Transgenic potato plants expressing cysteine protease inhibitors are resistant to the Colorado potato beetle, a notorious pest in temperate regions.

   » **Maize:** Cystatin-expressing maize disrupts the digestive processes of rootworms, reducing damage to maize roots and improving crop resilience.

3. **Legumes and Vegetables**

   » **Soybeans and Beans:** These crops benefit from Bowman-Birk inhibitors, which protect them against aphids and other sucking pests.

   » **Tomatoes and Peppers:** Kunitz inhibitors are introduced to reduce damage from caterpillars and beetle larvae.

Advantages of Genetic Engineering for Protease Inhibitors

The use of genetic engineering to enhance pest resistance through protease inhibitors offers several advantages over conventional pest management strategies.

1. **Environmentally Friendly Pest Control**

   » Transgenic plants expressing protease inhibitors reduce the need for chemical insecticides, lowering environmental contamination and protecting non-target species.

   » The inhibitors degrade naturally in the environment, leaving no toxic residues behind.

2. **Specificity and Safety**

   » Protease inhibitors target pest-specific enzymes, minimizing unintended effects on non-target organisms, including beneficial insects and soil microbes.

   » Their specificity ensures that ecosystem biodiversity is preserved, aligning with sustainable farming practices.

3. **Cost-Effective Pest Management**

   » Once established, transgenic plants provide continuous protection against pests without the need for repeated pesticide applications.

   » This reduces input costs for farmers and enhances profitability, particularly in resource-limited settings.

4. **Integration with Other Technologies**

   » Protease inhibitor genes can be stacked with other pest-resistant traits, such as *Bt* proteins or RNAi, to create crops with multi-layered defenses against diverse pest populations.

Challenges and Considerations

While genetic engineering for protease inhibitor-based resistance holds significant promise, it is not without challenges. Addressing these issues is crucial to ensure the long-term success and acceptance of this technology.

1. **Potential for Resistance Development**

   » Prolonged exposure to protease inhibitors can lead to the evolution of resistant pest populations, which may produce alternative proteases or suppress inhibitor activity.

   » Resistance management strategies, such as rotating crops and combining multiple resistance traits, are essential to mitigate this risk.

2. **Regulatory Hurdles**

   » Transgenic crops face stringent regulatory requirements in many countries, increasing the cost and time needed for approval.

   » Public concerns about genetically modified organisms (GMOs) add another layer of complexity, requiring transparent communication and robust safety assessments.

3. **Unintended Ecological Effects**

   » While protease inhibitors are generally specific, their broader ecological impact is not fully understood. For example, effects on soil microbial communities or unintended interactions with non-target insects require further study.

4. **Economic Accessibility**

   » Developing and deploying transgenic crops with protease inhibitors can be expensive, potentially limiting their availability to smallholder farmers in developing regions.

   » Efforts to reduce production costs and improve accessibility are critical for ensuring equitable benefits from this technology.

Future Prospects for Genetic Engineering of Protease Inhibitors

The continued advancement of genetic engineering offers new opportunities to enhance the efficacy and sustainability of protease inhibitor-based pest resistance.

1. **Gene Editing Technologies**

   » Tools like CRISPR-Cas9 enable precise modifications to protease inhibitor genes, improving their specificity and potency.

   » These technologies also allow the rapid development of new resistance traits to address emerging pest threats.

2. **Development of Multi-Trait Crops**

   » Combining protease inhibitors with other genetic traits, such as drought tolerance or nutrient use efficiency, can create crops that are more resilient to both biotic and abiotic stresses.

3. **Expansion to Emerging Pests**

   » As climate change shifts pest distributions, protease inhibitors can be tailored to target newly invasive species, ensuring continued crop protection.

4. **Sustainable Pest Management Systems**

   » Integrating protease inhibitor-based resistance with other sustainable practices, such as conservation tillage and biological control, can enhance the overall resilience and sustainability of farming systems.

2.1.3.3 Applications

The applications of protease inhibitor-transgenic plants highlight the versatility and efficacy of this advanced biotechnology in addressing critical agricultural challenges. By equipping plants with genes encoding protease inhibitors, scientists have developed crops that are inherently resistant to a range of destructive pests. These innovations are transforming pest management in agriculture, reducing reliance on chemical pesticides, and promoting sustainability.

Applications in Protecting Legumes and Cereals

Protease inhibitors have been successfully introduced into legumes and cereals, two of the most widely cultivated crop groups, to enhance their resistance to pests. These crops play a crucial role in global food security, making the development of pest-resistant varieties a priority for scientists and farmers alike.

1. **Legumes**

   » **Soybeans and Beans:**

   - Protease inhibitors such as Bowman-Birk and Kunitz inhibitors have been integrated into soybean and common bean varieties to combat pests like aphids and pod borers.
   - These pests are known for reducing crop yield and quality, particularly in regions where legumes serve as staple foods or cash crops.

   » **Peas and Lentils:**

   - Transgenic peas expressing protease inhibitors have demonstrated increased resistance to weevils and moth larvae, reducing yield losses significantly.

Impact on Legume Farming:

» By reducing pest damage, protease inhibitor-transgenic legumes contribute to higher yields and improved food quality.

» Reduced pesticide use lowers production costs and minimizes the environmental impact of legume farming.

1. **Cereals**

   » **Rice:**

   - Transgenic rice varieties expressing cysteine protease inhibitors have been developed to resist pests such as rice stem borers and brown planthoppers. These pests are among the most damaging to rice crops in Asia and Africa.

   » **Maize:**

   - Cysteine protease inhibitors have been introduced into maize to protect against rootworms, a pest that causes significant root damage and yield reductions.

   » **Wheat and Barley**:

   - Wheat and barley expressing protease inhibitors have shown resistance to aphids and caterpillars, which feed on the leaves and stems, reducing photosynthetic capacity.

Impact on Cereal Farming:

» Protease inhibitor-transgenic cereals improve yield stability in pest-prone regions, ensuring consistent food supply and farmer incomes.

» These innovations are particularly valuable in developing countries, where cereals form the dietary staple for millions of people.

Example: Potato Plants and Colorado Potato Beetle Resistance

One of the most notable applications of protease inhibitors in transgenic plants is the development of potato varieties resistant to the Colorado potato beetle (*Leptinotarsa decemlineata*). This pest is a significant threat to potato crops in temperate regions, capable of causing severe yield losses if not controlled.

1. **Mechanism of Resistance**

   » Potato plants have been engineered to express cysteine protease inhibitors, which specifically target the digestive enzymes of the Colorado potato beetle.

   » When the beetles feed on these transgenic potatoes, the protease inhibitors bind to the active sites of their digestive enzymes, disrupting protein digestion and nutrient absorption.

   » The resulting malnutrition leads to reduced growth, reproductive failure, and mortality in the beetles.

2. **Field Performance of Transgenic Potatoes**

   » Studies have demonstrated that potato plants expressing cysteine protease inhibitors exhibit significantly lower pest damage compared to non-transgenic varieties.

   » In addition to reducing beetle populations, these transgenic potatoes maintain higher yields and require fewer pesticide applications.

3. **Economic and Environmental Benefits**

   » The introduction of protease inhibitor-transgenic potatoes has reduced the reliance on chemical insecticides, lowering production costs for farmers.

» By minimizing pesticide use, these potatoes also contribute to environmental health, reducing contamination of soil and water resources.

» The technology has been particularly beneficial for smallholder farmers in regions where the Colorado potato beetle is endemic, improving food security and livelihoods.

Broader Applications of Protease Inhibitors in Transgenic Plants

The use of protease inhibitors is not limited to legumes, cereals, or potatoes. The technology has been applied to a wide range of crops, demonstrating its adaptability and effectiveness in various agricultural systems.

1. **Vegetables**

» **Tomatoes and Peppers:**

- Protease inhibitor-transgenic varieties have been developed to resist pests like caterpillars and beetles, which commonly infest solanaceous crops.

» **Eggplants:**

- Protease inhibitors have been combined with other pest-resistant traits to create eggplants with enhanced protection against fruit borers.

2. **Fruits**

» **Bananas and Apples:**

- Transgenic fruit crops expressing protease inhibitors are being developed to combat pests like fruit flies and aphids, which damage fruit quality and reduce marketability.

3. **Industrial Crops**

» **Cotton:**

- Protease inhibitors have been introduced into cotton plants to complement *Bt* technology, providing a broader range of protection against chewing and sucking pests.

» **Oilseeds:**

- Transgenic oilseeds, such as canola, expressing protease inhibitors help protect against pests like aphids and beetles, which can reduce oil yield and quality.

Advantages of Protease Inhibitor Applications

The application of protease inhibitors in transgenic plants offers several advantages, making it a valuable tool for sustainable agriculture.

1. **Targeted Pest Control**

» Protease inhibitors provide pest-specific resistance, minimizing the impact on non-target organisms and preserving beneficial insects, such as pollinators and natural predators.

2. **Reduced Reliance on Chemical Pesticides**

» Transgenic plants expressing protease inhibitors reduce the need for chemical insecticides, lowering production costs and minimizing environmental harm.

3. **Enhanced Yield Stability**
   » By protecting crops from pest damage, protease inhibitors contribute to stable yields, particularly in regions with high pest pressures.
4. **Compatibility with Integrated Pest Management (IPM)**
   » Protease inhibitors can be combined with other pest management strategies, such as crop rotation and biological control, to create comprehensive and sustainable farming systems.

Challenges and Considerations

While protease inhibitors offer significant benefits, their application also presents challenges that must be addressed to ensure their sustainability and effectiveness.

1. **Potential for Resistance Development**
   » Pests exposed to protease inhibitors over prolonged periods may develop resistance, either through the production of alternative proteases or other adaptive mechanisms.
   » Implementing resistance management strategies, such as rotating crops or stacking traits, is essential to mitigate this risk.
2. **Regulatory and Public Acceptance**
   » As with all genetically modified crops, protease inhibitor-transgenic plants face regulatory scrutiny and public skepticism. Transparent communication and rigorous safety assessments are critical for fostering acceptance.
3. **Economic Accessibility**
   » The high costs associated with developing and deploying transgenic crops may limit their adoption by smallholder farmers. Efforts to reduce production costs and provide subsidies or support programs are necessary to ensure equitable access.

2.1.4 Secondary Metabolite Pathways

2.1.4.1 Role of Secondary Metabolites in Pest Resistance

Secondary metabolites play a crucial role in the natural defense mechanisms of plants, offering a sustainable and efficient method of pest resistance. These compounds, which include alkaloids, tannins, terpenoids, flavonoids, and phenolics, are not directly involved in the primary metabolic processes essential for plant growth and development. Instead, they function as chemical defenses against herbivores, pathogens, and environmental stresses. By deterring pests through toxicity, anti-digestive properties, or feeding deterrence, secondary metabolites form an integral part of a plant's arsenal against biotic threats. Advances in genetic engineering have amplified the effectiveness of these compounds, allowing scientists to enhance or introduce secondary metabolite pathways in crops to bolster pest resistance.

The Natural Role of Secondary Metabolites in Plant Defense

1. **Chemical Deterrence of Herbivores**
   » Secondary metabolites act as natural pesticides, making plants less palatable or toxic to herbivores.

- They interfere with the physiological or neurological systems of pests, reducing feeding activity, reproductive success, or survival rates.

Examples:

- **Alkaloids:** Compounds such as nicotine in tobacco and caffeine in coffee act as neurotoxins, disrupting the nervous systems of insects and other pests.
- **Terpenoids:** Compounds like pyrethrins in chrysanthemum flowers and limonoids in citrus plants deter pests through direct toxicity or anti-feedant properties.
- **Tannins:** Found in plants like oak and tea, tannins interfere with protein digestion in herbivores, leading to reduced growth and development.

2. **Anti-Digestive Properties**
   - Many secondary metabolites inhibit the digestive enzymes of herbivores, reducing their ability to process nutrients from plant tissues.
   - This mode of action often results in malnutrition and starvation for the pests.
3. **Attraction of Natural Enemies**
   - Certain secondary metabolites, such as volatile organic compounds (VOCs), act as signaling molecules that attract natural predators or parasitoids of herbivores.
   - For example, terpenoids released by maize under caterpillar attack attract parasitic wasps that prey on the caterpillars.

Types of Secondary Metabolites in Pest Resistance

Secondary metabolites vary widely in their chemical structure and function, enabling plants to target different pest species and adapt to diverse environmental conditions. The primary classes include:

1. **Alkaloids**
   - **Function:** Alkaloids are nitrogen-containing compounds that disrupt the nervous systems of pests or interfere with their metabolism.
   - Examples in Pest Resistance:
     - ◊ Nicotine in tobacco deters aphids and caterpillars.
     - ◊ Quinine in cinchona plants inhibits feeding in certain insect pests.
2. **Terpenoids**
   - **Function:** Terpenoids are hydrocarbons that act as toxins, feeding deterrents, or signaling molecules.
   - Examples in Pest Resistance:
     - ◊ Pyrethrins in chrysanthemum act as neurotoxins for a wide range of insects.
     - ◊ Limonene in citrus plants repels aphids and other soft-bodied pests.

3. **Tannins**
   - **Function:** Tannins bind to proteins in the herbivore's digestive system, reducing nutrient absorption and causing malnutrition.
   - Examples in Pest Resistance:
     - Tannins in oak leaves deter caterpillars and beetles.
4. **Phenolics**
   - **Function:** Phenolics disrupt pest development and act as feeding deterrents.
   - Examples in Pest Resistance:
     - Salicylic acid in willow trees repels sap-feeding insects.
5. **Flavonoids**
   - **Function:** Flavonoids provide UV protection and act as antioxidants, while some also deter herbivores.
   - Examples in Pest Resistance:
     - Isoflavonoids in legumes reduce the feeding of nematodes and insects.

Genetic Engineering to Enhance Secondary Metabolite Pathways

The natural production of secondary metabolites in plants often occurs at suboptimal levels for effective pest resistance. Genetic engineering allows scientists to enhance or introduce these metabolic pathways, increasing the quantity and diversity of secondary metabolites produced by crops.

1. **Enhancement of Existing Pathways**
   - Genes encoding enzymes that regulate secondary metabolite synthesis can be overexpressed to boost production.
   - For example, increasing the expression of key enzymes in the terpenoid biosynthetic pathway can amplify the production of volatile terpenoids, improving pest deterrence.
2. **Introduction of New Pathways**
   - Genetic engineering can introduce secondary metabolite pathways from one species into another, conferring new defensive capabilities.
   - For instance, genes for nicotine biosynthesis have been transferred from tobacco to non-related crops, providing protection against aphids and other pests.
3. **Targeted Expression in Vulnerable Tissues**
   - By using tissue-specific promoters, scientists can ensure that secondary metabolites are produced in parts of the plant most susceptible to pest attack, such as leaves, stems, or roots.
   - This targeted approach minimizes the metabolic cost to the plant while maximizing pest resistance.

Examples of Genetically Engineered Crops Utilizing Secondary Metabolites

Several crops have been engineered to enhance pest resistance through secondary metabolite pathways, with promising results.

1. **Rice**
   - **Enhanced Pathway:** Overexpression of genes involved in phenolic acid synthesis increases resistance to rice stem borers and leafhoppers.
   - **Outcome:** Reduced pest damage and improved yield stability in pest-prone regions of Asia.
2. **Maize**
   - **Enhanced Pathway:** Introduction of terpenoid biosynthesis genes boosts the production of VOCs that attract parasitic wasps to control caterpillar infestations.
   - **Outcome:** Improved natural pest control and reduced need for insecticides.
3. **Tomatoes**
   - **Enhanced Pathway:** Overexpression of genes involved in alkaloid synthesis increases resistance to whiteflies and aphids.
   - **Outcome:** Better crop protection and higher marketable yields.
4. **Soybeans**
   - **Enhanced Pathway:** Isoflavonoid biosynthesis genes have been upregulated to deter nematodes and weevils.
   - **Outcome:** Reduced pest pressure and improved root health.

Advantages of Secondary Metabolite-Based Pest Resistance

1. **Environmentally Friendly Pest Control**
   - Secondary metabolites offer a natural alternative to synthetic pesticides, reducing environmental contamination and promoting sustainable farming.
2. **Multi-Pest Resistance**
   - Many secondary metabolites have broad-spectrum activity, protecting crops from a wide range of pests, including insects, nematodes, and pathogens.
3. **Reduced Pesticide Use**
   - Crops enhanced with secondary metabolites require fewer chemical inputs, lowering production costs and reducing health risks to farmers.
4. **Compatibility with Integrated Pest Management (IPM)**
   - Secondary metabolite-based resistance can be combined with other pest control strategies, such as biological control and crop rotation, for more comprehensive pest management.

Challenges and Future Directions

Despite their potential, the application of secondary metabolites in pest resistance faces several challenges that must be addressed to maximize their impact.

1. **Metabolic Trade-Offs**
   » The production of secondary metabolites can impose metabolic costs on plants, potentially affecting growth and yield.
   » Research is focused on optimizing metabolic pathways to balance pest resistance with crop productivity.
2. **Pest Adaptation**
   » Prolonged exposure to secondary metabolites may lead to the evolution of pest resistance.
   » Strategies such as rotating resistance traits and combining multiple metabolites are being developed to mitigate this risk.
3. **Regulatory and Public Acceptance**
   » Genetically engineered crops face regulatory hurdles and public skepticism, particularly in regions with strict GMO policies.
   » Transparent communication and rigorous safety assessments are critical to building trust in this technology.
4. **Scaling Up Production**
   » Large-scale implementation of secondary metabolite-enhanced crops requires significant investment in research, development, and infrastructure.

2.1.4.2 Engineering Plants for Enhanced Secondary Metabolite Production

Engineering plants to enhance the production of secondary metabolites represents a significant advancement in agricultural biotechnology. Secondary metabolites, such as alkaloids, terpenoids, and phenolics, are natural plant compounds that deter pests through toxicity, anti-feeding properties, or by attracting pest predators. By modifying plant genomes to amplify these compounds or introduce new biosynthetic pathways, scientists are creating crops with enhanced pest resistance. This strategy not only reduces crop losses but also minimizes the environmental impact of traditional pest control methods, such as chemical pesticides.

Engineering Plants for Enhanced Terpenoid Pathways

Terpenoids are a diverse class of secondary metabolites with significant roles in plant defense. These compounds are produced via complex biosynthetic pathways, making them an ideal target for genetic engineering to enhance pest resistance.

1. **Understanding Terpenoid Pathways in Plants**
   » Terpenoids are synthesized through the mevalonate (MVA) pathway in the cytoplasm and the methylerythritol phosphate (MEP) pathway in plastids.
   » These pathways generate a wide variety of terpenoid compounds, including monoterpenes,

sesquiterpenes, and diterpenes, which serve as insect deterrents, toxins, or attractants for natural pest predators.

» For example, volatile terpenoids released by maize and rice can repel pests or attract parasitic wasps that attack herbivorous larvae.

2. **Genetic Modifications for Terpenoid Production**

   » **Overexpression of Native Genes:** Engineering plants to overexpress genes involved in terpenoid biosynthesis can increase the production of naturally occurring defense compounds.

   » **Introduction of Foreign Genes:** Genes encoding enzymes from other species can be introduced to create new terpenoid pathways, providing crops with resistance to a broader range of pests.

   » **Regulation of Expression:** Tissue-specific promoters ensure that terpenoids are produced where they are most needed, such as in leaves or stems vulnerable to pest attacks.

Example: Transgenic Rice Expressing Terpenoid Pathways

Rice (*Oryza sativa*) is one of the world's most important staple crops, feeding billions of people annually. However, rice production is severely threatened by pests such as stem borers, which cause significant yield losses in many regions. By engineering rice plants to enhance terpenoid production, scientists have developed a novel solution to this persistent problem.

1. **Stem Borers: A Major Pest of Rice**

   » Stem borers are lepidopteran pests whose larvae tunnel into rice stems, disrupting nutrient transport and weakening plant structure.

   » Chemical pesticides have been the primary method of control, but they are costly, environmentally harmful, and can lead to pest resistance.

2. **Engineering Terpenoid Pathways in Rice**

   » **Gene Introduction:** Genes encoding key enzymes for terpenoid biosynthesis, such as terpene synthases (TPS), are introduced into the rice genome.

   » **Pathway Optimization:** By modifying the metabolic flux in rice plants, scientists enhance the production of specific terpenoids that repel or kill stem borers.

   » **Volatile Organic Compounds (VOCs):** Engineered rice emits higher levels of terpenoid VOCs, which act as both repellents for stem borers and attractants for natural predators, such as parasitic wasps.

3. **Results and Benefits**

   » Transgenic rice expressing terpenoid pathways has shown significant resistance to stem borers in field trials.

   » The engineered plants experience reduced pest damage, leading to higher yields and improved economic outcomes for farmers.

   » This approach minimizes the need for chemical pesticides, reducing environmental contamination and promoting biodiversity.

Enhancing Pest Deterrence Without Harming Beneficial Organisms

One of the most significant advantages of secondary metabolite engineering is the ability to target pests selectively without affecting beneficial organisms, such as pollinators or natural predators. This precision aligns with the goals of sustainable agriculture.

1. **Selective Pest Targeting**
   » Terpenoids and other secondary metabolites are highly specific in their mode of action. For example:
      ◊ Certain terpenoids target the nervous systems of herbivorous insects while leaving other organisms unaffected.
      ◊ Alkaloids such as nicotine disrupt pest neurotransmission without harming mammals or birds in the ecosystem.
2. **Preservation of Natural Pest Predators**
   » VOCs produced by engineered plants not only repel pests but also attract predators and parasitoids that naturally control pest populations.
   » For example, rice engineered to emit terpenoid VOCs attracts parasitic wasps that attack stem borer larvae, enhancing biological pest control.
3. **Reduction in Pesticide Residues**
   » By reducing the reliance on chemical pesticides, engineered plants lower the risk of pesticide residues in the environment, safeguarding soil and water quality and protecting non-target organisms.

Broader Applications of Secondary Metabolite Engineering

The principles used to enhance terpenoid pathways in rice can be applied to a wide range of crops, providing pest resistance across diverse agricultural systems.

1. **Cereal Crops**
   » **Maize:** Engineering maize to produce volatile terpenoids has improved resistance to rootworms and caterpillars.
   » **Wheat:** Enhanced production of phenolic compounds in wheat deters aphids and other sap-feeding pests.
2. **Vegetables and Fruits**
   » **Tomatoes:** Overexpression of alkaloid biosynthetic pathways in tomatoes increases resistance to whiteflies and thrips.
   » **Citrus Fruits:** Enhanced production of limonoids in citrus plants protects against aphids and mites.
3. **Industrial Crops**
   » **Cotton:** Engineering cotton plants to produce secondary metabolites like gossypol has increased resistance to bollworms.

» **Oilseeds:** Enhanced secondary metabolite production in oilseeds like canola reduces damage from weevils and beetles.

Challenges and Considerations

Despite the promise of secondary metabolite engineering, there are challenges that must be addressed to ensure the sustainability and effectiveness of these approaches.

1. **Metabolic Trade-Offs**
   » Overproduction of secondary metabolites can divert resources from primary metabolic processes, potentially affecting plant growth and yield.
   » Balancing pest resistance with crop productivity is a critical area of research.
2. **Pest Adaptation**
   » Prolonged exposure to secondary metabolites may lead to pest resistance, requiring the development of multi-target approaches to maintain effectiveness.
3. **Regulatory and Public Perception Issues**
   » Genetically engineered crops face strict regulatory scrutiny and public skepticism, particularly in regions with restrictive GMO policies.
   » Transparent communication and robust safety assessments are necessary to build public trust and acceptance.

2.1.4.3 Ecological and Agricultural Benefits

The ecological and agricultural benefits of secondary metabolites are profound, particularly in the context of pest resistance and sustainable farming practices. These naturally occurring compounds, including alkaloids, tannins, terpenoids, and flavonoids, offer plants an intrinsic mechanism to deter pests, reducing reliance on synthetic chemical pesticides. The ability of secondary metabolites to provide broad-spectrum pest resistance, coupled with their alignment with the principles of sustainable agriculture, makes them a cornerstone of modern, eco-friendly farming practices.

Broad-Spectrum Pest Resistance

1. **Mechanisms of Pest Resistance**

Secondary metabolites function as natural defense compounds that protect plants from a wide variety of pests, including insects, nematodes, fungi, and even some bacterial pathogens. Their modes of action are diverse, allowing them to target pests through multiple pathways:

» **Toxicity:** Certain metabolites, such as alkaloids and terpenoids, are directly toxic to pests, interfering with their physiological processes. For example:
   ◊ Alkaloids like nicotine and caffeine disrupt the nervous systems of herbivorous insects, leading to paralysis or death.
   ◊ Terpenoids such as limonoids inhibit growth and reproduction in aphids and whiteflies.

- **Feeding Deterrence:** Many secondary metabolites make plants unpalatable to pests. Tannins, for instance, bind to proteins in the digestive systems of herbivores, reducing nutrient absorption and causing malnutrition.
- **Reproductive Disruption:** Some metabolites interfere with the hormonal systems of pests, reducing their reproductive success and limiting population growth.

2. Wide Applicability Across Pest Types

The broad-spectrum activity of secondary metabolites allows them to protect crops from diverse pest species, including:

- **Insects:** Terpenoids and phenolics effectively deter lepidopteran caterpillars, coleopteran beetles, and hemipteran sap-suckers.
- **Nematodes:** Flavonoids and saponins in legumes act as repellents or toxins to plant-parasitic nematodes, such as root-knot nematodes.
- **Fungi and Bacteria:** Phenolics and phytoalexins inhibit the growth of pathogenic fungi and bacteria, reducing disease incidence in crops like wheat and soybeans.

3. Examples of Broad-Spectrum Resistance

- **Maize:** Produces phenolic compounds that deter both insects and fungal pathogens, contributing to reduced crop losses.
- **Tomatoes:** Enhanced production of alkaloids and flavonoids in genetically modified varieties protects against whiteflies and leaf miners, while also reducing fungal infections.

Alignment with Sustainable Farming Practices

1. Reduced Dependency on Chemical Pesticides

Secondary metabolites serve as natural alternatives to synthetic pesticides, enabling farmers to reduce chemical inputs. This shift offers several benefits:

- **Environmental Health:** Reduced pesticide application minimizes soil and water contamination, protecting ecosystems from the harmful effects of chemical residues.
- **Biodiversity Conservation:** Natural pest resistance preserves beneficial organisms, such as pollinators (e.g., bees) and predatory insects (e.g., ladybugs), which play vital roles in maintaining ecosystem balance.
- **Reduced Carbon Footprint:** Lower pesticide production and transportation requirements contribute to decreased greenhouse gas emissions, supporting global efforts to mitigate climate change.

2. Improved Soil Health

- Secondary metabolites contribute to healthier soils by reducing the need for chemical treatments that can disrupt microbial communities.
- Some metabolites, such as flavonoids, promote beneficial interactions between plants and soil microorganisms, enhancing nutrient cycling and root health.

3. **Enhanced Resilience to Climate Change**

   » Secondary metabolites can provide dual benefits by protecting crops from pests and enhancing tolerance to abiotic stresses, such as drought and heat.

   » This adaptability is crucial for ensuring food security in the face of changing climatic conditions.

Agricultural Productivity Benefits

1. **Higher Yields and Crop Quality**

   » By reducing pest damage, secondary metabolites help ensure consistent yields and improve the quality of harvested crops.

   » For example:

   ◊ Rice varieties engineered to produce higher levels of terpenoids show increased resistance to stem borers, resulting in higher grain yields.

   ◊ Tomatoes with enhanced alkaloid production are less prone to pest infestations, ensuring better fruit quality for markets.

2. **Economic Advantages for Farmers**

   » Reduced reliance on chemical pesticides lowers production costs, increasing profitability for farmers.

   » Smallholder farmers in developing regions particularly benefit from crops with built-in pest resistance, as they often lack access to costly chemical inputs.

3. **Extended Shelf Life and Market Value**

   » Secondary metabolites with antifungal properties reduce post-harvest losses caused by pathogens, extending the shelf life of crops like wheat and maize.

   » Higher-quality produce fetches better prices in markets, boosting farmer incomes.

Environmental Conservation

1. **Support for Biodiversity**

   » The use of secondary metabolites in pest resistance ensures that non-target organisms, including beneficial insects, are not harmed.

   » Crops engineered to produce volatile organic compounds (VOCs) attract natural enemies of pests, such as parasitic wasps and predatory beetles, enhancing biological control.

2. **Reduction in Pesticide Residues**

   » By reducing chemical pesticide use, secondary metabolites help prevent pesticide residues from entering food chains and water systems. This aligns with global goals for food safety and environmental sustainability.

3. **Promotion of Sustainable Ecosystems**

   » Natural pest resistance reduces the disruption of ecosystems caused by intensive pesticide application, promoting ecological balance.

- » Secondary metabolites also support soil health by fostering symbiotic relationships with beneficial microorganisms, such as nitrogen-fixing bacteria in legumes.

Integration into Modern Agricultural Systems

1. **Compatibility with Integrated Pest Management (IPM)**
   - » Secondary metabolite-based resistance is easily incorporated into IPM strategies, providing a complementary approach to cultural, mechanical, and biological pest control methods.
   - » This integration enhances the overall effectiveness of pest management while reducing reliance on any single strategy, such as chemical pesticides.
2. **Potential for Multi-Trait Crops**
   - » Advances in genetic engineering allow for the development of crops with multiple resistance traits, combining secondary metabolite pathways with other biotechnological innovations, such as RNAi or *Bt* proteins.
   - » These multi-trait crops offer enhanced resilience to both biotic and abiotic stresses, ensuring sustainable agricultural production.

Challenges and Future Directions

1. **Balancing Pest Resistance and Crop Growth**
   - » Overproduction of secondary metabolites can divert resources from primary metabolic processes, potentially affecting plant growth and yield.
   - » Research is focused on optimizing metabolic pathways to achieve a balance between pest resistance and productivity.
2. **Pest Adaptation**
   - » Prolonged exposure to secondary metabolites may lead to the evolution of resistant pest populations.
   - » Strategies such as combining multiple metabolites, rotating crops, and stacking resistance traits are essential to mitigate this risk.
3. **Public Perception and Regulatory Hurdles**
   - » Genetically modified crops that enhance secondary metabolite production face regulatory scrutiny and public skepticism.
   - » Transparent communication and rigorous safety assessments are necessary to build trust and acceptance.

Future Prospects

1. **Synthetic Biology and Secondary Metabolites**
   - » Synthetic biology offers opportunities to create entirely new secondary metabolite pathways, providing novel solutions for pest resistance.
   - » For example, designing metabolites that specifically target emerging pest species or pathogens could address challenges posed by climate change and global trade.

2. **Gene Editing and Precision Agriculture**
   » CRISPR-Cas9 technology enables precise modifications to secondary metabolite pathways, enhancing their effectiveness while minimizing off-target effects.
3. **Scaling Up for Global Agriculture**
   » Efforts are underway to scale up the production of secondary metabolite-enhanced crops, ensuring their accessibility to smallholder farmers worldwide.
   » Partnerships between governments, research institutions, and private companies are critical for achieving this goal.

2.1.5 Chitinase and Glucanase Genes

2.1.5.1 Role in Pest Defense

Chitinases and glucanases are pivotal enzymes in plant defense mechanisms, targeting two critical components of pest and pathogen structures: the chitin in insect exoskeletons and fungal cell walls, and the glucans in fungal and microbial cell walls. By degrading these essential structural polysaccharides, these enzymes weaken and disrupt the integrity of pests and pathogens, rendering them vulnerable and inhibiting their ability to cause damage. While these enzymes are naturally present in many plants, advancements in genetic engineering have enabled their overexpression, enhancing their pest defense capabilities.

Natural Role of Chitinases and Glucanases in Plants

1. **Chitinases: Targeting Chitin in Insect Exoskeletons and Fungal Cell Walls**
   » **Chitin Structure and Function:**
      - Chitin is a long-chain polymer of N-acetylglucosamine, forming a key structural component of insect exoskeletons and fungal cell walls.
      - It provides rigidity and protection, enabling pests and pathogens to survive environmental stresses and immune responses.

   » **Chitinase Action:**
      - Chitinases are enzymes that hydrolyze the β-1,4 linkages in chitin, breaking it down into smaller fragments.
      - In insects, this disrupts the exoskeleton, making them more susceptible to desiccation, predation, or environmental damage.
      - In fungi, chitinase activity compromises the integrity of the cell wall, inhibiting growth and reproduction.
2. **Glucanases: Breaking Down Glucans in Fungal and Microbial Cell Walls**
   » **Glucan Structure and Function:**
      - Glucans are polysaccharides found in the cell walls of fungi and some bacteria. They contribute to cell wall strength and structural stability.

» **Glucanase Action:**

- Glucanases hydrolyze β-1,3 or β-1,4 linkages in glucans, degrading the cell wall and rendering pathogens unable to invade plant tissues or reproduce effectively.

3. **Natural Plant Defense Mechanism**

» Chitinases and glucanases are often produced as part of the plant's innate immune response, triggered by pest or pathogen attacks.

» These enzymes are synthesized in response to signaling molecules such as jasmonic acid and salicylic acid, which activate the plant's defense pathways.

» Their production is localized to sites of attack, ensuring a targeted and energy-efficient response.

Genetic Engineering to Enhance Chitinase and Glucanase Activity

The natural production of chitinases and glucanases, while effective, is often insufficient to combat severe pest and pathogen infestations. Genetic engineering offers a solution by enabling plants to overexpress these enzymes, providing heightened and sustained defense capabilities.

1. **Overexpression of Chitinase and Glucanase Genes**

» **Gene Selection and Cloning:**

- Genes encoding potent chitinases and glucanases are isolated from naturally resistant plants or microorganisms. For example:

◊ Chitinase genes from *Trichoderma* fungi are known for their strong antifungal properties.

◊ Glucanase genes from barley and wheat have been widely studied for their role in pathogen resistance.

» **Gene Insertion:**

- These genes are inserted into the genomes of target crops using genetic engineering techniques, such as *Agrobacterium tumefaciens*-mediated transformation or biolistic methods.

» **Promoter Optimization:**

- Tissue-specific or inducible promoters are used to ensure that enzyme production occurs at the right time and location, such as in leaves or root tissues prone to pest attacks.

2. **Enhanced Pest and Pathogen Resistance**

» Transgenic plants with overexpressed chitinases exhibit increased resistance to insect pests like caterpillars and beetles, as well as fungal pathogens like *Fusarium* and *Rhizoctonia*.

» Glucanase-overexpressing plants show improved defense against fungal and bacterial pathogens, reducing disease incidence and crop losses.

Examples of Chitinase and Glucanase Applications in Crops

1. **Rice**

   » **Chitinase Overexpression:**

   - Transgenic rice plants expressing chitinase genes are resistant to rice blast, a fungal disease caused by *Magnaporthe oryzae*.
   - These plants experience reduced lesion formation and higher yields in pathogen-prone regions.

2. **Tomatoes**

   » **Combined Chitinase and Glucanase Expression:**

   - Tomato plants engineered to express both chitinase and glucanase genes show enhanced resistance to fungal pathogens like *Alternaria* and *Botrytis*.
   - The dual action of these enzymes disrupts the fungal cell wall more effectively than either enzyme alone.

3. **Wheat**

   » **Glucanase Overexpression:**

   - Wheat varieties with enhanced glucanase activity are resistant to fungal diseases like powdery mildew and rust, which are significant threats to global wheat production.

4. **Cotton**

   » **Chitinase Expression for Insect Resistance:**

   - Cotton plants engineered to express chitinase genes exhibit resistance to bollworms, reducing damage to cotton bolls and improving fiber quality.

Advantages of Chitinases and Glucanases in Pest Defense

1. **Broad-Spectrum Activity**

   » These enzymes target a wide range of pests and pathogens, including insects, fungi, and some bacterial species.

   » Their broad-spectrum activity makes them versatile tools in integrated pest management systems.

2. **Environmentally Friendly**

   » Chitinases and glucanases provide a natural alternative to chemical pesticides, reducing environmental contamination and pesticide residues in food products.

   » They are biodegradable and do not accumulate in the environment, aligning with the principles of sustainable agriculture.

3. **Minimal Non-Target Effects**

   » These enzymes specifically target structural components unique to pests and pathogens, such as chitin and glucans, minimizing unintended effects on beneficial organisms, such as pollinators and soil microbes.

4. **Improved Yield Stability**

   » By enhancing pest and pathogen resistance, chitinase- and glucanase-expressing crops experience reduced damage and improved yield stability, particularly in regions with high pest pressures.

2.1.5.2 Transgenic Plants with Enhanced Enzyme Activity

The development of transgenic plants with enhanced chitinase and glucanase enzyme activity marks a significant breakthrough in agricultural biotechnology. By engineering plants to overexpress these enzymes, researchers have created crops with dual protection against biotic stresses, including insect pests and fungal pathogens. These transgenic plants provide a sustainable alternative to chemical pesticides, reducing environmental impact while improving crop yields and resilience.

Engineering Plants for Enhanced Chitinase and Glucanase Activity

1. **The Role of Chitinases and Glucanases in Pest Resistance**

   » Chitinases break down chitin, a key component of insect exoskeletons and fungal cell walls, while glucanases degrade glucans, another structural polysaccharide in fungal cell walls.

   » These enzymes disrupt the structural integrity of pests and pathogens, inhibiting their growth, reproduction, and survival.

2. **Genetic Engineering for Overexpression**

   » **Gene Selection:**

   - Chitinase and glucanase genes are isolated from plants, fungi, or bacteria with strong antifungal and insecticidal properties.
   - For example, chitinase genes from *Trichoderma* fungi and glucanase genes from barley have been widely used in transgenic crops.

   » **Gene Insertion and Expression:**

   - These genes are introduced into crop plants using genetic engineering techniques, such as *Agrobacterium*-mediated transformation or particle bombardment.
   - Promoters are used to regulate expression, ensuring that the enzymes are produced in specific tissues or in response to pest attacks.

3. **Enhanced Enzyme Activity in Transgenic Plants**

   » Transgenic plants with enhanced chitinase and glucanase activity produce higher levels of these enzymes, providing continuous and effective protection against pests and pathogens.

Example: Tobacco and Tomato Plants Engineered with Chitinase Genes

1. **Tobacco Plants**

   » **Engineering for Fungal Resistance:**

   - Tobacco plants have been engineered to overexpress chitinase genes from *Trichoderma* fungi.

  - These transgenic plants exhibit increased resistance to fungal pathogens like *Alternaria alternata* and *Fusarium oxysporum*, which cause leaf spots and vascular wilt, respectively.

» **Insect Pest Resistance:**

  - Enhanced chitinase activity also targets the chitin in insect exoskeletons, providing protection against pests like aphids and caterpillars.

2. **Tomato Plants**

» **Dual Resistance to Fungi and Insects:**

  - Tomato plants expressing chitinase genes demonstrate resistance to fungal diseases such as early blight (*Alternaria solani*) and gray mold (*Botrytis cinerea*), as well as insect pests like whiteflies and thrips.

» **Improved Yield and Quality:**

  - By reducing pest and pathogen damage, these transgenic tomatoes produce higher yields and better-quality fruits, with fewer blemishes and post-harvest losses.

Dual Benefits of Enhanced Enzyme Activity

1. **Protection Against Fungal Pathogens**

» Chitinases and glucanases degrade the structural components of fungal cell walls, compromising their integrity and inhibiting growth.

» This provides effective resistance against a wide range of fungal diseases that threaten global crop production.

2. **Resistance to Insect Pests**

» Chitinases disrupt the chitin in insect exoskeletons, leading to structural weakness, desiccation, and death.

» This dual functionality enables transgenic plants to combat both fungal pathogens and herbivorous insects, reducing the need for multiple pest control strategies.

3. **Sustainability and Environmental Benefits**

» Transgenic plants with enhanced enzyme activity reduce reliance on chemical fungicides and insecticides, lowering environmental contamination and pesticide residues in food.

» These crops also support biodiversity by minimizing the impact on non-target organisms, including pollinators and natural pest predators.

4. **Economic Benefits for Farmers**

» Reduced pesticide costs and improved yields enhance the profitability of farming.

» These benefits are particularly valuable for smallholder farmers in developing regions, where access to chemical inputs is limited.

2.1.5.3 Broader Applications

Chitinase and glucanase transgenes have wide-ranging applications in modern agriculture, extending their utility beyond pest and pathogen resistance during plant growth to post-harvest preservation. Their effectiveness in degrading the structural components of pests and pathogens makes these transgenes particularly valuable across various crop types, including cereals, fruits, and vegetables. By enhancing crop durability and reducing post-harvest losses, these genes contribute to increased food security, improved marketability, and reduced reliance on chemical treatments.

Application in Cereal Crops

Cereal crops are critical to global food security, providing staple nutrition for billions of people. However, cereals are highly susceptible to fungal pathogens and post-harvest pests, resulting in significant yield and storage losses.

1. **Enhancing Field Resistance**
   - » **Target Pests and Pathogens:**
     - Chitinase and glucanase transgenes help protect cereals from fungal diseases such as rusts, smuts, and blights, as well as insect pests like stem borers and aphids.
   - » **Examples of Transgenic Cereals:**
     - **Rice:**
       - ◊ Chitinase-expressing rice varieties demonstrate resistance to *Magnaporthe oryzae* (rice blast) and *Rhizoctonia solani* (sheath blight).
       - ◊ These transgenic plants maintain higher yields in regions with severe fungal outbreaks.
     - **Wheat:**
       - ◊ Glucanase-transgenic wheat varieties exhibit resistance to fungal pathogens like powdery mildew and rust, reducing the need for fungicide applications.
2. **Reducing Post-Harvest Losses**
   - » Chitinase and glucanase activity continues to protect cereal grains during storage by inhibiting fungal growth and deterring storage pests such as weevils and moths.
   - » Enhanced resistance reduces spoilage, improving the shelf life and quality of stored grains.

Application in Fruit Crops

Fruits are highly perishable commodities that are particularly vulnerable to fungal and insect damage both in the field and during storage. Chitinase and glucanase transgenes offer significant benefits for fruit production and preservation.

1. **Protection During Cultivation**
   - » **Target Pathogens:**
     - Fungal pathogens like *Botrytis cinerea* (gray mold), *Colletotrichum* (anthracnose), and *Fusarium* species cause significant pre-harvest and post-harvest losses in fruit crops.
     -

- » **Examples of Transgenic Fruits:**
  - **Tomatoes:**
    - ◊ Tomato plants engineered with chitinase genes exhibit resistance to gray mold and anthracnose, improving fruit yield and quality.
  - **Bananas:**
    - ◊ Glucanase-transgenic bananas show enhanced resistance to *Fusarium oxysporum* (Panama disease), a devastating fungal pathogen that threatens global banana production.

2. **Post-Harvest Benefits**
   - » Enhanced enzyme activity in fruits delays fungal spoilage, reducing waste during transportation and storage.
   - » The improved shelf life of transgenic fruits allows for longer distribution chains, benefiting farmers and consumers in distant markets.

Application in Vegetable Crops

Vegetables are essential components of human diets and are often consumed fresh, making pest and pathogen resistance critical to maintaining quality and minimizing losses.

1. **Field Protection**
   - » **Target Pests and Pathogens:**
     - Leafy vegetables, root crops, and solanaceous vegetables are frequently attacked by fungal pathogens such as *Alternaria*, *Sclerotinia*, and *Fusarium*.
   - » **Examples of Transgenic Vegetables:**
     - **Potatoes:**
       - ◊ Chitinase- and glucanase-expressing potatoes show resistance to *Phytophthora infestans* (late blight), reducing the need for fungicide applications.
     - **Peppers and Eggplants:**
       - ◊ These crops engineered with chitinase genes exhibit improved resistance to fungal pathogens and aphids, leading to better crop health and higher yields.

2. **Reduction in Post-Harvest Losses**
   - » For perishable vegetables, transgenic modifications prevent fungal spoilage during storage and transport.
   - » Enhanced durability ensures that vegetables retain their freshness and nutritional value for longer periods.

Cross-Cutting Benefits of Chitinase and Glucanase Transgenes

1. **Enhanced Durability**
   - Crops expressing chitinase and glucanase genes are better equipped to withstand pest and pathogen pressures during cultivation and storage.
   - Improved durability translates to lower spoilage rates, reduced waste, and increased availability of fresh produce.
2. **Economic Gains for Farmers**
   - Reduced crop losses and lower reliance on chemical treatments increase profitability for farmers.
   - Transgenic crops are particularly beneficial for smallholder farmers in developing regions, where post-harvest losses account for a significant portion of agricultural waste.
3. **Environmental Sustainability**
   - By reducing the need for chemical pesticides and fungicides, these transgenic crops minimize environmental contamination and promote biodiversity.
   - Enhanced durability also reduces the carbon footprint associated with food production and transportation, contributing to more sustainable agricultural systems.
4. **Food Security and Market Access**
   - By improving yield stability and reducing post-harvest losses, chitinase- and glucanase-expressing crops contribute to global food security.
   - Extended shelf life and improved quality allow farmers to access larger and more distant markets, increasing their economic opportunities.

2.2 Pest Resistance Mechanisms

Pest resistance mechanisms in agriculture represent a cornerstone of modern crop protection strategies. With the increasing demand for sustainable food production, innovative approaches to pest resistance have become essential. These mechanisms range from integrating multiple resistance genes into crops to leveraging cutting-edge technologies like CRISPR-Cas9 for targeted pest management. Each method contributes uniquely to reducing crop losses, minimizing reliance on chemical pesticides, and improving environmental health.

2.2.1 Integration of Multiple Resistance Genes

The integration of multiple resistance genes, commonly referred to as "gene stacking," is a transformative approach in genetic engineering. By combining different transgenes with complementary modes of action, scientists create crops capable of resisting multiple pests or addressing pest resistance more effectively. This strategy is particularly valuable for mitigating the risk of pests evolving resistance to single traits, ensuring long-term efficacy and broader protection.

Stacking Multiple Transgenes for Enhanced Resistance

1. **Concept and Advantages of Gene Stacking**

   » **Broader Spectrum of Protection:**

   - Gene stacking enables crops to target a wider range of pests, addressing both primary and secondary threats. For example, combining *Bt* proteins with RNAi technology allows plants to simultaneously protect against lepidopteran and coleopteran pests.

   » **Delay in Resistance Development:**

   - Pests exposed to multiple modes of action face greater challenges in evolving resistance, as they would need simultaneous mutations in multiple target genes.

   » **Synergistic Effects:**

   - Stacked traits can work synergistically to enhance overall pest control. For instance, *Bt* proteins disrupt gut integrity, while RNAi silences genes critical for pest survival, leading to compounded effects.

2. **Examples of Stacked Traits in Crops**

   » ***Bt* + Protease Inhibitors:**

   - Crops combining *Bt* proteins with protease inhibitors provide dual protection by targeting different physiological systems in pests.
   - Example: Cotton with stacked traits resists both chewing pests like bollworms and sap-feeding insects like aphids.

   » ***Bt* + RNAi Traits in Maize:**

   - *Maize engineered with Bt proteins targets lepidopteran pests like corn borers, while RNAi silences genes specific to coleopteran pests such as rootworms.*
   - *This dual mechanism reduces crop losses and minimizes the need for chemic*al insecticides, particularly in regions with diverse pest pressures.

3. **Scientific Principles Behind Gene Stacking**

   » **Selection of Compatible Traits:**

   - Traits are chosen based on their compatibility and complementary modes of action. For example:
     - ◊ *Bt* proteins disrupt pest midgut membranes, causing cell lysis.
     - ◊ RNAi interferes with gene expression in vital metabolic pathways.

   » **Molecular Engineering:**

   - Gene stacking involves the use of molecular tools to integrate multiple transgenes into a single locus, ensuring stable expression without disrupting the plant's genome.
   - Promoters and regulatory sequences are carefully selected to optimize expression levels and tissue-specific activity.

4. **Challenges in Gene Stacking**
   » **Metabolic Trade-Offs:**
      - Expressing multiple resistance traits can impose metabolic costs on plants, potentially affecting growth and yield.
   » **Regulatory Complexity:**
      - Approval processes for stacked traits are more stringent, requiring comprehensive testing for safety and efficacy.
   » **Cost of Development**:
      - Gene stacking is resource-intensive, necessitating significant investment in research, development, and commercialization.

Broader Implications of Gene Stacking

1. **Impact on Sustainable Agriculture**
   » Stacked traits reduce the reliance on chemical pesticides, promoting environmental sustainability and reducing pesticide runoff.
   » Enhanced pest resistance contributes to yield stability, ensuring food security in pest-prone regions.
2. **Economic Benefits for Farmers**
   » Crops with stacked traits offer cost savings through reduced pesticide applications and higher yields.
   » These benefits are particularly significant for smallholder farmers in developing countries, where pest pressures often lead to substantial losses.
3. **Future Prospects**
   » **Triple-Stack Traits:**
      - Advances in genetic engineering are enabling the development of crops with three or more stacked traits, targeting multiple pests and diseases simultaneously.
   » **Integration with Abiotic Stress Resistance:**
      - Combining pest resistance traits with tolerance to abiotic stresses like drought and salinity could create crops resilient to both biotic and environmental challenges.

2.2.2 Prospects for New Technologies

The advent of genome-editing technologies, particularly CRISPR-Cas9, has opened new frontiers in pest resistance. Unlike traditional genetic engineering, which involves introducing foreign genes, CRISPR-Cas9 enables precise editing of existing genes to enhance pest resistance. This approach offers unparalleled specificity, efficiency, and flexibility, making it a promising tool for developing next-generation pest-resistant crops.

Advancements in Genome Editing for Pest Resistance

1. **Overview of CRISPR-Cas9 Technology**

   » **Mechanism:**

   - CRISPR-Cas9 is a gene-editing tool that uses a guide RNA (gRNA) to direct the Cas9 protein to a specific DNA sequence. Cas9 acts as molecular scissors, cutting the DNA at the target site.
   - The cut DNA is repaired through natural cellular processes, allowing for the introduction, deletion, or modification of specific genes.

   » **Advantages:**

   - High precision and reduced off-target effects.
   - Customizability to target specific pests or plant traits.
   - Faster and more cost-effective compared to traditional genetic engineering.

2. **Applications in Pest Resistance**

   » **Direct Editing of Pest Genes**:

   - CRISPR-Cas9 can be used to target and disrupt essential genes in pest populations, such as those involved in reproduction, digestion, or detoxification of plant defenses.
   - Example: Editing genes in aphids to reduce their ability to feed on crops.

   » **Enhancing Plant Defenses:**

   - Editing plant genes to overexpress natural defense compounds, such as alkaloids or phenolics, strengthens resistance to pests.
   - Example: Modifying rice to produce higher levels of terpenoids for enhanced resistance to stem borers.

3. **Examples of CRISPR Applications in Pest-Resistant Crops**

   » **Tomatoes:**

   - CRISPR has been used to knock out susceptibility genes in tomatoes, providing resistance to *Botrytis cinerea* and other fungal pathogens.

   » **Maize:**

   - Genome editing has enhanced maize resistance to fall armyworms by targeting genes involved in pest detoxification pathways.

   » **Soybeans:**

   - CRISPR-modified soybeans with enhanced resistance to nematodes have shown promise in reducing yield losses.

Customizable Solutions and Ecological Benefits

1. **Reduced Ecological Impact**
   - » Genome editing allows for highly specific targeting, minimizing unintended effects on non-target organisms.
   - » Unlike chemical pesticides, CRISPR-based solutions do not leave harmful residues in the environment, supporting biodiversity and ecosystem health.
2. **Adaptability to Regional Needs**
   - » CRISPR-Cas9 enables the development of tailored solutions for specific pests and regions.
   - » For example, genome-edited crops can be customized to address emerging pests or those resistant to existing control methods.
3. **Integration with Other Technologies**
   - » CRISPR can be combined with other pest resistance strategies, such as RNAi or protease inhibitors, to create multi-layered defenses.
   - » This integration enhances overall effectiveness and delays the evolution of pest resistance.

Broader Implications of New Technologies

1. **Transforming Pest Management**
   - » Genome-editing technologies have the potential to revolutionize pest management by providing precise, efficient, and sustainable solutions.
   - » These innovations can reduce reliance on chemical pesticides, lower production costs, and support environmental conservation.
2. **Addressing Climate Change Challenges**
   - » By enabling rapid responses to emerging pests and diseases, CRISPR and other technologies can help crops adapt to the challenges posed by climate change.
3. **Fostering Global Food Security**
   - » The widespread adoption of genome-edited crops can enhance agricultural productivity, reduce food waste, and ensure a stable food supply for growing populations.

Chapter 6

Insect Gene Transformation

Introduction to Insect Gene Transformation and Biotechnology

Insect gene transformation represents a groundbreaking frontier in the field of biotechnology, opening up a wealth of possibilities for addressing critical challenges in agriculture, medicine, and environmental management. This transformative approach involves modifying the genetic makeup of insects to enhance desirable traits, suppress harmful ones, or introduce entirely new characteristics. From controlling pest populations to curbing the spread of insect-borne diseases, gene transformation offers innovative solutions that were once inconceivable.

The history of insect gene transformation is rooted in the broader field of genetic engineering, which began gaining momentum in the late 20th century. Early advancements in molecular biology and the development of tools like recombinant DNA technology paved the way for the manipulation of insect genomes. Techniques such as microinjection, electroporation, and the use of viral vectors were refined over the years, enabling scientists to target specific genes with precision. In recent years, revolutionary tools like CRISPR-Cas9 have accelerated progress, offering unparalleled accuracy and efficiency in genetic modifications.

The scope of insect gene transformation is vast and multifaceted. In agriculture, genetically engineered insects play a critical role in pest management and crop protection. For instance, the release of sterile male insects or genetically modified pests that suppress their populations has significantly reduced the reliance on chemical pesticides. This not only minimizes environmental damage but also enhances the sustainability of agricultural practices. Additionally, advances in genetic engineering have been extended to beneficial insects such as silkworms and honey bees, yielding remarkable benefits in silk production and pollination services.

Silkworms (*Bombyx mori*), historically valued for their silk, have emerged as a model organism for insect biotechnology. Genetic modifications in silkworms have been used to improve the quality and quantity of silk production, as well as to produce silk with novel properties, such as increased strength or elasticity. Moreover, transgenic silkworms have been developed to produce pharmaceutical proteins and other biomolecules, demonstrating their potential as biofactories for industrial and medical applications.

DOI: 10.1201/9781003757405-6

Honey bees, essential pollinators in natural ecosystems and agriculture, are another focus of insect gene transformation. Genetic engineering in honey bees aims to address critical challenges such as colony collapse disorder and the devastating effects of parasites like Varroa mites. By introducing genes that enhance disease resistance or improve their adaptability to environmental stressors, researchers seek to safeguard these vital pollinators and ensure the stability of food production systems worldwide.Another promising area of insect gene transformation is the introduction of lectin genes for pest suppression. Lectins are naturally occurring carbohydrate-binding proteins found in many plants, known for their ability to interfere with the digestive processes of insect pests. By integrating lectin genes into pest insects or crops, scientists aim to disrupt the life cycle of pests, reducing their populations and protecting crops from damage. This approach offers a sustainable and targeted alternative to traditional pest control methods, reducing the ecological footprint of agriculture.

Despite its immense potential, insect gene transformation is not without challenges. Ethical concerns, ecological risks, and regulatory hurdles must be addressed to ensure the responsible development and deployment of genetically modified insects. Public perception also plays a significant role in determining the acceptance of this technology, necessitating transparent communication and rigorous safety assessments.As biotechnology continues to advance, the possibilities for insect gene transformation are expanding rapidly. From enhancing silk production in transgenic silkworms to safeguarding honey bee populations and introducing innovative pest control strategies, the applications of this technology are as diverse as they are impactful.The integration of synthetic biology, genomics, and molecular tools holds promise for unlocking new frontiers, making insect gene transformation a cornerstone of sustainable solutions for global challenges.

1.1 Definition and Evolution of Gene Transformation in Insects

Gene transformation in insects involves the precise modification of an insect's genetic material through the introduction, deletion, or alteration of specific genes. This process enables researchers to study gene functions, develop innovative pest control strategies, and address challenges in disease vector management. The evolution of gene transformation in insects spans decades of scientific advancements, from early transposable element-based systems to state-of-the-art genome-editing techniques such as CRISPR-Cas9.

Definition of Gene Transformation in Insects

Gene transformation in insects refers to the deliberate introduction of foreign or modified genetic material into an insect genome to achieve a specific outcome. The goal is to manipulate genetic traits to study biological processes, control pest populations, or alter insect-pathogen interactions. This can involve:

1. **Gene Introduction:** Adding exogenous genes (transgenes) to express new traits, such as resistance to diseases or markers for tracking.
2. **Gene Editing:** Altering existing genes to suppress or enhance specific functions.
3. **Gene Silencing:** Temporarily disabling genes using RNA interference (RNAi) to understand their roles or reduce pest viability.

Key tools and techniques include:

- » **Transposon Systems:** Mobile genetic elements like *P-element* and *piggyBac* that integrate transgenes into the genome.
- » **Microinjection:** A mechanical method to deliver genetic material into insect embryos.
- » **CRISPR-Cas9 Technology:** A genome-editing tool that allows precise and targeted modifications to the DNA sequence.
- » **Viral Vectors:** Engineered viruses used to deliver genes into insect cells efficiently.

Historical Evolution of Gene Transformation in Insects

The field of gene transformation in insects has undergone remarkable progress since its inception. Each phase of development has contributed to refining techniques and expanding applications.

Early Foundations (1950s–1980s)

1. **Discovery of Transposable Elements:**
   - Transposable elements, or "jumping genes," were first discovered by Barbara McClintock, who identified their ability to move within genomes.
   - In insects, the *P-element* transposon in *Drosophila melanogaster* became a model tool for genetic studies.
2. **First Transformations in Model Organisms:**
   - The first successful transformation of *Drosophila* in the early 1980s involved introducing transgenes using the *P-element*.
   - This marked a breakthrough in genetic engineering, demonstrating that exogenous DNA could be stably integrated into insect genomes.
3. **Applications in Basic Research:**
   - Early gene transformation studies focused on fundamental questions, such as understanding developmental pathways and genetic regulation in insects.

Expansion and Refinement (1990s–2000s)

1. **Development of Additional Transposon Systems:**
   - New transposons, such as *piggyBac*, *Hermes*, and *Minos*, were developed to broaden the range of species amenable to genetic manipulation.
   - These systems increased transformation efficiency and enabled genetic studies in non-model insects, including mosquitoes and agricultural pests.
2. **Marker Genes and Visualization Techniques:**
   - The incorporation of reporter genes, such as green fluorescent protein (GFP), allowed researchers to identify successfully transformed insects visually.
   - This innovation improved the precision and reproducibility of transformation experiments.

3. **Pest Control Applications:**
   - Transgenic techniques were applied to suppress pest populations. For instance, genetically modified sterile male insects were introduced into populations to control pests like the Mediterranean fruit fly.
   - In mosquitoes, transgenes were designed to reduce their ability to transmit diseases such as malaria and dengue.

Modern Advancements (2010s–Present)

1. **CRISPR-Cas9 Revolution:**
   - The introduction of CRISPR-Cas9 as a gene-editing tool transformed the field, enabling precise modifications of specific genes.
   - This method allowed the creation of gene drives, which promote the inheritance of modified genes, ensuring their rapid spread through wild populations.
2. **Increased Versatility with RNAi:**
   - RNA interference emerged as a complementary tool for silencing specific genes without permanent modifications.
   - This method has been widely used to reduce the fitness of agricultural pests and to study gene function in insect biology.
3. **Synthetic Biology and Designer Genes**:
   - Advances in synthetic biology have enabled the creation of highly specific genetic constructs tailored to address particular challenges, such as disease vector control or pest suppression.
   - For example, genes encoding anti-pathogen peptides were introduced into mosquitoes to disrupt malaria transmission.
4. **High-Throughput Methods:**
   - Automation and high-throughput screening techniques have made large-scale transformations more efficient, facilitating population-level studies and field applications.

Applications of Gene Transformation in Insects

Gene transformation in insects serves a variety of purposes across scientific and practical domains.

1. **Pest Population Suppression:**
   - Genetically engineered insects, such as sterile males, are released to reduce pest populations through incompatible matings.
   - Example: Modified *Aedes aegypti* mosquitoes are used to control dengue and Zika virus vectors.
2. **Disease Vector Managemen**t:
   - Transformed insects are designed to interrupt disease transmission. For instance:

◊ *Anopheles* mosquitoes carrying anti-malarial genes have reduced vector competency.

◊ Gene drives have been developed to spread anti-disease traits through mosquito populations.

3. **Agricultural Pest Control:**
   - Transgenic agricultural pests, such as bollworms and fruit flies, have been modified to enhance control strategies, protecting crops from damage.
   - Resistance genes introduced into insects disrupt their ability to metabolize pesticides, making them more susceptible to control measures.
4. **Developmental and Behavioral Studies:**
   - Gene transformation has provided insights into insect development, physiology, and behavior.
   - By tagging specific genes with fluorescent markers, researchers can track developmental processes in real time.
5. **Bioproduction:**
   - Transgenic insects are used as bioreactors to produce valuable compounds, such as pharmaceuticals, enzymes, or biofuels.

Methodologies for Gene Transformation

The methods used for gene transformation in insects have become increasingly sophisticated, each offering distinct advantages depending on the target species and research goals.

1. **Microinjection:**
   - DNA is injected directly into embryos during early developmental stages.
   - This method remains the gold standard for transforming many insect species.
2. **Transposon-Based Systems:**
   - Transposons like *piggyBac* integrate transgenes into the genome with high efficiency.
   - These systems are particularly effective for non-model insects.
3. **CRISPR-Cas9:**
   - CRISPR allows for precise editing of DNA at target sites, enabling the modification, deletion, or insertion of specific genes.
   - It is widely used for creating functional knockouts and studying gene roles in development and disease transmission.
4. **Electroporation and Protoplast Transformation:**
   - These methods are used to deliver genetic material into cells using electrical pulses or enzymatically isolated cells.

5. **Viral Vectors:**
    - Engineered viruses serve as delivery vehicles for genetic material, ensuring efficient transformation of large populations.

Key Examples of Gene Transformation

1. **Mosquitoes for Malaria Control:**
    - *Anopheles* mosquitoes were engineered to express genes that block the malaria parasite's development in the gut.
    - These mosquitoes have shown promise in reducing malaria transmission in controlled trials.
2. **Fruit Flies in Agriculture:**
    - Sterile male fruit flies were developed using transgenic approaches, helping control populations of invasive agricultural pests.
3. **Lepidopteran Pests:**
    - Genes encoding insecticidal proteins, such as *Bt* toxins, have been introduced into pests like cotton bollworms to improve pest management strategies.

1.2 Innovations and Milestones in Genetic Engineering Techniques for Insects

Genetic engineering techniques for insects have evolved significantly over the past several decades, revolutionizing research and applications in pest control, disease vector management, and insect biology. These advancements have transitioned from foundational discoveries to cutting-edge technologies that allow precise, efficient, and large-scale modifications of insect genomes.

1.2.1 Foundational Discoveries in Genetic Engineering

1. **Discovery of Transposable Elements (1950s–1980s):**
    - The discovery of transposable elements, often called "jumping genes," laid the groundwork for insect genetic engineering.
    - In the 1950s, Barbara McClintock's work on maize revealed the existence of mobile genetic elements that can move within a genome.
    - In the 1980s, the *P-element* transposon was identified in *Drosophila melanogaster* and adapted for use as a tool to introduce genetic material into insect genomes.
2. **First Genetic Transformation in Insects (1982):**
    - The first successful genetic transformation in insects was achieved in *Drosophila melanogaster* using microinjection of DNA and the *P-element*.
    - This breakthrough demonstrated that transgenes could be stably integrated into insect genomes and expressed.
3. **Development of Fluorescent Markers (1990s):**
    - The introduction of green fluorescent protein (GFP) and other marker genes allowed researchers to visualize and track genetic modifications in living insects.

- This innovation significantly improved the efficiency of genetic transformation experiments.

1.2.2 Expanding the Toolbox for Genetic Engineering

1. **Advancements in Transposon Systems (1990s–2000s):**
    - Transposons, such as *piggyBac*, *Hermes*, and *Minos*, emerged as versatile tools for genetic transformation in non-model insects.
    - These systems expanded the range of species amenable to genetic manipulation, including mosquitoes, agricultural pests, and beneficial insects like bees.
    - *PiggyBac* gained popularity due to its ability to integrate transgenes into a wide variety of insect genomes with high efficiency.
2. **Gene Targeting with Homologous Recombination (2000s):**
    - Techniques leveraging homologous recombination enabled targeted modifications in specific genomic loci.
    - While initially challenging to apply in insects, this method paved the way for more precise genetic engineering techniques.
3. **RNA Interference (RNAi) (2000s):**
    - RNAi became a powerful tool for gene silencing in insects, allowing researchers to suppress specific genes temporarily.
    - This technique has been widely applied to study gene functions, disrupt pest biology, and develop RNAi-based pest control products.

1.2.3 Precision with Genome Editing

1. **Emergence of CRISPR-Cas9 (2010s):**
    - CRISPR-Cas9 revolutionized genetic engineering by enabling precise and efficient editing of specific DNA sequences.
    - The technology uses a guide RNA (gRNA) to direct the Cas9 protein to a target site in the genome, where it introduces a double-strand break.
    - This break is repaired by cellular mechanisms, allowing for gene knockouts, insertions, or replacements.

Applications of CRISPR-Cas9 in Insects:

- Disruption of disease transmission genes in *Anopheles* mosquitoes to combat malaria.
- Knockout of pesticide resistance genes in agricultural pests like *Helicoverpaarmigera* (cotton bollworm).
- Modification of genes affecting behavior and reproduction to control pest populations.

2. **Gene Drives (Mid-2010s):**
    - CRISPR-Cas9 enabled the development of gene drives, which promote the rapid spread of genetic modifications through wild insect populations.
    - Gene drives have been particularly impactful in efforts to control disease vectors, such as *Aedes aegypti* mosquitoes, by spreading traits like infertility or pathogen resistance.
3. **Base Editing and Prime Editing (Late 2010s):**
    - These next-generation genome-editing tools allow for precise nucleotide changes without introducing double-strand breaks.
    - Base editing has been used to modify insect genes with unparalleled accuracy, opening new avenues for studying gene functions and developing targeted control strategies.

1.2.4 Innovations in Genetic Transformation Methods

1. **Microinjection Techniques (1980s–Present):**
    - Microinjection remains a foundational technique for introducing genetic material into insect embryos.
    - Modern advancements in automation and precision tools have improved the efficiency and scalability of this method.
2. **Electroporation and Liposome-Based Delivery (2000s):**
    - Electroporation uses electrical pulses to create temporary pores in cell membranes, allowing DNA or RNA to enter.
    - Liposome-based systems facilitate the delivery of genetic material encapsulated in lipid vesicles, improving transformation efficiency in certain insect species.
3. **Viral Vectors for Gene Delivery (2000s):**
    - Engineered viruses, such as baculoviruses, have been used to introduce genetic material into insect cells or tissues.
    - Viral vectors are particularly useful for large-scale transformations in industrial or agricultural applications.
4. Aut**omated High-Throughput Systems (2010s):**
    - Robotics and automation have enabled large-scale transformations, allowing for population-level studies and the production of transgenic insects for field release.

1.2.5 Applications of Advanced Genetic Engineering

The innovations in genetic engineering have enabled transformative applications in various fields, including:

1. **Pest Control and Management:**
    - **Sterile Insect Technique (SIT):**
        - ◊ Genetically modified sterile male insects are released to mate with wild populations, reducing pest numbers.

- ◊ Example: Transgenic *Aedes aegypti* mosquitoes used to control dengue vectors.
- **Gene Drives:**
  - ◊ Rapidly spread traits like infertility or pesticide susceptibility through pest populations to suppress numbers.

2. **Disease Vector Control:**
   - **Pathogen Resistance:**
     - ◊ Mosquitoes genetically modified to resist malaria parasites or dengue viruses reduce transmission rates.
     - ◊ Example: *Anopheles gambiae* with modified genes disrupting malaria parasite development.
   - **Behavioral Modifications:**
     - ◊ Engineering mosquitoes to avoid biting humans reduces disease transmission.
3. **Improved Agricultural Productivity:**
   - **Enhanced Pest Control:**
     - ◊ Agricultural pests like fruit flies and bollworms have been genetically modified to improve control strategies.
   - **Pollinator Protection:**
     - ◊ Genetic engineering ensures pollinator resilience to diseases, preserving their ecological and agricultural roles.
4. **Insect-Based Bioproduction:**
   - Transgenic insects are used as biological factories to produce pharmaceuticals, enzymes, or biomaterials.
   - Example: Silkworms engineered to produce spider silk proteins for high-strength fibers.

1.2.6 Overcoming Technical Challenges

1. **Species-Specific Barriers:**
   - Techniques like CRISPR-Cas9 and transposon systems have been adapted to work across diverse insect species, overcoming the limitations of earlier methods.
2. **Stability of Genetic Transformations**:
   - Advances in integration systems, such as site-specific recombination, ensure stable and long-term expression of transgenes in transformed insects.
3. **Precision and Off-Target Effects:**
   - Tools like CRISPR-Cas9 have been refined to minimize unintended genetic modifications, improving safety and reliability.

1.3 The Significance of Insect Gene Transformation in Agriculture, Medicine, and Industrial Biotechnology

Insect gene transformation has emerged as a powerful tool with transformative implications across multiple domains, including agriculture, medicine, and industrial biotechnology. By enabling precise genetic modifications in insects, this technology addresses challenges such as pest control, disease vector management, and the production of valuable biomaterials.

1.3.1 Significance in Agriculture

The agricultural sector has been one of the primary beneficiaries of insect gene transformation, leveraging the technology to enhance crop productivity and sustainability. Insects play a dual role in agriculture: as destructive pests and as essential pollinators. Gene transformation enables targeted interventions to control harmful pests while protecting beneficial species.

1.3.1.1 Pest Control and Population Suppression

1. **Sterile Insect Technique (SIT):**
   - Genetically transformed sterile male insects are released into wild populations to mate with females, producing no viable offspring and thereby reducing pest populations.
   - Example: *Aedes aegypti* mosquitoes engineered with self-limiting genes have been deployed to control dengue and Zika vectors.
2. **Gene Drives for Pest Suppression:**
   - Gene drives engineered using CRISPR-Cas9 enable the rapid spread of genetic modifications, such as traits that render pests infertile or susceptible to pesticides, throughout populations.
   - Example: *Anopheles gambiae* mosquitoes have been modified to suppress malaria vectors.
3. **Resistance Management:**
   - Insects that are genetically modified to lose their resistance to pesticides or to become susceptible to biological control agents help mitigate resistance issues in pest management.

1.3.1.2 Protection of Crops and Yields

1. **Disruption of Pest Behavior:**
   - Genetic engineering alters pest behavior to minimize damage to crops.
   - Example: Engineered moths and bollworms that fail to reproduce effectively reduce crop infestations.
2. **Targeting Specific Pest Genes:**
   - Techniques like RNAi are used to silence essential genes in pest species, reducing their survival or reproductive capacity without harming non-target organisms.
3. **Reduction in Chemical Pesticide Use:**
   - By reducing pest populations genetically, farmers can decrease their reliance on harmful chemical pesticides, leading to lower environmental contamination and production costs.

1.3.1.3 Pollinator Conservation

1. **Enhancing Disease Resistance in Pollinators:**
   - Bees and other pollinators are critical for crop pollination but face threats from diseases and environmental stressors.
   - Gene transformation is used to enhance pollinators' resistance to diseases, such as viruses or mites, ensuring their survival and continued ecological and agricultural roles.
2. **Behavioral Modifications to Improve Efficiency:**
   - Genetically modified pollinators can be designed to target specific crops more effectively, optimizing pollination rates and yields.

1.3.2 Significance in Medicine

In the field of medicine, insect gene transformation has been instrumental in combating vector-borne diseases, developing therapeutic agents, and producing medical-grade biomaterials.

1.3.2.1 Disease Vector Management

1. **Reduction of Disease Transmission:**
   - Gene transformation reduces the ability of disease vectors to transmit pathogens.
   - Example:
     - ◊ *Anopheles* mosquitoes engineered to resist *Plasmodium* parasites interrupt the malaria transmission cycle.
     - ◊ *Aedes aegypti* mosquitoes with altered genes fail to transmit dengue, Zika, and chikungunya viruses.
2. **Pathogen Resistance in Vectors:**
   - Genetic engineering introduces genes into insects that block the life cycle of pathogens within the vector.
   - Example: Transformed mosquitoes expressing anti-malaria peptides prevent the development of the parasite in their gut.
3. **Tracking and Monitoring Disease Vectors:**
   - Fluorescent markers and other reporter genes help researchers monitor the spread and behavior of genetically modified vectors in the wild, improving control strategies.

1.3.2.2 Therapeutic Agent Production

1. **Insect Bioreactors for Pharmaceuticals:**
   - Transgenic insects are engineered to produce pharmaceutical proteins, such as vaccines, antibodies, and enzymes.
   - Example: Engineered silkworms produce recombinant proteins for therapeutic use, including anticoagulants and growth factors.

2. **Insect-Based Gene Delivery Systems:**
    - Engineered insects or insect-derived systems are used for delivering therapeutic genes to specific tissues or for producing vaccines.

1.3.2.3 Modeling Human Diseases

1. **Functional Genomics Studies:**
    - Gene transformation in model insects, such as *Drosophila melanogaster*, enables the study of genes involved in human diseases, including cancer, neurodegenerative disorders, and metabolic diseases.
2. **Drug Screening and Discovery:**
    - Genetically modified insects are used to test the efficacy of drugs targeting specific pathways relevant to human diseases.

1.3.3 Significance in Industrial Biotechnology

Insects serve as efficient bioreactors for producing high-value materials and compounds in industrial biotechnology. Gene transformation enhances their capacity for these applications, improving production efficiency and scalability.

1.3.3.1 Bioproduction of Materials

1. **Production of Silk and Other Fibers:**
    - Transgenic silkworms are engineered to produce high-strength fibers, such as spider silk, which have applications in medical sutures, military gear, and high-performance textiles.
    - Example: Silkworms expressing spider silk genes produce biomaterials with enhanced tensile strength and elasticity.
2. **Enzyme Production:**
    - Insects are genetically modified to produce industrial enzymes for applications in biofuels, food processing, and waste management.

1.3.3.2 Sustainable Resource Development

1. **Insects as Protein Sources**:
    - Gene transformation improves the nutritional profile of insects bred for human consumption or animal feed, enhancing their sustainability as a protein source.
2. **Biofuel Production:**
    - Engineered insects produce lipids or other biofuel precursors, contributing to renewable energy development.

1.3.3.3 Waste Bioconversion

1. **Improved Waste Decomposition:**
    - Transgenic insects, such as black soldier flies, are modified to enhance their efficiency in breaking down organic waste and converting it into high-value compost or feed.

2. **Pollutant Bioremediation:**

   - Engineered insects can metabolize and detoxify environmental pollutants, offering eco-friendly solutions for waste management.

2: Biotechnology in Relation to Silkworms and Honey Bees

Biotechnology has revolutionized our understanding and utilization of insects, particularly silkworms (*Bombyx mori*) and honey bees (*Apis mellifera*). These two species are not only integral to agriculture and industry but also crucial to ecological balance and biodiversity. Genetic engineering has enhanced their productivity, disease resistance, and adaptability, addressing global challenges in agriculture, textile production, and environmental conservation.

2.1 Silkworms: Genetic Engineering to Enhance Silk Production

Silkworms are the backbone of sericulture, producing silk that is not only a luxury commodity but also a high-demand material for medical, industrial, and technological applications. Advances in biotechnology have transformed silkworm production, enabling researchers to modify genetic pathways for higher yield, improved fiber properties, and enhanced disease resistance.

2.1.1 Enhancing Silk Yield and Quality

The advancements in biotechnology have transformed silkworm (*Bombyx mori*) sericulture, enabling unprecedented improvements in silk yield and quality. By manipulating the genetic pathways governing silk production, researchers have optimized both the quantity and structural properties of silk fibers. These innovations not only address the growing demand for high-quality silk in industries ranging from textiles to medicine but also provide sustainable solutions to improve the economics of sericulture.

2.1.1.1 Silk Yield Optimization Through Gene Overexpression

Silkworm silk production relies on two primary proteins, fibroin and sericin, which work synergistically to create the structural framework and adhesive matrix of silk fibers. Genetic engineering approaches focus on increasing the expression of these proteins to enhance overall silk yield.

1. Fibroin: The Structural Backbone of Silk

Fibroin constitutes about 75% of the silk thread and is primarily responsible for its strength and luster. Advances in genetic engineering have enabled the overexpression of fibroin genes, resulting in increased production of this key structural protein.

» **Gene Overexpression:**

- By introducing regulatory elements that upregulate fibroin synthesis, researchers have developed silkworm strains capable of producing up to 50% more fibroin.
- Example: Transgenic silkworm lines engineered with enhanced heavy chain fibroin (*H-Fib*) expression exhibit a significant increase in cocoon weight and silk thread length.

» **Impact on Silk Yield:**

- Traditional silkworm strains produce approximately 300–400 meters of silk thread per cocoon. Engineered strains exceed this benchmark, producing up to 600 meters per cocoon, thereby doubling productivity.

2. Sericin: Optimizing Adhesive Properties

Sericin, which constitutes the outer coating of silk fibers, acts as an adhesive, holding fibroin filaments together during cocoon formation.

- » **Modulation of Sericin Pathways:**
  - Researchers have manipulated sericin production pathways to enhance its adhesive properties, ensuring better thread formation during reeling and improved efficiency in the silk-reeling process.
  - Optimized sericin reduces thread breakage during processing, resulting in higher-quality silk.

3. Economic and Environmental Benefits

- » Enhanced silk yield reduces the number of silkworms required to meet production demands, lowering the environmental footprint of sericulture.
- » Higher productivity also improves profitability for silk farmers, particularly in regions where sericulture forms the backbone of rural economies.

2.1.1.2 Improving Silk Fiber Strength and Elasticity

The tensile strength and elasticity of silk fibers make them invaluable not only in the textile industry but also in fields such as medicine and aerospace. Advances in biotechnology have focused on improving these properties through targeted genetic modifications.

1. Enhancing Tensile Strength

Silk's tensile strength is attributed to the β-sheet crystalline structures formed by fibroin proteins. Genetic engineering has further enhanced these molecular arrangements.

- » **Modifications to Fibroin Genes:**
  - By altering the amino acid composition of fibroin, researchers have increased hydrogen bonding within the protein structure, creating fibers that are stronger and more durable.
  - Example: Transgenic silkworm lines with a modified fibroin sequence produce silk with 20% greater tensile strength, suitable for heavy-duty applications.

2. Increasing Elasticity

Elasticity is critical for silk's adaptability in applications requiring flexibility, such as medical sutures and composite materials.

- » **Targeted Gene Editing:**
  - Gene editing tools like CRISPR-Cas9 have been employed to modify fibroin genes, enhancing the polymer's flexibility without compromising strength.

- The resulting silk exhibits improved elasticity, making it ideal for applications in biocompatible materials.

3. Applications of Improved Silk

» **Medical Applications:**

- High-strength, elastic silk fibers are used in sutures, wound dressings, and scaffolds for tissue engineering.
- Example: Modified silk threads provide superior knot security and tensile strength, ensuring better healing outcomes.

» **Industrial Applications:**

- Enhanced silk fibers are utilized in the manufacture of parachutes, bulletproof vests, and flexible yet durable composite materials for aerospace engineering.

4. Real-World Case Studies

» **Hybrid Silk Development:**

- Collaboration between research institutions and textile industries has led to the development of hybrid silk fibers tailored to specific industrial requirements.
- Example: A Japanese biotech firm has commercialized high-strength silk for use in medical sutures and high-performance sportswear.

Impact of Enhanced Silk Yield and Quality

The genetic engineering of silkworms to optimize silk yield and improve fiber properties represents a paradigm shift in sericulture.

1. **Economic Gains:**
   - Enhanced productivity translates into higher profitability for silk farmers and manufacturers.
2. **Sustainability:**
   - Reduced reliance on chemical treatments and lower resource input requirements align with global sustainability goals.
3. **Technological Advancements:**
   - The development of tailored silk fibers meets the needs of high-tech industries, expanding silk's applications far beyond traditional textiles.

2.1.2 Incorporation of Spider Silk Genes

Spider silk, known for its extraordinary mechanical properties, is often regarded as nature's "super fiber." It possesses exceptional tensile strength, elasticity, and durability, surpassing many synthetic materials such as Kevlar. Despite these advantages, spider silk production at scale has been hindered by spiders' territorial and cannibalistic behavior, making traditional farming infeasible. Advances in biotechnology have resolved this challenge by introducing spider silk genes into silkworms, a well-established silk-producing organism. This genetic modification enables the mass production of hybrid silk fibers that retain the desirable qualities of spider silk while leveraging the efficiency of silkworm sericulture.

2.1.2.1 Production of Hybrid Silk Fibers

The incorporation of spider silk genes into silkworms has been a significant breakthrough in genetic engineering, transforming the silk production landscape.

1. **Key Spider Silk Proteins: MaSp1 and MaSp2**
    - The dragline silk of spiders, primarily composed of major ampullate spidroins MaSp1 and MaSp2, is responsible for its unique strength and elasticity.
    - These proteins form β-sheet nanocrystals and amorphous regions, giving spider silk its impressive mechanical properties.
2. **Genetic Engineering of Silkworms for Spider Silk Production**
    - Using advanced molecular techniques, researchers have successfully transferred spider silk genes into the genomes of silkworms.
    - Transgenic silkworms are engineered to express MaSp1 and MaSp2 proteins in their silk glands, enabling them to produce fibers with properties similar to spider silk.

Process:

» Spider silk genes are synthesized and optimized for expression in silkworms.

» These genes are then inserted into silkworm genomes using techniques such as transposon-based transformation or CRISPR-Cas9.

» The transgenic silkworms produce silk fibers enriched with spider silk proteins during cocoon formation.

3. **Enhanced Mechanical Properties of Hybrid Silk**
    - Fibers produced by these genetically engineered silkworms exhibit tensile strengths up to five times greater than regular silkworm silk.
    - The incorporation of spider silk proteins also improves elasticity and toughness, making the fibers suitable for high-performance applications.
4. **Scalability of Production**
    - Silkworms, already domesticated for large-scale silk production, serve as an efficient bioreactor for spider silk protein synthesis.
    - This approach overcomes the challenges of spider farming, providing a sustainable and economically viable solution for spider silk production.

2.1.2.2 Applications of Spider Silk-Enhanced Fibers

The unique properties of spider silk-enhanced fibers have opened new avenues for their use in diverse fields, from medicine to industrial manufacturing.

1. Medical Applications

Spider silk's biocompatibility and exceptional mechanical properties make it an ideal material for various medical applications:

» **Bio-Compatible Sutures:**

- Hybrid silk fibers are used to create surgical sutures that are strong, elastic, and compatible with biological tissues.
- These sutures exhibit minimal tissue reaction and biodegrade naturally, reducing the risk of complications.

» **Tissue Engineering and Scaffolds:**

- Spider silk-enhanced fibers are used to fabricate scaffolds for tissue engineering, supporting cell growth and tissue regeneration.
- Example: Hybrid silk scaffolds are being explored for regenerating cartilage, skin, and nerve tissues.

» **Wound Dressings:**

- The antimicrobial properties of spider silk make it suitable for advanced wound dressings, promoting healing and preventing infections.

2. Industrial Applications

The superior strength and elasticity of spider silk fibers have made them invaluable in industrial sectors requiring high-performance materials:

» **Bulletproof Vests and Military Gear:**

- Hybrid silk fibers are used in lightweight bulletproof vests and protective clothing, offering enhanced protection without compromising mobility.
- Example: Military research programs are exploring spider silk for creating next-generation armor.

» **Climbing Ropes and Cables:**

- High-strength spider silk fibers are ideal for manufacturing climbing ropes, cables, and other load-bearing materials, offering a combination of durability and flexibility.

» **Aerospace Components**:

- The lightweight and high-strength properties of spider silk make it suitable for aerospace applications, such as composite materials for aircraft and space vehicles.

3. Specialized Applications

» **Sportswear and Performance Fabrics:**

- A Japanese biotechnology company has developed hybrid silk fibers for creating sportswear and high-performance fabrics.

- These materials are lightweight, durable, and moisture-wicking, catering to the needs of athletes and outdoor enthusiasts.

» **Surgical Mesh:**

- Hybrid silk is used in manufacturing surgical mesh for hernia repair and other procedures, offering superior strength and compatibility with human tissues.

Example of Successful Development: Japanese Biotechnology Company

A notable example of the successful application of spider silk-enhanced fibers is a Japanese biotech company that has pioneered the use of transgenic silkworms to produce hybrid silk.

» **Innovations:**

- The company integrated MaSp1 and MaSp2 genes into silkworm genomes, creating a scalable production system for spider silk fibers.
- These fibers have been commercialized for various applications, including sports apparel and medical devices.

» **Impact:**

- The technology has demonstrated the feasibility of using silkworms as bioreactors for high-performance silk production, bridging the gap between laboratory research and industrial application.

Broader Implications of Spider Silk Gene Incorporation

1. **Economic Benefits:**
   - Hybrid silk production reduces reliance on synthetic materials, lowering production costs and environmental impact.
   - The commercialization of spider silk fibers creates new revenue streams for sericulture industries.
2. **Environmental Sustainability:**
   - Spider silk production through genetically modified silkworms is eco-friendly, requiring fewer resources than synthetic fiber manufacturing.
   - Biodegradable hybrid silk fibers align with global sustainability goals.
3. **Advancements in Genetic Engineering:**
   - The success of spider silk gene incorporation showcases the potential of genetic engineering in creating novel biomaterials.
   - This breakthrough has set the stage for future innovations in silk production and beyond.

2.1.3 Disease Resistance in Silkworms

Silkworm diseases represent a significant challenge in sericulture, causing widespread economic losses and disrupting the global silk supply chain. Viruses, bacteria, fungi, and microsporidia are the major pathogens that threaten silkworm populations, leading to reduced cocoon production and compromised silk quality. Advances in genetic engineering, including CRISPR-Cas9 and RNA interference (RNAi), have provided innovative solutions to enhance disease resistance in silkworms. These breakthroughs ensure stable production, reduce reliance on chemical treatments, and safeguard the livelihoods of sericulture-dependent communities.

2.1.3.1 Economic Losses from Silkworm Diseases

1. **Impact of Silkworm Diseases on Sericulture**
   - Silkworm diseases are a persistent threat, causing up to 30–50% losses in cocoon yield in severely affected areas.

» Key pathogens include:

- **Viral Infections:**
  - ◊ The *Bombyx mori* nuclear polyhedrosis virus (BmNPV) is one of the most devastating viral pathogens, causing high mortality rates and disrupting silk production.
- **Bacterial Infections:**
  - ◊ *Streptococcus* species and *Bacillus thuringiensis* can lead to septicemia and gut infections, weakening silkworms and reducing cocoon quality.
- **Fungal Pathogens:**
  - ◊ Fungal infections, such as muscardine caused by *Beauveria bassiana*, result in mummified larvae and significant economic losses.
- **Microsporidia:**
  - ◊ *Nosema bombycis* is a microsporidian parasite that causes pebrine disease, leading to high mortality rates and production failures.

2. **Economic Impact**
   - In countries heavily reliant on sericulture, such as India and China, silkworm diseases disrupt rural livelihoods and national silk production.
   - Annual losses from diseases can amount to millions of dollars, making disease management a priority for sericulture industries.

3. **Challenges with Traditional Disease Control**
   - Chemical treatments and antibiotics used to manage silkworm diseases can lead to residues in silk products and environmental contamination.
   - Preventative measures, such as sterilization of rearing environments, are labor-intensive and often insufficient to control outbreaks.

4. **Need for Genetic Solutions**
   - Genetic engineering offers a sustainable alternative by developing silkworm strains with innate resistance to pathogens.
   - These innovations reduce the need for chemical treatments, ensure consistent production, and protect the environment.

2.1.3.2 CRISPR and RNAi Technologies

Modern genetic engineering tools, including CRISPR-Cas9 and RNAi, have revolutionized the development of disease-resistant silkworms. These technologies provide precise and effective methods to enhance silkworm immunity and reduce susceptibility to pathogens.

1. CRISPR-Cas9 for Silkworm Disease Resistance

The CRISPR-Cas9 system has emerged as a powerful genome-editing tool for silkworms, enabling precise modifications to genes involved in susceptibility or immunity.

» **Knocking Out Susceptibility Genes:**

- Researchers use CRISPR-Cas9 to disrupt genes that make silkworms vulnerable to pathogens.
- Example: By knocking out the *BmNPV receptor gene*, scientists have developed silkworm strains resistant to this lethal virus.
- These modifications ensure that viruses cannot bind to host cells, effectively blocking infection.

» **Enhancing Immune Genes:**

- CRISPR-Cas9 has also been used to upregulate genes involved in the silkworm immune response, boosting resistance to bacterial and fungal pathogens.

» **Advantages of CRISPR-Cas9:**

- Precision: Edits are made at specific genomic loci, minimizing off-target effects.
- Durability: Modifications are heritable, ensuring long-term resistance in breeding programs.
- Efficiency: The tool allows rapid development of resistant silkworm strains, addressing outbreaks in a timely manner.

2. RNAi for Pathogen Control

RNA interference (RNAi) is another transformative technology for silkworm disease management. Unlike CRISPR, which makes permanent genetic changes, RNAi silences genes temporarily by degrading messenger RNA (mRNA) transcripts.

» **Silencing Pathogen-Associated Genes:**

- RNAi is used to silence genes essential for pathogen survival or replication.
- Example: By targeting BmNPV replication genes, RNAi effectively reduces viral loads in infected silkworms.

» **Enhancing Silkworm Immunity:**

- RNAi can also target silkworm genes involved in pathogen susceptibility, temporarily boosting resistance during infection.
- Example: RNAi-mediated silencing of gut receptor genes reduces susceptibility to bacterial infections.

» **Advantages of RNAi:**

- Flexibility: RNAi can be applied as needed, providing a non-permanent resistance mechanism.
- Specificity: The technology targets specific pathogen genes, reducing non-target effects.
- Application: RNAi can be delivered through feeding, simplifying its use in large-scale sericulture operations.

3. Example of Successful Application

Transgenic silkworm lines resistant to microsporidia have been developed using RNAi technology:

» **Mechanism:**

- RNAi targets *Nosema bombycis* genes required for spore germination and infection.
- Silkworms ingest RNAi-coated feed, silencing microsporidian genes and preventing infection.

» **Outcomes:**

- Cocoon yield increased by 30% in treated populations, reducing economic losses significantly.
- The method is environmentally friendly and scalable, offering a practical solution for sericulture.

Broader Implications of Disease-Resistant Silkworms

The development of disease-resistant silkworms has far-reaching implications for the sericulture industry, rural economies, and global sustainability.

1. **Economic Stability**
   - Disease-resistant silkworms ensure consistent cocoon production, stabilizing income for farmers and sericulture industries.
   - Reduced losses from outbreaks enhance profitability and make sericulture a viable livelihood in disease-prone regions.
2. **Environmental Benefits**
   - Genetic solutions reduce the need for chemical treatments, lowering environmental contamination and promoting eco-friendly practices.
3. **Improved Silk Quality**
   - Healthier silkworms produce higher-quality silk, meeting the demands of luxury and industrial markets.
4. **Global Competitiveness**
   - Countries adopting advanced biotechnologies in sericulture gain a competitive edge in the global silk market.

2.2 Honey Bees: Genetic Modification for Improved Resistance

Honey bees are essential pollinators, responsible for pollinating over 70% of the crops consumed globally. However, they face significant challenges, including colony collapse disorder (CCD), Varroa mite infestations, and viral diseases. Biotechnology offers innovative solutions to improve honey bee health and safeguard their role in ecosystems.

2.2.1 Resistance to Varroa Mites

Varroa mites (*Varroa destructor*) are among the most destructive threats to honey bee (*Apis mellifera*) colonies, posing a global challenge to apiculture and agriculture. These ectoparasitic mites feed on the hemolymph of bees, weakening their immune systems, reducing colony vigor, and transmitting harmful pathogens such as deformed wing virus (DWV) and Israeli acute paralysis virus (IAPV). The resulting damage leads to weakened colonies, reduced productivity, and in many cases, complete colony collapse. The scale of the problem is immense, with infestations resulting in losses exceeding $500 million annually in the apiculture industry. Addressing this challenge has been a priority for researchers and beekeepers alike. Advances in genetic engineering and RNA interference (RNAi) have introduced groundbreaking approaches to enhance honey bee resilience and directly target the Varroa mites, offering sustainable solutions to mitigate their impact.

2.2.1.1 Impact of Varroa Destructor on Honey Bees

The detrimental effects of Varroa mites on honey bee colonies are multifaceted, impacting individual bee health, colony dynamics, and the broader agricultural ecosystem.

1. Physical Damage and Nutritional Stress

- Varroa mites attach to honey bees, puncturing their exoskeleton to feed on hemolymph (a bee's equivalent of blood).
- This feeding weakens individual bees by depleting their nutritional reserves and compromising their immune defenses.
- Larvae and pupae are particularly vulnerable, as mite infestations during their developmental stages lead to stunted growth, deformities (such as deformed wings), and shortened lifespans.

2. Vector for Pathogens

- Varroa mites are efficient vectors for viruses, exacerbating their impact on bee health.
  - **Deformed Wing Virus (DWV):** Bees infected with DWV often exhibit wing deformities, reduced mobility, and early mortality.
  - **Israeli Acute Paralysis Virus (IAPV):** This virus causes paralysis and death, further accelerating colony decline.
- Viral transmission within colonies is amplified by the social nature of bees, leading to rapid spread among individuals.

3. Economic Consequences

- The economic toll of Varroa mite infestations is staggering:

- Annual losses in the apiculture industry exceed $500 million globally, driven by colony mortality, reduced honey yields, and increased costs for mite control measures.
- Farmers relying on pollination services also face significant economic risks due to declining bee populations.

4. Environmental Impact

» Colony losses affect not only the production of honey and other hive products but also the pollination of crops and wild plants, disrupting ecosystems and reducing biodiversity.

» A decline in pollination services leads to reduced yields of fruits, nuts, and vegetables, compounding food security concerns.

2.2.1.2 Genetic Engineering for Mite Resistance

Genetic engineering has emerged as a powerful tool to enhance honey bees' natural defenses against Varroa mites. By introducing genes associated with behaviors or physiological traits that counter mite infestations, researchers aim to create more resilient bee populations.

1. Enhancing Hygienic Behavior

» Hygienic behavior refers to the ability of honey bees to detect and remove infested or diseased brood from their hives, interrupting the life cycle of mites.

» Researchers have identified and introduced genes that enhance this behavior in honey bees:

- **Odor Detection Genes:** Enhanced expression of odorant receptor genes improves bees' ability to detect chemical signals associated with mite infestations.
- **Behavioral Response Genes:** Modifications to genes linked to grooming and brood inspection behaviors lead to more effective removal of mites.

2. Transgenic Bees with Enhanced Grooming Behavior

» Grooming behavior is another critical defense mechanism, where bees remove mites from their own bodies or the bodies of their hive-mates.

» Transgenic bees have been developed to exhibit increased grooming frequency and efficiency:

- **Outcomes in Experimental Colonies:**
  - ◊ Studies have shown that colonies with enhanced grooming behaviors experience significantly reduced mite loads, improving overall colony health.
  - ◊ Example: In one trial, transgenic bees demonstrated a 40–50% reduction in mite populations compared to non-modified controls.

3. Advantages of Genetic Engineering

» **Durability:** Genetically engineered traits are heritable, ensuring long-term resistance across bee generations.

» **Sustainability:** These innovations reduce reliance on chemical miticides, which can harm bees and lead to environmental contamination.

- » **Compatibility:** Genetically engineered bees can integrate into existing hives without disrupting colony dynamics.

2.2.1.3 RNAi-Based Approaches for Mite Control

RNA interference (RNAi) represents a complementary approach to genetic engineering, targeting the Varroa mites directly rather than modifying the bees themselves. RNAi silences specific genes in the mites, disrupting their vital processes and reducing their ability to harm bees.

1. Mechanism of RNAi Technology

- » RNAi works by introducing double-stranded RNA (dsRNA) molecules that match the sequence of target genes in Varroa mites.
- » When mites ingest or come into contact with these dsRNA molecules, the target genes are silenced, impairing essential physiological functions such as reproduction or feeding.

2. Target Genes in Varroa Mites

- » **Reproductive Genes:** Silencing genes involved in egg production reduces mite fertility, limiting population growth within hives.
- » **Feeding-Associated Genes:** RNAi targeting genes essential for mite feeding weakens their ability to extract hemolymph from bees, ultimately leading to mite starvation.

3. Delivery Methods for RNAi

- » RNAi treatments are delivered to bees through their food or water:
  - Bees ingest dsRNA-treated feed, which is absorbed by the mites when they feed on the bees' hemolymph.
  - The dsRNA molecules then activate the RNAi pathway in mites, silencing the target genes.

4. Applications and Effectiveness

- » **Experimental Outcomes:**
  - RNAi-treated colonies show a 70–80% reduction in mite populations, significantly alleviating stress on the bees.
  - These reductions translate into improved colony health, higher survival rates, and increased productivity.
- » **Practical Applications:**
  - RNAi-based treatments are easy to administer and compatible with existing beekeeping practices, making them accessible for widespread use.

Broader Implications of Mite Resistance

The development of mite-resistant honey bees and RNAi-based mite control measures has far-reaching implications for apiculture, agriculture, and environmental sustainability.

1. Economic Benefits

» Enhanced resistance to Varroa mites reduces colony losses, ensuring consistent honey production and pollination services.

» Beekeepers experience lower costs for mite management, improving the profitability of apiculture operations.

2. Improved Crop Pollination

» Healthier bee populations ensure reliable pollination for crops, boosting yields of fruits, nuts, and vegetables.

» This contributes to food security and supports the agricultural economy.

3. Environmental Sustainability

» Genetic and RNAi-based approaches minimize the use of chemical miticides, reducing environmental contamination and preserving biodiversity.

» These methods align with sustainable beekeeping practices, promoting harmony between agriculture and natural ecosystems.

2.2.2 Disease Resistance

Honey bees (*Apis mellifera*), vital pollinators in agriculture and ecosystems, face significant threats from various pathogens, including viruses, bacteria, and fungi. These pathogens exacerbate colony collapse disorder (CCD), a global phenomenon characterized by the sudden and unexplained loss of bee populations. Among these threats, viral infections such as deformed wing virus (DWV) and Israeli acute paralysis virus (IAPV) stand out as major contributors to colony declines, particularly when coupled with other stressors like Varroa mite infestations and environmental changes. Advanced biotechnological tools, such as genetic engineering and CRISPR-Cas9, offer groundbreaking solutions to enhance honey bee immunity and resilience, providing a sustainable path to protect bee populations and ensure their continued role in agriculture and biodiversity.

Combating Viral Infections

Viral diseases represent a particularly destructive class of pathogens affecting honey bee colonies. These infections, often compounded by other factors like parasitic infestations, weaken individual bees, reduce colony productivity, and frequently lead to widespread mortality. Addressing these viral threats has become a critical focus of modern apiculture research.

Major Viral Threats

1. **Deformed Wing Virus (DWV):**
   - DWV is one of the most prevalent and damaging viruses affecting honey bees.
   - It is closely associated with Varroa mite infestations, as the mites act as vectors for the virus, transmitting it directly into the bees' hemolymph.
   - Bees infected with DWV exhibit visible deformities, such as shriveled wings, reduced

flight ability, and shortened lifespans. These impairments significantly diminish the productivity and survival of the colony.

2. **Israeli Acute Paralysis Virus (IAPV):**
   - IAPV causes rapid paralysis and death in honey bees, leading to large-scale colony losses in a matter of weeks.
   - This virus is particularly problematic in regions with high Varroa mite prevalence, as the mites facilitate its spread within and between colonies.
3. **Black Queen Cell Virus (BQCV):**
   - BQCV primarily affects queen larvae, disrupting the reproductive cycle of colonies and undermining their ability to sustain population growth.

Genetic Engineering Solutions for Antiviral Resistance

Modern genetic engineering approaches offer novel solutions to combat these viral threats by directly enhancing honey bees' innate immunity.

1. **Introduction of Antiviral Genes:**
   - Researchers have engineered honey bees to express antiviral proteins capable of inhibiting viral replication.
   - These proteins target key viral enzymes or structural components, preventing the spread of infection within the colony.
2. **Enhancing Innate Immune Pathways:**
   - Genetic modifications focus on strengthening the bees' natural immune responses, particularly the Toll and Imdsignaling pathways, which play crucial roles in antiviral defense.
   - Example: Transgenic bees with enhanced immune signaling exhibit significantly lower viral loads and improved survival rates during viral outbreaks.

Impact of Enhanced Antiviral Resistance

1. **Colony Stability and Productivity**:
   - Colonies with engineered antiviral traits demonstrate reduced mortality during viral outbreaks, maintaining stability and productivity.
   - Healthier bees are better equipped to perform essential tasks, such as foraging and brood care, ensuring the colony's long-term success.
2. **Sustainability of Pollination Services:**
   - By reducing the impact of viral diseases, these innovations help maintain the availability of honey bees for critical pollination services, supporting agricultural productivity and food security.

Improving Colony Resilience

Beyond viral infections, honey bee colonies are also vulnerable to bacterial and fungal pathogens, which further threaten their health and productivity. Genetic engineering, particularly CRISPR-Cas9, has enabled precise modifications to honey bee genomes, improving their ability to resist these threats and enhancing overall colony resilience.

CRISPR-Based Modifications

CRISPR-Cas9, a revolutionary genome-editing tool, allows researchers to make precise alterations to honey bee DNA, targeting genes associated with susceptibility to pathogens or enhancing immune system function.

1. **Targeting Susceptibility Genes**:
   - CRISPR is used to knock out genes that increase bees' vulnerability to bacterial and fungal infections.
   - Example: Genes associated with susceptibility to *Nosema* fungi, a common cause of microsporidian infections, have been successfully silenced using CRISPR, reducing the severity of these infections in experimental colonies.
2. **Enhancing Immune Responses:**
   - Researchers employ CRISPR to upregulate genes linked to key immune pathways, such as prophenoloxidase activation and antimicrobial peptide production.
   - These modifications strengthen the bees' defenses against a wide range of pathogens, ensuring better colony health.

Enhancing Antifungal Activity

Fungal pathogens such as *Ascosphaeraapis*, which causes chalkbrood disease, pose a significant threat to honey bee larvae. Genetic engineering has been instrumental in developing bees with enhanced antifungal properties.

1. **Improved Disease Resistance:**
   - Transgenic bees engineered to produce high levels of antifungal peptides are better equipped to eliminate fungal spores from the hive environment.
2. **Experimental Success:**
   - Example: In controlled studies, transgenic bees with enhanced antifungal activity exhibited survival rates 25–30% higher than non-modified colonies in pathogen-rich environments.
   - These colonies also maintained healthier brood patterns and produced higher yields of honey and other hive products.

Benefits of Improved Resilience

1. **Healthier Colonies:**
    - Enhanced immunity reduces the prevalence of bacterial and fungal pathogens within hives, leading to stronger and more productive colonies.
2. **Reduced Chemical Dependency:**
    - Genetically engineered bees require fewer chemical treatments for disease management, lowering costs for beekeepers and minimizing environmental contamination.
3. **Economic and Environmental Impact:**
    - Disease-resistant colonies improve the sustainability of apiculture, supporting rural livelihoods and contributing to the preservation of ecosystems.

Implications of Disease-Resistant Honey Bees

The development of disease-resistant honey bees has far-reaching implications for agriculture, apiculture, and environmental conservation.

1. **Ensuring Pollination Services:**
    - Healthy bee populations are essential for the pollination of crops, ensuring stable agricultural yields and food security.
    - Disease-resistant colonies contribute to reliable pollination services, reducing the risk of crop failures caused by declining bee populations.
2. **Economic Stability for Beekeepers:**
    - Reduced colony losses and lower treatment costs improve the profitability of apiculture, benefiting beekeepers and associated industries.
3. **Sustainability of Apiculture Practices:**
    - By reducing dependence on chemical treatments, genetic solutions align with environmentally sustainable practices, preserving biodiversity and ecosystem health.

2.3 Role of Transgenic Insects in Pollination, Ecosystem Health, and Commercial Applications

The contributions of transgenic silkworms and honey bees extend beyond immediate benefits to sericulture and pollination, influencing broader ecological, agricultural, and industrial domains.

2.3.1 Enhanced Pollination and Ecosystem Stability

Honey bees (*Apis mellifera*) are indispensable pollinators, playing a critical role in the agricultural economy and maintaining ecosystem stability. They facilitate the pollination of over 70% of the world's crop species, including fruits, vegetables, nuts, and oilseeds, contributing to global food security. Beyond agriculture, honey bees are essential for the survival of wild plants and the ecosystems that depend on them. Advances in genetic engineering have opened new possibilities to enhance pollination efficiency and improve ecosystem stability by addressing challenges such as declining bee populations, resource competition, and suboptimal crop pollination. These innovations not only boost agricultural productivity but also support biodiversity and ecological health.

2.3.1.1 Optimizing Pollination Efficiency

One of the key applications of biotechnology in honey bees is the development of genetically modified bees that exhibit enhanced pollination behaviors, tailored to meet the specific needs of agricultural crops. These modifications aim to maximize crop yields while minimizing the inefficiencies caused by resource competition and suboptimal foraging behaviors.

1. Enhancing Crop-Specific Pollination

1. **Selective Foraging Behavior:**
    - Genetic engineering enables the modification of honey bee foraging preferences, directing them toward specific crops during pollination.
    - This targeted approach improves pollination rates for high-value crops while reducing the time and resources spent on non-target plants.
2. **Example: Almond Blossoms**
    - In regions where almond cultivation is a key agricultural activity, genetically modified bees have been engineered to prefer almond blossoms.
    - These bees exhibit a heightened sensitivity to the chemical cues of almond flowers, such as scent and color, prompting them to prioritize almonds over competing wild plants.
    - **Impact:**
        - ◊ Studies have shown that colonies of modified bees increased almond yields by 20%, demonstrating the potential of this approach to enhance agricultural productivity.

2. Benefits of Optimized Pollination

1. **Improved Crop Yields:**
    - Crops that receive consistent and efficient pollination produce higher yields and better-quality fruits, seeds, and nuts.
    - Enhanced pollination efficiency ensures uniform flowering and fruit set, reducing the risks of poor harvests caused by inadequate pollinator activity.
2. **Reduced Resource Competition:**
    - By prioritizing target crops, genetically modified bees reduce resource competition with wild plants, ensuring more efficient use of available floral resources.
    - This also minimizes the need for artificial pollination techniques, such as hand-pollination, which are labor-intensive and costly.
3. **Economic Gains for Farmers:**
    - Enhanced pollination translates directly into higher revenues for farmers, particularly for crops with high market value, such as almonds, apples, and berries.

2.3.1.2 Supporting Biodiversity and Ecosystem Health

The health and stability of ecosystems are intricately linked to the well-being of pollinator populations.

Honey bees, as keystone species, play a vital role in maintaining biodiversity by supporting the reproduction of a wide variety of plants. By enhancing the health and efficiency of honey bees, biotechnology also contributes to the preservation of ecosystems and the wildlife that depends on them.

1. Role of Healthy Bee Populations in Biodiversity

1. **Pollination of Wild Plants:**
   - Many wild plant species rely on honey bees for pollination, as they lack access to alternative pollinators.
   - Healthy honey bee populations ensure the reproduction and genetic diversity of these plants, which form the foundation of terrestrial ecosystems.
2. **Indirect Support for Wildlife Habitats:**
   - The survival of plant species pollinated by honey bees has a cascading effect on wildlife habitats.
   - Animals, including birds, mammals, and insects, depend on these plants for food, shelter, and nesting sites.
3. **Preservation of Rare and Endangered Plants:**
   - In areas where rare or endangered plant species exist, healthy pollinator populations are crucial for their conservation.
   - Enhanced honey bee health and efficiency directly support the survival of these vulnerable plant species.

2. Biotechnological Contributions to Ecosystem Health

1. **Improved Bee Resilience:**
   - Genetically engineered bees with enhanced disease resistance and environmental adaptability are better equipped to withstand stressors such as pathogens, pesticides, and habitat loss.
   - These resilient populations ensure consistent pollination services, maintaining the balance of ecosystems.
2. **Reducing Pollinator Declines:**
   - Biotechnology addresses the root causes of pollinator declines, such as susceptibility to diseases and parasites, contributing to the recovery of bee populations.
   - Restoring bee populations has a ripple effect on ecosystems, stabilizing plant pollinator networks.

3. Broader Ecological Benefits

1. **Ecosystem Stability:**
   - Enhanced pollination efficiency reduces the risks of ecosystem disruption caused by declining bee populations.

- Consistent pollination supports the stability of food webs and nutrient cycles, promoting ecosystem resilience.

2. **Biodiversity Conservation:**
   - By supporting the reproductive success of diverse plant species, honey bees contribute to the preservation of biodiversity across habitats.
   - This biodiversity, in turn, provides ecosystem services such as carbon sequestration, soil stabilization, and water regulation.

Implications of Enhanced Pollination and Ecosystem Stability

1. **Agricultural Sustainability**:
   - Optimized pollination efficiency supports sustainable agriculture by reducing the need for artificial inputs and increasing crop yields.
2. **Environmental Protection:**
   - Healthier bee populations reduce the reliance on chemical pesticides and fertilizers, mitigating their impact on ecosystems.
3. **Global Food Security:**
   - Enhanced pollination services ensure the consistent production of food crops, contributing to global food security in the face of rising population demands.
4. **Ecosystem Restoration**:
   - Biotechnological interventions in honey bee populations play a crucial role in restoring degraded ecosystems and reversing biodiversity loss.

2.3.2 Commercial Applications

The commercial applications of biotechnology in silkworms and honey bees have revolutionized industries reliant on these species, creating transformative opportunities in silk production and agriculture. Through genetic engineering, silkworms have been optimized to produce high-value hybrid silk fibers with enhanced properties, catering to luxury and industrial markets. Similarly, honey bees have been genetically enhanced to improve pollination efficiency and ensure stable agricultural yields. These advancements have addressed longstanding challenges in sustainability, efficiency, and productivity, paving the way for eco-friendly innovations and economic growth.

2.3.2.1 Silk Industry

The silk industry has always been associated with high-quality textiles and luxury goods. With the advent of transgenic silkworms, the production of silk has entered a new era, offering materials that meet the demands of both traditional markets and modern industrial applications. By incorporating biotechnology, silkworms now produce hybrid silk fibers with superior properties such as greater tensile strength, elasticity, and biodegradability.

1. Hybrid Silk for Luxury Markets

The production of hybrid silk through genetic engineering has elevated the status of silk in luxury markets, offering innovative materials that surpass traditional silk in quality and versatility.

1. **Enhanced Fiber Quality and Customization:**
   - Transgenic silkworms are engineered to produce hybrid silk fibers that combine the structural proteins of traditional silkworm silk (fibroin and sericin) with spider silk proteins such as MaSp1 and MaSp2.
   - These modifications result in silk fibers with enhanced tensile strength, elasticity, and durability, meeting the rigorous demands of high-end fashion and design.
   - Example: Hybrid silk garments exhibit improved resistance to wear and tear, ensuring longer-lasting textiles with superior aesthetic appeal.
2. **Impact on Luxury Fashion:**
   - Designers and luxury brands are increasingly adopting hybrid silk for its unique properties, such as its ability to hold intricate patterns and its soft yet durable texture.
   - Enhanced dye absorption and color retention make hybrid silk a preferred material for vibrant, long-lasting designs.
   - Example: Luxury fashion houses have incorporated hybrid silk into their collections, marketing it as a sustainable and innovative alternative to traditional fabrics.
3. **Economic Benefits for High-Value Markets:**

» The premium pricing of hybrid silk generates higher revenues for producers and manufacturers.

» By appealing to environmentally conscious consumers, hybrid silk aligns with the growing demand for sustainable luxury goods, further expanding its market potential.

2. Industrial Applications of Hybrid Silk

The exceptional properties of hybrid silk extend its utility beyond textiles, making it a valuable material for various industrial and technical applications.

1. **Medical Applications:**
   - Hybrid silk is increasingly used in biomedical fields due to its biocompatibility, biodegradability, and mechanical strength.
   - Applications include:
     - ◊ **Surgical Sutures:** Strong and flexible hybrid silk sutures provide better knot security and wound healing outcomes.
     - ◊ **Tissue Scaffolds:** Hybrid silk scaffolds support cell growth and tissue regeneration, offering innovative solutions for tissue engineering.
     - ◊ **Drug Delivery Systems:** The controlled biodegradability of hybrid silk makes it an ideal material for sustained-release drug delivery platforms.

2. **Aerospace and Defense Applications:**
   - Hybrid silk's lightweight and durable properties make it suitable for high-performance materials in aerospace and defense sectors.
   - Applications include:
     - ◊ **Composite Materials:** Hybrid silk fibers are integrated into lightweight composites for use in aircraft and spacecraft.
     - ◊ **Bulletproof Clothing:** The superior tensile strength of hybrid silk is being explored for protective gear, such as bulletproof vests.
3. **Eco-Friendly Packaging and Industrial Materials:**
   - As an alternative to plastic, hybrid silk is being developed for biodegradable packaging materials, aligning with global sustainability initiatives.
   - Its use in flexible electronics and environmentally friendly adhesives further diversifies its industrial applications.

3. Boosting the Sericulture Industry

1. **Economic Impact on Silk-Producing Regions:**
   - The adoption of transgenic silkworms enhances productivity and profitability for silk farmers, particularly in silk-producing countries like China, India, and Japan.
   - Improved cocoon yields and higher-quality fibers contribute to the growth of local and national economies.
2. **Sustainability and Resource Efficiency:**
   - Hybrid silk production requires fewer resources compared to synthetic fibers, reducing the environmental footprint of the silk industry.
   - Transgenic technologies minimize waste and promote sustainable sericulture practices, benefiting both producers and the environment.

2.3.2.2 Agriculture

Honey bees are indispensable for agricultural productivity, playing a vital role in the pollination of crops that sustain the global food supply. Genetic advancements have enhanced honey bee efficiency and resilience, offering significant commercial benefits in agriculture by ensuring stable food production and reducing reliance on artificial pollination methods.

1. Ensuring Stable Food Production

The enhancement of honey bee pollination efficiency through genetic engineering addresses key challenges in modern agriculture, such as inconsistent yields and pollinator decline.

1. **Optimized Pollination Behaviors:**
   - Genetically engineered honey bees are designed to exhibit targeted foraging behaviors, ensuring efficient pollination of high-value crops such as almonds, apples, and berries.

- Example: Modified bees that preferentially visit almond blossoms have increased almond yields by up to 20%, demonstrating their potential to stabilize production in monoculture farming systems.

2. **Improved Yield and Quality:**
   - Enhanced pollination results in higher yields, larger fruits, and better-quality produce.
   - Example: Crops like sunflowers and tomatoes benefit from consistent pollination, producing more uniform and marketable harvests.

3. **Economic Impact:**
   - Improved pollination services directly increase farm profitability, reducing losses associated with poor pollination or crop failure.
   - The economic value of honey bee pollination in global agriculture is estimated to exceed $200 billion annually, with enhanced pollination further boosting this figure.

2. Reducing Reliance on Artificial Pollination

1. **Challenges of Artificial Pollination:**
   - Hand-pollination and other artificial methods are labor-intensive, costly, and often inefficient, making them impractical for large-scale farming.

2. **Benefits of Natural Pollination:**
   - Genetically enhanced honey bees reduce the need for artificial pollination techniques, offering a more cost-effective and sustainable alternative.
   - Example: Greenhouses employing enhanced bees for pollination report increased productivity and reduced labor costs compared to those using manual pollination methods.

3. Supporting Agricultural Sustainability

1. **Resilience to Environmental Stressors:**
   - Enhanced honey bees are more resistant to diseases, parasites, and environmental changes, ensuring consistent pollination services even in challenging conditions.
   - These resilient populations contribute to agricultural sustainability by reducing crop losses and maintaining productivity.

2. **Securing Food Supply Chains:**
   - Reliable pollination services provided by enhanced honey bees ensure the stability of food supply chains, supporting global food security in the face of rising population demands.

Broader Implications of Commercial Applications

1. **Economic Growth and Global Trade:**
   - Biotechnology-driven advancements in silkworms and honey bees contribute to the growth of industries such as sericulture and agriculture, fostering innovation and expanding global trade opportunities.

- These innovations strengthen the competitiveness of silk and agricultural products in international markets, benefiting producers and consumers alike.

2. **Sustainability and Environmental Benefits:**
   - Hybrid silk fibers and enhanced pollination services align with global sustainability goals, reducing reliance on synthetic materials, chemical treatments, and artificial inputs.
   - These eco-friendly solutions promote biodiversity conservation and environmental stewardship, supporting long-term ecological balance.

3. **Innovation and Product Diversification:**
   - Continued advancements in biotechnology create new avenues for product development, expanding the commercial potential of silkworms and honey bees in industries beyond their traditional roles.
   - Example: Hybrid silk is being explored for use in flexible electronics, while enhanced honey bees are being studied for their potential to pollinate novel crop varieties.

3: Introduction of Lectin Genes for Pest Suppression

3.1 The Role of Lectins (Carbohydrate-Binding Proteins) in Pest Resistance

Lectins, known for their precise carbohydrate-binding properties, are pivotal in the natural defense mechanisms of plants and other organisms. Their ability to interact specifically with sugar moieties on glycoproteins and glycolipids has positioned them as powerful agents in pest management strategies. Understanding the multifaceted roles of lectins in pest resistance reveals their potential to transform sustainable agriculture.

Plants synthesize lectins as part of their innate defense system against biotic stressors, including insects, nematodes, and microbial pathogens. These proteins accumulate in tissues vulnerable to attack, such as leaves, seeds, and roots. The binding of lectins to carbohydrate residues on pest gut epithelial cells disrupts digestion by interfering with nutrient absorption. This interaction leads to malnutrition, developmental delays, or mortality in pests. Furthermore, lectins can bind to specific glycan structures on microbial pathogens, thereby inhibiting their attachment or entry into plant cells.One of the most well-known examples of lectin-mediated pest resistance is the role of *Phaseolus vulgaris* (common bean) lectin. This lectin has demonstrated significant anti-insect activity against pests like aphids and weevils. Upon ingestion, it binds to the epithelial cells lining the insect gut, impairing digestive functions and causing structural damage. Similarly, the pea lectin from *Pisum sativum* has shown promise in combating nematode infestations by altering their development and mobility. The versatility of lectins lies in their broad spectrum of activity. While they are naturally present in many plant species, their potential as a pest management tool is amplified through biotechnological advancements. For instance, lectins from non-plant sources, such as fungal and bacterial lectins, also exhibit strong insecticidal properties. These lectins often target glycan structures unique to certain insect pests, reducing the risk of off-target effects on non-target organisms.

The ecological benefits of using lectins for pest resistance are significant. Unlike chemical pesticides, lectins are biodegradable and less likely to persist in the environment, reducing the risk of pollution and harm to beneficial organisms. Additionally, the use of lectin genes in transgenic crops minimizes reliance on synthetic chemicals, contributing to the development of sustainable agricultural practices.However, deploying lectins in pest management is not without challenges. The specificity of lectin activity, while advantageous, may limit their efficacy against pests with different glycan profiles. Additionally, some insect species have evolved mechanisms to counteract the effects of lectins, such as gut proteases that degrade the protein or altered gut epithelium glycan structures that reduce lectin binding. Understanding these mechanisms is critical for optimizing the use of lectins in pest control.

Research into lectins continues to expand, focusing on identifying novel lectins with unique binding properties and improved insecticidal activities. The integration of lectins into pest management strategies requires a multidisciplinary approach, combining molecular biology, agricultural science, and ecology. By leveraging the natural roles of lectins in plant defense, it is possible to enhance crop resilience and reduce dependency on chemical inputs, paving the way for a more sustainable and environmentally friendly agricultural system.

3.2 Genetic Engineering of Insects and Plants to Produce Lectins for Controlling Pest Populations

The advent of genetic engineering has revolutionized the field of pest management by enabling the introduction of lectin genes into plants and, in some cases, insect populations. This biotechnological approach enhances the ability of plants to defend themselves against pests while offering a novel strategy to suppress pest populations directly. The genetic manipulation of organisms to express lectins has emerged as a promising tool for integrated pest management.

Transgenic plants engineered to produce lectins are at the forefront of this approach. By introducing lectin genes from various sources into the genomes of crop plants, scientists have been able to develop varieties with enhanced resistance to a wide range of insect pests. These genetically modified crops express lectins in specific tissues, such as leaves or seeds, where pest attacks are most likely to occur. Upon ingestion, the lectins interfere with the pests' digestive processes, causing mortality or reduced fecundity.

One notable example is the introduction of snowdrop lectin (*Galanthus nivalis agglutinin* or GNA) into crops such as rice, maize, and potatoes. GNA has demonstrated potent insecticidal activity against sap-sucking pests like aphids and planthoppers. These pests are notorious for their role in transmitting plant viruses, and their control through lectin-producing transgenic plants has the dual benefit of reducing crop damage and curbing the spread of viral diseases.In addition to plants, genetic engineering has been applied to pest populations themselves. One strategy involves modifying insect pests to produce lectins that act against their own species. For example, genetically engineered mosquitoes expressing lectins that target the midgut epithelium of larvae can disrupt their development, thereby reducing population densities over time. While this approach is still in the experimental stages, it represents a novel avenue for pest suppression.

The use of lectin genes in genetic engineering also extends to biopesticides. Microorganisms, such as bacteria and fungi, can be engineered to produce lectins that are toxic to specific pests. When applied to crops, these microorganisms act as bioinoculants, delivering lectins directly to the target pests. This approach combines the specificity of lectin activity with the versatility of microbial application, offering an environmentally friendly alternative to chemical pesticides.Despite the promising results, several challenges must be addressed to ensure the success and sustainability of lectin-based genetic engineering. One concern is the potential for off-target effects on non-pest organisms, including beneficial insects like pollinators and natural predators. While lectins are generally selective in their binding, careful assessment of their impact on non-target species is crucial to avoid unintended ecological consequences.

Another challenge is the risk of pests developing resistance to lectins, similar to how they develop resistance to chemical pesticides. This issue underscores the need for integrated pest management strategies that combine lectin-producing crops with other control measures, such as crop rotation, biological control, and habitat management. Additionally, public acceptance of genetically modified organisms (GMOs) remains a significant hurdle, requiring transparent communication of the benefits and risks associated with lectin-based genetic engineering.

Regulatory frameworks also play a critical role in the deployment of genetically engineered plants and insects. The development and commercialization of these technologies must adhere to stringent biosafety and environmental impact assessments to ensure their safe use. Collaborative efforts between researchers, policymakers, and stakeholders are essential to navigate these challenges and unlock the full potential of lectin genes in pest management.By integrating genetic engineering with the natural pest-resistance properties of lectins, it is possible to develop innovative solutions for agriculture. These approaches not only enhance crop protection but also contribute to the sustainability of farming practices by reducing the reliance on chemical inputs and promoting ecological balance. The continued exploration of lectin genes and their applications promises to revolutionize pest management in the years to come.

3.3 Case Studies of Lectin-Based Pest Suppression Strategies and Their Impact on Agriculture

The application of lectin-based pest suppression strategies has been extensively explored through various case studies that highlight their effectiveness, feasibility, and impact on agriculture. These examples provide valuable insights into how lectins can be integrated into pest management programs to address both pest-related challenges and broader agricultural issues, such as food security and environmental sustainability.

One of the most prominent examples of lectin-based pest control involves the use of transgenic crops expressing the snowdrop lectin (*Galanthus nivalis agglutinin* or GNA). Research has demonstrated the efficacy of GNA against several sap-sucking insects, including aphids, whiteflies, and planthoppers, which are responsible for significant crop losses worldwide. In a case study conducted on rice plants, GNA-expressing transgenic lines showed a marked reduction in damage caused by the brown planthopper (*Nilaparvatalugens*), a major pest in Asia. These crops exhibited increased resistance to feeding damage, higher yields, and lower incidences of virus transmission by the pests, underscoring the multifaceted benefits of lectin expression.Another case study focused on the incorporation of bean (*Phaseolus vulgaris*) lectins into potato plants to combat the Colorado potato beetle (*Leptinotarsa decemlineata*). This pest poses a significant threat to potato production due to its ability to develop resistance to chemical insecticides. Transgenic potato plants expressing bean lectins demonstrated reduced feeding and reproduction rates in the beetles, leading to enhanced crop protection and decreased reliance on chemical pesticides.

In Africa, a pilot project utilizing lectin-based pest control targeted the maize stem borer (*Busseolafusca*), a devastating pest of maize crops. Researchers engineered maize varieties to express wheat germ agglutinin (WGA), a lectin known for its broad-spectrum insecticidal activity. Field trials revealed that WGA-expressing maize experienced significantly lower infestation rates and yield losses compared to non-transgenic controls. This success has spurred interest in expanding the use of lectin genes in staple crops across regions heavily impacted by pests.

Lectins have also been utilized in biopesticide formulations as part of integrated pest management (IPM) programs. For instance, Bacillus thuringiensis (*Bt*) strains engineered to produce insecticidal lectins have been developed to target caterpillar pests such as the cotton bollworm (*Helicoverpaarmigera*). These formulations combine the advantages of microbial biopesticides with the specificity of lectin activity, offering a sustainable and targeted approach to pest control. Field trials have shown that these biopesticides are effective in reducing pest populations without harming beneficial insects, such as pollinators and natural predators.

Beyond their direct impact on pest suppression, lectin-based strategies have been found to contribute to broader agricultural benefits. Reduced pest damage translates into improved crop yields and quality, enhancing food security and farmer income. Moreover, the decreased reliance on chemical insecticides helps mitigate environmental pollution, safeguard biodiversity, and promote soil health. These outcomes align with the goals of sustainable agriculture and highlight the potential of lectin-based approaches to address global challenges in food production.

However, the case studies also reveal certain limitations and challenges associated with lectin-based pest control. For example, field trials with transgenic crops expressing high levels of lectins have occasionally shown variable performance under different environmental conditions. Factors such as pest population dynamics, climate, and soil composition can influence the effectiveness of lectin-based resistance. Furthermore, some studies have reported the development of tolerance in pest populations exposed to lectins over multiple generations, emphasizing the need for integrated strategies to delay resistance development.Another concern raised in some case studies is the potential non-target effects of lectins. Although most lectins are highly specific in their binding properties, there is a possibility of unintended impacts on beneficial organisms, such as pollinators or soil microbes. Comprehensive ecological risk assessments are necessary to ensure that lectin-based approaches do not disrupt agroecosystem balance.

In summary, case studies of lectin-based pest suppression strategies underscore their potential to revolutionize agricultural practices. These approaches have demonstrated effectiveness in controlling a wide range of pests, improving crop productivity, and reducing the environmental footprint of pest management. By addressing the limitations identified in these case studies and fostering collaboration among researchers, policymakers, and farmers, lectin-based pest control can be further optimized to meet the demands of sustainable agriculture.

3.4 Challenges and Future Perspectives in Utilizing Lectin Genes

While lectin genes offer immense potential for pest suppression, their application is accompanied by various challenges that must be addressed to ensure their sustainable and effective use. These challenges span scientific, ecological, socio-economic, and regulatory domains, necessitating a comprehensive and multidisciplinary approach to fully realize the benefits of lectin-based pest control strategies.

One of the primary scientific challenges in utilizing lectin genes is ensuring their efficacy across diverse pest populations. Pests often exhibit considerable genetic and physiological diversity, which can affect their susceptibility to lectins. For instance, some pests possess gut proteases capable of degrading lectins before they can exert their toxic effects. Others may exhibit glycan structures on their gut epithelia that reduce lectin binding affinity. Addressing these challenges requires ongoing research to identify and engineer lectins with enhanced stability, specificity, and potency.

Another significant concern is the potential for pests to develop resistance to lectin-based control measures. Similar to the resistance observed with chemical insecticides, prolonged exposure to lectins can lead to the selection of resistant pest populations. For example, some pests may evolve mechanisms to neutralize lectins or bypass their effects. To mitigate this risk, lectin-based strategies must be integrated into broader pest management programs that incorporate crop rotation, natural predators, and other complementary approaches. The development of lectins with multiple modes of action may also help reduce the likelihood of resistance.

Ecological challenges associated with lectin genes primarily involve their impact on non-target organisms. While lectins are generally selective for specific pests, their effects on beneficial insects, such as pollinators and natural enemies of pests, remain a concern. Studies have shown that some lectins can interact with the gut microbiota of non-target organisms, potentially affecting their health and behavior. Comprehensive risk assessments and ecological monitoring are essential to minimize these unintended consequences and ensure that lectin-based strategies do not disrupt agroecosystem balance.Socio-economic factors also play a critical role in the adoption and implementation of lectin-based pest control. Public perception of genetically modified organisms (GMOs) remains a significant barrier in many regions. Concerns about the safety of consuming crops expressing lectins and their potential environmental impacts have led to resistance from consumers, farmers, and policymakers. Transparent communication of the scientific evidence supporting the safety and benefits of lectin-based technologies is crucial to building public trust and acceptance.

Regulatory frameworks for lectin-based pest control present additional challenges. The development and commercialization of transgenic crops expressing lectin genes must navigate complex regulatory processes to ensure their safety for human health and the environment. These processes often involve rigorous testing, extensive documentation, and lengthy approval timelines, which can increase costs and delay the deployment of these technologies. Harmonizing regulatory standards across countries and regions could help streamline the approval process and facilitate the global adoption of lectin-based solutions.

Despite these challenges, the future perspectives for utilizing lectin genes in pest suppression are promising. Advances in synthetic biology and molecular engineering are enabling the design of novel lectins with improved properties, such as enhanced binding specificity, increased stability, and reduced off-target effects. The integration of lectin genes into multi-trait transgenic crops, which combine pest resistance with other desirable traits like drought tolerance or nutrient efficiency, represents a potential game-changer for agriculture.

Another exciting development is the exploration of lectins in combination with other biotechnological tools, such as RNA interference (RNAi) and CRISPR-based gene editing. These tools offer opportunities to create synergistic pest control strategies that leverage the unique strengths of each technology. For example, RNAi could be used to silence genes critical for pest survival, while lectins target their digestive systems, providing a dual mechanism of action.The adoption of lectin-based pest control strategies will also benefit from increased collaboration between researchers, industry stakeholders, and policymakers. By fostering partnerships and investing in education and outreach programs, it is possible to overcome the socio-economic and regulatory barriers associated with these technologies. Initiatives that prioritize smallholder farmers in developing regions, where pest pressures are often highest, can ensure that the benefits of lectin-based approaches are equitably distributed.

4: Advanced Methods and Applications of Insect Gene Transformation

4.1 Overview of Cutting-Edge Technologies: CRISPR-Cas9, RNA Interference, and Transposon-Based Gene Delivery

The advent of advanced molecular biology tools has revolutionized the field of insect gene transformation, allowing researchers to manipulate insect genomes with unprecedented precision and efficiency. Three key technologies—CRISPR-Cas9, RNA interference (RNAi), and transposon-based gene delivery—are at the forefront of this transformative wave, offering unique mechanisms to modify insect traits for applications in pest management, disease control, and beyond.

CRISPR-Cas9 is arguably the most groundbreaking tool in genetic engineering, enabling precise editing of DNA sequences in virtually any organism, including insects. This system leverages a guide RNA (gRNA) to direct the Cas9 enzyme to a specific target sequence in the genome, where it induces a double-strand break. The break can be repaired through non-homologous end joining (NHEJ) or homology-directed repair (HDR), leading to gene knockouts, insertions, or replacements. In insects, CRISPR-Cas9 has been employed to knock out genes responsible for pest resilience, insecticide resistance, and disease vector competence. For example, researchers have used CRISPR to edit genes in mosquitoes that are essential for transmitting malaria-causing *Plasmodium* parasites, significantly reducing the insects' ability to spread the disease.

RNA interference (RNAi) is another powerful technology for gene silencing. It works by introducing double-stranded RNA (dsRNA) molecules corresponding to a target gene, which are processed into small interfering RNAs (siRNAs) by the insect's cellular machinery. These siRNAs guide the degradation of the target mRNA, effectively silencing the gene. RNAi has been widely adopted in pest management to suppress genes critical for insect survival, reproduction, or feeding behavior. For instance, RNAi has been used to target genes in the western corn rootworm (*Diabroticavirgiferavirgifera*), a major pest of maize, resulting in reduced larval feeding and crop damage.

Transposon-based gene delivery, a method predating CRISPR and RNAi, remains a robust and widely used technique for introducing genetic material into insect genomes. Transposons, also known as "jumping genes," are DNA sequences capable of moving from one location in the genome to another. By harnessing this natural property, scientists can deliver transgenes or other genetic constructs into insect genomes. Transposon systems like *piggyBac* and *Hermes* are particularly effective in insects, enabling stable and heritable integration of genes. This technology has been instrumental in creating transgenic insect lines for research, pest control, and disease vector studies.

Each of these technologies has unique strengths and applications. CRISPR-Cas9 excels in precise, targeted editing and creating permanent genetic changes. RNAi is highly effective for transient gene silencing, especially when long-term effects are not required. Transposon-based delivery is well-suited for the stable insertion of large genetic constructs and creating transgenic lines.Despite their transformative potential, these technologies face challenges. Off-target effects remain a concern with CRISPR-Cas9, though advances in gRNA design and Cas9 variants have mitigated these issues. RNAi can be limited by variability in dsRNA uptake and degradation in different insect species. Transposon-based methods, while efficient, sometimes result in unintended genomic insertions. Addressing these challenges through continuous refinement of these tools will enhance their reliability and expand their applicability.

4.2 Applications in Pest Management, Disease Vector Control, and Environmental Monitoring

The applications of advanced insect gene transformation technologies span diverse domains, from pest management to disease control and environmental monitoring. These innovations are reshaping how we address pressing challenges in agriculture, public health, and ecosystem management.

Pest Management:

Gene transformation technologies have been harnessed to create insect strains with reduced capacity for pestilence. RNAi has shown particular promise in targeting agricultural pests like the cotton bollworm (*Helicoverpaarmigera*) and the Colorado potato beetle (*Leptinotarsa decemlineata*). By silencing genes critical for pest survival or reproduction, these technologies offer an eco-friendly alternative to chemical pesticides, reducing the environmental impact of pest control practices.

CRISPR-Cas9 has further expanded the possibilities by enabling gene drive systems in pests. Gene drives bias the inheritance of a particular genetic trait, ensuring its rapid spread through a population. For instance, a CRISPR-based gene drive in fruit flies has been used to suppress populations by propagating sterility genes. This approach holds significant potential for controlling invasive pest species and reducing their impact on crops and ecosystems.

Disease Vector Control:

Insects such as mosquitoes and tsetse flies serve as vectors for devastating diseases like malaria, dengue, and sleeping sickness. Gene transformation technologies have opened new avenues for controlling these vectors by targeting the genes responsible for disease transmission. For example, CRISPR-Cas9 has been used to create malaria-resistant mosquitoes by disrupting genes required for *Plasmodium* parasite development within the mosquito.

Another promising application is the use of RNAi to block virus replication in mosquitoes. By silencing genes critical for the replication of dengue and Zika viruses, RNAi can reduce the capacity of mosquitoes to transmit these diseases. Transposon-based systems have also contributed by introducing genes that render mosquitoes incapable of hosting pathogens, effectively breaking the disease transmission cycle.

Environmental Monitoring:

Gene transformation in insects is not limited to pest and disease control; it also plays a role in environmental monitoring and conservation. For instance, transgenic insects can be engineered to serve as biosensors for detecting environmental pollutants. Fluorescent markers or stress-responsive genes can be integrated into insect genomes to signal the presence of specific chemicals or heavy metals in the environment. This approach offers a cost-effective and real-time method for monitoring ecosystem health.

Moreover, genetic tools are being developed to protect beneficial insect populations, such as pollinators, from environmental stressors. By enhancing the resilience of bees and other pollinators to pathogens, pesticides, and climate change, gene transformation technologies can safeguard pollination services essential for global food production.The integration of gene transformation tools into these applications represents a paradigm shift in how we approach complex problems. These technologies provide highly targeted, efficient, and sustainable solutions, reducing reliance on chemical interventions and preserving ecological balance. However, their implementation must be accompanied by rigorous ethical and ecological considerations to ensure their long-term safety and efficacy.

4.3 Integration of Synthetic Biology to Design Insects with Novel Traits for Industrial and Ecological Benefits

Synthetic biology has emerged as a powerful interdisciplinary field, enabling the design of insects with novel traits for industrial and ecological applications. By combining principles of biology, engineering, and computational modeling, synthetic biology allows for the creation of custom genetic circuits and pathways within insects, transforming them into "living machines" capable of performing specific functions.

One of the most exciting applications of synthetic biology in insects is the development of biofactory insects. These engineered insects are designed to produce high-value products, such as pharmaceuticals, enzymes, and biomaterials. For example, transgenic silkworms have been engineered to produce recombinant spider silk proteins, which are highly sought after for their exceptional strength and flexibility. Similarly, insects like black soldier flies (*Hermetiaillucens*) are being explored as biofactories for producing antimicrobial peptides and enzymes used in industrial processes.

In agriculture, synthetic biology is being applied to design insects that enhance crop productivity and resilience. For instance, pollinator insects like bees can be engineered to resist pathogens, pesticides, or adverse environmental conditions. Additionally, synthetic biology can be used to create insects that deliver beneficial microbes or nutrients to crops, promoting growth and reducing dependency on chemical fertilizers.

Ecologically, synthetic biology enables the creation of insects with traits designed to restore or maintain ecosystem balance. Invasive species, which often disrupt local ecosystems, can be targeted using synthetic biology tools. Gene drives, powered by CRISPR-Cas9, can be used to spread traits that suppress invasive populations or make them less harmful to native species. Alternatively, engineered insects can be deployed to control pest populations without the need for chemical pesticides, reducing their environmental footprint.

Another innovative application is the use of engineered insects for bioremediation. By incorporating genes that allow insects to metabolize pollutants or heavy metals, synthetic biology can create insect species capable of cleaning up contaminated environments. For instance, insects engineered to break down hydrocarbons could be deployed to sites affected by oil spills, providing a natural and cost-effective remediation strategy.The integration of synthetic biology in designing insects also extends to public health. Engineered mosquitoes that are incapable of transmitting diseases, such as malaria and dengue, have already been developed using synthetic biology tools. These innovations have the potential to significantly reduce the global burden of vector-borne diseases while minimizing the use of chemical insecticides.

Despite these promising developments, the integration of synthetic biology into insect design raises important ethical, ecological, and regulatory questions. Concerns about the unintended consequences of releasing genetically engineered insects into the wild, such as their potential impact on non-target species or ecosystems, must be carefully addressed. Public acceptance and trust in these technologies are also critical, requiring transparent communication and stakeholder engagement.To ensure the responsible use of synthetic biology in insect design, robust risk assessment frameworks and regulatory guidelines must be established. These measures should balance innovation with precaution, enabling the safe deployment of engineered insects while minimizing potential risks. Collaboration between scientists, policymakers, and the public will be key to achieving this balance.

Chapter 7

DNA Fingerprinting for Taxonomy and Phylogeny

Introduction

The fields of taxonomy and phylogeny have undergone transformative changes with the advent of molecular biology, particularly through the application of DNA fingerprinting techniques. As traditional methods of classifying and understanding evolutionary relationships relied heavily on morphological characteristics and fossil records, these approaches often encountered limitations due to convergent evolution, incomplete fossil evidence, and subjective interpretations. DNA fingerprinting, a robust and precise molecular tool, has revolutionized the way scientists delineate species boundaries, resolve taxonomic ambiguities, and reconstruct evolutionary histories. DNA fingerprinting refers to a set of molecular techniques used to identify genetic differences among individuals, populations, or species. By analyzing unique genetic markers, researchers can generate "fingerprints" that serve as reliable identifiers, much like a barcode for organisms. These fingerprints provide invaluable data for studying biodiversity, species identification, and phylogenetic relationships. Unlike traditional methods, DNA fingerprinting leverages genetic material that is inherently more stable and less susceptible to environmental influences, offering unparalleled accuracy in classification.

The application of DNA fingerprinting in taxonomy, the science of naming, describing, and classifying organisms, has addressed long-standing challenges in biological research. For centuries, taxonomy was based on observable traits, a practice that often led to misidentifications due to phenotypic plasticity, hybridization, or cryptic speciation. Cryptic species, for example, are groups of organisms that are morphologically indistinguishable but genetically distinct. Identifying such species is critical for conservation biology, as they often represent unique genetic lineages that require protection. DNA fingerprinting provides a molecular lens to uncover these hidden diversities, ensuring more accurate classification and effective biodiversity management.

In phylogenetics, the study of evolutionary relationships among species, DNA fingerprinting has proven equally transformative. The molecular data o*Bt*ained from DNA fingerprints enable the construction of phylogenetic trees, which depict the evolutionary pathways of species. These trees are essential for understanding the origins and diversification of life forms, as well as for addressing questions about the timing and mechanisms of evolutionary events. For instance, the use of DNA

DOI: 10.1201/9781003757405-7

fingerprinting has elucidated relationships between closely related species, resolved debates over evolutionary lineages, and even identified previously unknown connections between distant taxa. The precision offered by DNA fingerprinting has reshaped the understanding of evolutionary biology, providing insights into speciation, gene flow, and adaptation.

One of the most remarkable aspects of DNA fingerprinting is its versatility. Various techniques are employed to generate genetic fingerprints, each with specific applications and advantages. Restriction Fragment Length Polymorphism (RFLP) was one of the earliest methods, relying on the digestion of DNA by restriction enzymes to produce fragments of varying lengths. Amplified Fragment Length Polymorphism (AFLP), Random Amplified Polymorphic DNA (RAPD), and microsatellite analysis followed, offering greater efficiency and resolution. More recently, high-throughput sequencing technologies have enabled researchers to analyze entire genomes, providing comprehensive insights into genetic variation and evolutionary history.

Beyond taxonomy and phylogeny, DNA fingerprinting plays a crucial role in addressing ecological and agricultural challenges, particularly in the genetic improvement of natural enemies. Natural enemies, including predators, parasitoids, and pathogens, are vital components of biological pest control, offering sustainable alternatives to chemical pesticides. However, the efficacy of natural enemies is often compromised by their susceptibility to inebriates such as insecticides, herbicides, and other environmental stressors. Enhancing the tolerance of these organisms through genetic improvement has become a critical area of research, where DNA fingerprinting serves as a foundational tool.

By identifying genetic markers associated with traits like pesticide resistance or environmental tolerance, scientists can selectively breed or genetically modify natural enemies to enhance their resilience. For instance, DNA fingerprinting has been instrumental in identifying pesticide-tolerant strains of beneficial insects such as ladybirds and parasitic wasps, which are deployed in integrated pest management programs. These genetically improved organisms not only survive in environments with residual pesticide presence but also maintain their effectiveness in suppressing pest populations. This approach aligns with the principles of sustainable agriculture, minimizing the ecological footprint of farming practices while ensuring crop protection.

Another notable application of DNA fingerprinting is in the study of genetic diversity within and among populations. Genetic diversity is a cornerstone of species resilience, enabling populations to adapt to changing environmental conditions and resist diseases. By assessing genetic variation, researchers can identify populations at risk of inbreeding, loss of heterozygosity, or genetic bottlenecks. This information is critical for designing conservation strategies and maintaining the health of ecosystems. DNA fingerprinting has, for example, revealed alarming levels of genetic erosion in some natural enemy populations, prompting efforts to reintroduce genetic diversity through crossbreeding or habitat restoration. In the broader context of biodiversity, DNA fingerprinting has also facilitated the discovery of new species and the reclassification of existing ones. In some cases, organisms previously thought to belong to a single species have been revealed to comprise multiple genetically distinct species. These findings have profound implications for conservation biology, as they often highlight the need to protect genetic lineages that were previously overlooked. Similarly, DNA fingerprinting has resolved taxonomic disputes, clarifying relationships among species that were once mired in controversy due to conflicting morphological or ecological data.

The introduction of DNA fingerprinting into the study of phylogeny has also enabled researchers to test and refine evolutionary hypotheses. Molecular data have challenged traditional phylogenetic models based on morphological traits, leading to a more nuanced understanding of evolutionary processes. For example, studies using DNA fingerprinting have shown that convergent evolution, where unrelated species develop similar traits due to similar environmental pressures, is more common than previously thought. This insight has reshaped the interpretation of evolutionary pathways, emphasizing the role of environmental factors in shaping biodiversity.

Despite its many advantages, the application of DNA fingerprinting in taxonomy and phylogeny is not without challenges. One significant issue is the reliance on high-quality DNA samples, which are not always available for rare or ancient specimens. Degraded DNA can compromise the accuracy of genetic analyses, necessitating the development of more robust techniques for sample preservation and analysis. Additionally, the sheer volume of data generated by modern sequencing technologies poses computational challenges, requiring advanced bioinformatics tools and expertise. Another concern is the potential for misinterpretation of genetic data. While DNA fingerprinting provides powerful insights, it is important to integrate molecular data with ecological and morphological information for a holistic understanding of biodiversity. Overreliance on genetic markers without considering ecological context can lead to erroneous conclusions about species boundaries or evolutionary relationships. Addressing these challenges requires a multidisciplinary approach, combining expertise in molecular biology, ecology, and evolutionary theory.

Looking to the future, the integration of DNA fingerprinting with emerging technologies such as artificial intelligence and machine learning holds promise for advancing taxonomy and phylogeny. These tools can analyze complex genetic data more efficiently, identifying patterns and relationships that might otherwise go unnoticed. Additionally, the advent of environmental DNA (eDNA) sampling has expanded the scope of DNA fingerprinting, allowing researchers to detect and monitor species using DNA shed into the environment. This non-invasive approach has significant implications for biodiversity assessment, especially in inaccessible or sensitive habitats. The role of DNA fingerprinting in the genetic improvement of natural enemies is also poised to expand, with ongoing research exploring the potential of CRISPR-Cas9 and other gene-editing tools. These technologies enable precise modifications of genetic traits, offering new possibilities for enhancing the effectiveness of biological control agents. For instance, gene-editing approaches could be used to confer resistance to multiple stressors or to tailor natural enemies for specific pest targets, further integrating them into sustainable agricultural practices.

1.1 Definition and Importance of DNA Fingerprinting in Biological Sciences

Definition of DNA Fingerprinting

DNA fingerprinting, also known as DNA profiling or genetic fingerprinting, is a highly specialized molecular technique that identifies and distinguishes individuals based on their unique genetic code. The process relies on analyzing specific regions of DNA, primarily non-coding sequences, which exhibit high variability among individuals. These regions include Variable Number Tandem Repeats (VNTRs) and Short Tandem Repeats (STRs). Their distinct patterns, akin to a biological barcode, allow the precise identification of individuals, even among closely related ones, except for identical twins.

DNA fingerprinting involves a series of carefully controlled steps to extract and analyze these genetic markers. These steps include:

1. **Sample Collection:** DNA is extracted from biological samples, such as blood, saliva, hair follicles, or tissue, which serve as reservoirs of genetic material. For forensic or environmental studies, samples might include degraded or ancient DNA.
2. **DNA Isolation and Purification:** Using specialized chemical and enzymatic techniques, DNA is separated from other cellular components to ensure a pure sample.
3. **Amplification of Target Sequences:** The polymerase chain reaction (PCR) is employed to amplify specific DNA regions of interest, creating millions of copies of these sequences for analysis.
4. **Fragment Analysis:** Amplified DNA is separated by size using electrophoresis methods, such as gel or capillary electrophoresis, producing distinct banding patterns that form the DNA fingerprint.
5. **Data Interpretation:** The banding patterns are compared with reference samples for identification, kinship analysis, or forensic matching.

The technique, introduced in 1984 by British geneticist Dr. Alec Jeffreys, revolutionized biology and its applied sciences. His groundbreaking work demonstrated that VNTRs exhibit patterns unique to each individual, laying the foundation for the development of DNA fingerprinting as a tool for identification.

Importance of DNA Fingerprinting in Biological Sciences

DNA fingerprinting is pivotal across numerous disciplines in biological sciences. Its precision, reliability, and versatility allow for applications ranging from solving criminal cases and tracing evolutionary lineages to conserving biodiversity and advancing personalized medicine. Each of these domains has benefited significantly from the transformative power of DNA fingerprinting.

1. DNA Fingerprinting in Forensic Science and Legal Applications

Forensic science was one of the first fields to adopt DNA fingerprinting extensively, owing to its unmatched accuracy in linking evidence to individuals. This technique has set a new standard for evidence-based justice systems worldwide. It has proven indispensable in:

- **Criminal Investigations:** DNA fingerprinting plays a crucial role in solving crimes by matching DNA found at crime scenes (e.g., hair, blood, or skin cells) with suspects. Even minuscule traces of DNA can produce definitive results, often becoming the cornerstone of a case.
- **Exonerating the Innocent:** Organizations such as The Innocence Project leverage DNA evidence to revisit old convictions, often overturning wrongful judgments and securing justice for those unjustly imprisoned.
- **Victim Identification in Disasters:** During natural disasters, terrorist attacks, or mass casualties, DNA fingerprinting aids in identifying victims by comparing their DNA with that of their relatives, providing closure to families.

DNA fingerprinting has profoundly transformed the legal landscape. Its evidentiary weight, coupled with its ability to establish guilt or innocence with near certainty, has redefined judicial processes. Moreover, the technique has sparked global conversations on ethical practices, particularly concerning privacy and data security in criminal databases.

2. Paternity and Kinship Testing

DNA fingerprinting serves as the cornerstone of paternity and kinship testing due to its unmatched accuracy. By comparing genetic profiles of individuals, the biological relationships between them can be determined with near certainty. Applications include:

» **Family Law Cases:** DNA fingerprinting is widely used in disputes involving child custody, inheritance, or parental rights. The method ensures fairness and accuracy, often serving as decisive evidence in court proceedings.

» **Immigration and Reunification Efforts:** DNA testing helps verify familial relationships in immigration cases or during humanitarian crises where displaced individuals seek to reunite with their families.

» **Tracing Genealogies:** Beyond legal contexts, DNA fingerprinting has sparked interest in personal ancestry testing, enabling individuals to explore their genetic heritage.

In paternity disputes, DNA fingerprinting is particularly significant as it can confirm or exclude biological parentage with an accuracy exceeding 99.99%. Its application extends beyond legal scenarios to personal and societal contexts, fostering deeper understanding and resolution of familial relationships.

3. Evolutionary Biology and Anthropology

In evolutionary biology, DNA fingerprinting illuminates the complexities of human origins, genetic diversity, and the migration of populations. Its contributions include:

» **Tracing Human Evolution:** Comparative analysis of DNA among humans, ancient hominids, and primates helps map evolutionary relationships, shedding light on our shared ancestry with other species.

» **Understanding Population Dynamics:** DNA analysis reveals how populations diverged, migrated, and adapted to different environments, providing insight into historical events such as the Out of Africa migration.

» **Cultural and Social Evolution:** DNA fingerprinting uncovers genetic linkages within and between ethnic groups, offering a window into the interplay between genetics, culture, and history.

The ability to analyze ancient DNA (aDNA) has expanded the horizons of anthropology, allowing scientists to study the genetic makeup of long-extinct species and uncover previously unknown details about our evolutionary past.

4. Conservation Biology and Wildlife Management

DNA fingerprinting is a cornerstone of conservation biology, offering solutions to challenges such as biodiversity loss, habitat degradation, and poaching. It is instrumental in:

- **Genetic Monitoring:** DNA analysis helps assess genetic diversity within and between populations, guiding conservation efforts to prevent inbreeding and maintain ecological balance.
- **Combating Wildlife Crime:** By analyzing DNA from confiscated animal products, such as ivory or scales, scientists can trace their origins, aiding law enforcement in cracking down on illegal poaching and trafficking.
- **Studying Migration Patterns:** DNA fingerprinting tracks animal movement, aiding in the development of policies that protect migratory species and their habitats.

Conservation efforts that leverage DNA fingerprinting emphasize its ecological significance, ensuring sustainable coexistence between humans and wildlife while safeguarding global biodiversity.

5. Medical Applications and Personalized Medicine

DNA fingerprinting has revolutionized the field of medicine, offering unprecedented insights into genetic predispositions, disease diagnosis, and therapeutic strategies. Its contributions include:

- **Genetic Disease Identification:** DNA profiling identifies mutations responsible for inherited conditions, enabling early diagnosis and intervention.
- **Pathogen Tracking:** In infectious disease outbreaks, DNA fingerprinting helps identify and track pathogens, facilitating containment and treatment efforts.
- **Personalized Treatment:** By analyzing an individual's genetic profile, DNA fingerprinting informs precision medicine, tailoring treatments to maximize efficacy while minimizing side effects.

The use of DNA fingerprinting in medical research extends to pharmacogenomics, where understanding genetic variations informs the development of drugs that cater to diverse genetic backgrounds.

6. Agriculture and Food Security

In agriculture, DNA fingerprinting addresses global challenges such as food security and sustainable farming practices. Applications include:

- **Crop Improvement:** DNA analysis identifies genetic traits associated with higher yield, pest resistance, and environmental adaptability, accelerating the development of improved crop varieties.
- **Livestock Breeding:** Genetic profiling ensures the selection of animals with desirable traits, enhancing productivity and disease resistance in livestock populations.
- **Food Traceability:** DNA fingerprinting authenticates the origin of food products, ensuring quality control and preventing fraud in supply chains.

The integration of DNA fingerprinting into agriculture promotes sustainable practices, contributing to global efforts to address hunger and climate resilience.

Ethical Considerations in DNA Fingerprinting

As DNA fingerprinting continues to influence diverse fields, ethical concerns have emerged regarding its use. Issues include:

» **Privacy Risks:** The storage of DNA profiles in databases poses risks of misuse, requiring robust regulatory frameworks to protect individuals' genetic information.

» **Discrimination Concerns:** Misuse of genetic data could lead to discrimination in areas like employment, insurance, or healthcare.

» **Informed Consent:** Ensuring that individuals understand and agree to how their genetic data will be used is a cornerstone of ethical DNA fingerprinting practices.

By addressing these challenges, DNA fingerprinting can continue to advance while respecting ethical principles and societal trust.

1.2 Overview of Taxonomy and Phylogeny and Their Significance in Evolutionary Studies

Definition of Taxonomy

Taxonomy is the scientific discipline of categorizing, naming, and organizing living organisms into hierarchical frameworks based on shared characteristics, genetic makeup, and evolutionary lineage. This system provides an essential structure for studying the vast diversity of life forms on Earth. The field is rooted in the work of Carl Linnaeus, who formalized the binomial nomenclature system in the 18th century, allowing for standardized naming of organisms using a genus and species format (e.g., *Homo sapiens*). Modern taxonomy, however, has evolved significantly with the advent of molecular biology and computational tools.

Taxonomy operates on several levels of classification, known as taxonomic ranks, which include:

1. **Domain**: The broadest classification, dividing life into three categories—Bacteria, Archaea, and Eukarya.
2. **Kingdom**: Subdivides domains into more specific groups such as Plantae, Animalia, and Fungi.
3. **Phylum**: Groups organisms within a kingdom based on shared structural or functional traits.
4. **Class, Order, Family, Genus, Species**: Successively narrower categories leading to the specific identity of an organism.

The classification relies on morphological, biochemical, and genetic data to draw distinctions between groups. For instance, earlier taxonomic systems relied heavily on observable traits, such as body structure and reproductive methods, but modern systems integrate genetic analysis for a more accurate representation of relationships.

Definition of Phylogeny

Phylogeny, on the other hand, delves deeper into the evolutionary history and relationships between organisms. It reconstructs the "tree of life," a metaphorical representation of how species diverged from common ancestors through evolutionary processes. Phylogenetic studies are built upon Darwin's theory of evolution, emphasizing descent with modification.

The study of phylogeny uses various tools and datasets:

1. **Morphological Data**: Observations of physical traits, such as skeletal structures, that suggest evolutionary linkages.

2. **Molecular Data**: DNA, RNA, and protein sequences are analyzed to uncover genetic similarities and divergences.
3. **Fossil Records**: Provide direct evidence of evolutionary transitions, aiding in the placement of extinct species within phylogenetic trees.
4. **Computational Methods**: Algorithms and software tools construct phylogenetic trees, utilizing methods such as Maximum Parsimony, Maximum Likelihood, and Bayesian Inference.

The result is often a phylogenetic tree, where:

- **Nodes** represent ancestors.
- **Branches** signify evolutionary paths.
- **Clades** indicate groups of organisms that share a common ancestor.

Interconnection Between Taxonomy and Phylogeny

Taxonomy and phylogeny, though distinct disciplines are inherently interlinked. Taxonomy provides a structured framework for naming and classifying organisms, while phylogeny contextualizes these groups within evolutionary history. Together, they form the backbone of evolutionary biology, enabling scientists to categorize life forms and trace their evolutionary pathways. Modern taxonomy increasingly incorporates phylogenetic insights to create classifications that reflect evolutionary relationships, a practice known as phylogenetic taxonomy. This approach ensures that taxa represent monophyletic groups (organisms descending from a common ancestor), which is a critical criterion for accuracy in evolutionary studies.

Significance of Taxonomy and Phylogeny in Evolutionary Studies

The integration of taxonomy and phylogeny has revolutionized our understanding of evolution, biodiversity, and ecological dynamics. Their significance is underscored by their applications across multiple domains:

1. Revealing Evolutionary Pathways

Phylogeny is central to understanding the evolutionary history of species. By analyzing genetic and morphological data, scientists can trace how traits have evolved and identify patterns of diversification. For instance:

- **Adaptive Traits:** Phylogenetic studies reveal how traits like feathers in birds or photosynthesis in plants have evolved as adaptations to environmental pressures.
- **Convergent Evolution**: Instances where unrelated species independently evolve similar traits (e.g., wings in bats and birds) can be studied through phylogenetic frameworks, highlighting the role of natural selection in shaping life.
- **Whale Evolution**: Phylogenetic analyses using fossil and molecular data have shown that whales share a common ancestor with modern hippopotamuses, redefining our understanding of cetacean origins.

2. Biodiversity Conservation

Taxonomy and phylogeny are indispensable tools for conserving biodiversity in the face of global challenges like habitat destruction and climate change. Their roles include:

» **Genetic Monitoring**: Phylogenetic data help identify populations with low genetic diversity, guiding breeding programs and habitat restoration to maintain healthy ecosystems.

» **Conservation Prioritization**: Evolutionary distinctiveness can prioritize species for conservation, focusing efforts on those with few close relatives or unique ecological roles.

» **Combatting Wildlife Trafficking**: DNA barcoding, a molecular taxonomy technique, allows identification of illegally trafficked wildlife products, aiding enforcement agencies.

For example, studies on coral reef biodiversity using phylogenetic trees have revealed critical hotspots for conservation, emphasizing the need to protect regions rich in evolutionary history.

3. Speciation and Extinction

Taxonomy and phylogeny shed light on the processes of speciation (formation of new species) and extinction (loss of species). They provide insights into:

» **Drivers of Speciation**: Factors like geographic isolation, ecological niches, and genetic mutations are better understood through phylogenetic comparisons.

» **Extinction Patterns**: Phylogenetic analysis of fossil records uncovers historical extinction events, such as the Permian-Triassic mass extinction, and their causes, offering lessons for modern conservation.

» **Human Impacts**: The phylogenetic consequences of anthropogenic actions, like habitat fragmentation, are studied to mitigate biodiversity loss.

4. Medical and Agricultural Applications

Taxonomy and phylogeny extend their relevance to practical fields like medicine and agriculture:

» **Pathogen Tracking**: In epidemiology, phylogenetic analysis tracks the evolution of pathogens, aiding in the development of treatments and vaccines. For example, studying the phylogeny of the SARS-CoV-2 virus has informed strategies to combat the COVID-19 pandemic.

» **Crop and Livestock Improvement**: Phylogenetic studies identify wild relatives of crops and livestock that possess traits for disease resistance, drought tolerance, or higher yields, informing breeding programs.

5. Ecosystem Functioning and Stability

Understanding the evolutionary relationships between species helps ecologists predict how ecosystems function and respond to changes:

» **Ecological Roles**: Taxonomy identifies species roles within ecosystems (e.g., pollinators, decomposers), while phylogeny reveals how these roles evolved.

» **Ecosystem Resilience**: Phylogenetic diversity is linked to ecosystem stability, with diverse lineages providing functional redundancy that buffers against environmental disruptions.

6. Advancing Evolutionary Theory and Research

Taxonomy and phylogeny are foundational to evolutionary biology. They enable scientists to test hypotheses about:

- **Speciation Rates**: Phylogenetic trees provide data on the rates at which new species arise and old ones disappear.
- **Trait Evolution**: Patterns of trait inheritance and innovation are analyzed through comparative phylogenetics.
- **Lineage Survival**: Factors influencing why some lineages thrive while others go extinct are explored using phylogenetic models.

Challenges and Innovations in Taxonomy and Phylogeny

Despite their transformative contributions, taxonomy and phylogeny face challenges:

- **Undiscovered Diversity**: An estimated 80% of Earth's species remain undiscovered, particularly in microbial communities and deep-sea ecosystems.
- **Integration of Data**: Combining traditional and molecular data requires advanced computational tools and interdisciplinary collaboration.
- **Ethical Concerns**: Balancing conservation goals with economic interests and indigenous rights poses ethical dilemmas.

Innovations such as high throughput sequencing, artificial intelligence, and machine learning are addressing these challenges, enabling faster and more accurate analyses of evolutionary relationships.

1.3 Historical Perspective on the Integration of Molecular Techniques in Taxonomy

Introduction to Molecular Techniques in Taxonomy

Taxonomy has traditionally relied on morphological traits—observable physical features such as body structure, size, and color—to classify organisms. While these traits provided the foundation for early classification systems, they often led to ambiguities, particularly in cases where species displayed convergent evolution, cryptic species, or phenotypic plasticity. The integration of molecular techniques into taxonomy marked a transformative era, enabling taxonomists to explore the genetic underpinnings of biodiversity with unprecedented precision. Molecular taxonomy utilizes genetic information, such as DNA and RNA sequences, to analyze evolutionary relationships and classify organisms. This approach emerged as a response to the limitations of traditional taxonomy, offering objective, quantifiable, and reproducible methods to resolve taxonomic complexities. The journey of molecular techniques in taxonomy spans several decades, reflecting the evolution of molecular biology itself.

The Pre-Molecular Era: Foundations of Taxonomy

Before the advent of molecular techniques, taxonomy was primarily morphological and descriptive. Carl Linnaeus, the father of modern taxonomy, introduced the binomial nomenclature system in the 18th century, standardizing the naming of organisms. This system relied heavily on observable traits to group organisms into hierarchical categories. While revolutionary for its time, Linnaean taxonomy faced challenges as naturalists encountered greater biodiversity and discovered anomalies in classification. By the late 19th and early 20th centuries, taxonomy began incorporating biochemical and cytological data. The examination of chromosome structures and protein profiles provided additional insights into species differences, but these methods lacked the resolution needed to uncover genetic relationships at the molecular level.

The Advent of Molecular Techniques in Taxonomy

The integration of molecular techniques into taxonomy began in the mid-20th century, catalyzed by advancements in molecular biology, biochemistry, and genetics. Key milestones include:

1. Protein-Based Methods

The first molecular approaches in taxonomy focused on proteins, particularly their structure and function. Techniques like electrophoresis and immunological assays were used to compare protein profiles among species. Notable advancements include:

» **Protein Electrophoresis (1950s-1960s):** By separating proteins based on size and charge, researchers could identify differences in protein composition among species. This method highlighted genetic variation underlying phenotypic traits.

» **Immunological Techniques:** Antigen-antibody reactions were used to assess similarities between proteins, offering a biochemical perspective on species relationships.

Although these methods were innovative, they were limited by the indirect nature of protein analysis and the influence of environmental factors on protein expression.

2. DNA Hybridization and Early Genomic Techniques

The focus shifted to DNA in the 1960s and 1970s, as scientists recognized the molecule's role as the blueprint of life. DNA hybridization, a technique measuring the degree of genetic similarity between DNA sequences of different organisms, emerged as a powerful tool.

» **DNA-DNA Hybridization:** This technique compared the thermal stability of hybridized DNA strands from different species. The higher the thermal stability, the greater the genetic similarity, providing a quantitative measure of relatedness. DNA hybridization was particularly influential in resolving higher-order taxonomic relationships, such as those between genera and families.

Despite its impact, DNA hybridization was labor-intensive and lacked the specificity required for fine-scale analyses, paving the way for more advanced molecular techniques.

3. The Polymerase Chain Reaction (PCR) Revolution

The development of the polymerase chain reaction (PCR) in the 1980s by Kary Mullis revolutionized molecular taxonomy. PCR enabled the amplification of specific DNA regions, even from minute or degraded samples, transforming how genetic data could be collected and analyzed.

» **Marker-Based Taxonomy:** PCR facilitated the use of molecular markers, such as ribosomal RNA (rRNA) genes, mitochondrial DNA (mtDNA), and chloroplast DNA (cpDNA), which provided reliable and species-specific genetic signatures.

- **DNA Barcoding:** The concept of DNA barcoding, introduced in the early 2000s, uses a standardized DNA sequence, such as the cytochrome c oxidase I (COI) gene in animals, to identify species. This approach democratized taxonomy, enabling rapid species identification across various domains of life.

PCR-based techniques continue to underpin modern molecular taxonomy, offering robust and reproducible methods for studying biodiversity.

Table: Historical integration of molecular techniques into taxonomy

S. No.	Time Period	Key Development	Description	Impact on Taxonomy
1	Pre-18th Century	Early Taxonomy	Taxonomy was based on morphological traits and observations without formal systems of classification.	Provided the foundation for the identification and description of organisms.
2	18th Century	Binomial Nomenclature (Carl Linnaeus)	Linnaeus introduced a standardized system for naming organisms using a two-part name (genus and species).	Enabled consistent naming and hierarchical classification of species.
3	Late 19th Century	Cytological Studies	Chromosome structures and counts were studied to differentiate species.	Added a genetic perspective to taxonomy, helping to distinguish species based on chromosomal differences.
4	Early 20th Century	Protein Analysis	Techniques like electrophoresis and immunological assays compared protein profiles among species.	Highlighted genetic differences indirectly through protein variations.
5	1960s	DNA Hybridization	DNA-DNA hybridization compared genetic similarity by measuring the stability of hybridized DNA strands.	Provided quantitative measures of genetic relatedness, aiding in higher-order classifications.
6	1970s	Restriction Fragment Length Polymorphism (RFLP)	DNA fragments produced by restriction enzymes were analyzed to detect genetic differences.	Allowed for the differentiation of species based on DNA patterns, marking the transition to molecular methods.
7	1980s	Polymerase Chain Reaction (PCR)	PCR amplified specific DNA sequences, enabling analysis from small or degraded samples.	Revolutionized molecular taxonomy by facilitating the study of molecular markers like rRNA and mitochondrial DNA.
8	Late 1980s	Use of Molecular Markers	Markers like mtDNA, cpDNA, and rRNA genes became standard for phylogenetic studies.	Enabled more accurate classification and understanding of evolutionary relationships.
9	Early 2000s	DNA Barcoding	Standardized genetic sequences, such as the COI gene, were used for species identification.	Democratized taxonomy by enabling rapid and consistent species identification across all taxa.
10	2000s	High-Throughput Sequencing	Next-generation sequencing allowed for large-scale analysis of genomes, transcriptomes, and metagenomes.	Provided unprecedented resolution in taxonomy, especially for microorganisms and cryptic species.
11	2010s	Environmental DNA (eDNA)	Genetic material from environmental samples (e.g., soil, water) was analyzed to detect and monitor biodiversity.	Enabled non-invasive biodiversity assessments and monitoring of rare or elusive species.

12	2010s-Present	Phylogenomics	Genomic data were integrated into phylogenetic analyses, reconstructing evolutionary histories with high resolution.	Enhanced understanding of evolutionary relationships, even among closely related species.
13	Present Day	Artificial Intelligence and Bioinformatics	AI and advanced computational tools analyze vast molecular datasets, facilitating faster and more accurate taxonomy.	Accelerates the discovery of new species and the resolution of complex evolutionary relationships.

The Genomics Era: High-Throughput Sequencing and Beyond

The 21st century ushered in the genomics era, characterized by high-throughput sequencing technologies and bioinformatics tools. These advancements have significantly expanded the scope and resolution of molecular taxonomy:

1. Whole-Genome Sequencing

Whole-genome sequencing (WGS) provides comprehensive insights into an organism's genetic makeup. By comparing entire genomes, taxonomists can:

» Resolve phylogenetic relationships at all taxonomic levels, from species to kingdoms.

» Identify horizontal gene transfer events and their impact on evolution.

» Uncover cryptic species and intraspecific genetic diversity.

WGS has been particularly impactful in microbial taxonomy, where traditional morphological methods were inadequate due to the simplicity of microbial structures.

2. Metagenomics and Environmental DNA (eDNA)

Metagenomics involves analyzing genetic material directly from environmental samples, such as soil, water, or air. This approach enables the study of complex microbial communities and uncultivable organisms.

» **eDNA Applications:** Environmental DNA allows for non-invasive biodiversity assessments, aiding conservation efforts and monitoring invasive species. For instance, eDNA sampling has been used to detect rare aquatic species without direct observation.

3. Phylogenomics

Phylogenomics integrates genomic data into phylogenetic analyses, providing unparalleled resolution in reconstructing evolutionary histories. By examining hundreds or thousands of genes, phylogenomics offers robust insights into the relationships among taxa.

Significance of Molecular Techniques in Taxonomy

The integration of molecular techniques has transformed taxonomy, offering numerous advantages:

1. **Increased Resolution:** Molecular data provide finer distinctions between closely related species, overcoming limitations of morphological traits.
2. **Objectivity and Reproducibility:** Genetic data are quantifiable and less subjective than morphological observations.

3. **Cryptic Species Discovery:** Molecular techniques have uncovered cryptic species, which are morphologically indistinguishable but genetically distinct.
4. **Biodiversity Assessment:** High-throughput sequencing enables comprehensive biodiversity surveys, contributing to conservation and ecological studies.
5. **Global Standardization:** DNA barcoding has facilitated the creation of global databases, such as the Barcode of Life Data System (BOLD), standardizing species identification.

2: Techniques of DNA Fingerprinting

2.1. Methods Used in DNA Fingerprinting: RFLP, AFLP, Microsatellites, and RAPD

DNA fingerprinting is a sophisticated molecular technique that identifies genetic variations among individuals or populations. This methodology relies on analyzing specific genetic markers or patterns in DNA sequences, providing critical applications in forensic science, paternity testing, population genetics, and evolutionary studies. Among the prominent methods used in DNA fingerprinting are Restriction Fragment Length Polymorphism (RFLP), Amplified Fragment Length Polymorphism (AFLP), microsatellites, and Random Amplified Polymorphic DNA (RAPD). Each method has unique principles, applications, advantages, and limitations.

1. Restriction Fragment Length Polymorphism (RFLP)

1.1 Principle

Restriction Fragment Length Polymorphism (RFLP) is a pioneering molecular technique used in DNA fingerprinting, founded on the principle that genetic variations between individuals influence the locations where restriction enzymes cut the DNA. These restriction enzymes act as molecular scissors, recognizing specific nucleotide sequences known as restriction sites and cleaving the DNA at these locations. Genetic differences such as insertions, deletions, point mutations, or sequence duplications can modify or eliminate restriction sites, resulting in DNA fragments of varying lengths when digested. These variations, which are unique to individuals or closely related groups, are termed polymorphisms. By analyzing these fragment patterns, RFLP provides a distinctive "fingerprint" of an individual's DNA.

For example:

- » If a mutation occurs within a restriction site, the restriction enzyme may fail to recognize and cut the DNA at that site, leading to a longer fragment.
- » Conversely, the creation of a new restriction site results in shorter fragments after enzymatic cleavage.

This principle laid the foundation for modern genetic analysis, enabling researchers to distinguish between individuals and study genetic inheritance with remarkable precision.

1.2 Methodology

The RFLP technique involves a series of carefully orchestrated steps to generate and analyze DNA fragments:

1. **DNA Extraction:**
   - High-quality genomic DNA is extracted from biological samples such as blood, saliva, or tissue. Ensuring the purity and integrity of the DNA is critical for successful digestion and subsequent analysis.
2. **Digestion with Restriction Enzymes:**
   - The extracted DNA is treated with specific restriction enzymes, such as *EcoRI* or *HindIII*. These enzymes cleave the DNA at their respective recognition sites, generating a mixture of fragments with varying lengths.
3. **Gel Electrophoresis:**
   - The digested DNA is loaded onto an agarose gel, where an electric current separates the fragments based on size. Smaller fragments migrate faster through the gel matrix, creating a pattern of bands.
4. **Southern Blotting:**
   - The DNA fragments from the gel are transferred onto a nylon or nitrocellulose membrane through capillary action or an electric current. This step ensures the DNA is immobilized and accessible for probe hybridization.
   - The membrane is treated with labeled probes, which are short, complementary DNA or RNA sequences that bind to specific regions of the DNA fragments.
5. **Visualization:**
   - After hybridization, the membrane is washed to remove unbound probes. The labeled probes are then visualized using techniques such as autoradiography (if radioactive probes are used) or chemiluminescence, revealing a unique pattern of bands representing the DNA fingerprint.

The final banding pattern reflects the individual's genetic profile and can be compared with other samples to determine similarities or differences.

1.3 Applications

The versatility of RFLP makes it a cornerstone technique in various scientific and practical fields:

» **Forensic Science:**
   - RFLP was historically used to match DNA evidence from crime scenes with potential suspects or to identify victims in criminal investigations. Its accuracy helped establish it as one of the earliest DNA-based forensic tools.

» **Paternity Testing:**
   - By comparing the DNA patterns of a child and potential parents, RFLP can determine biological relationships with high reliability. Each child inherits half of their DNA from each parent, creating recognizable patterns.

» **Genetic Mapping:**

- RFLP markers are instrumental in locating genes associated with specific traits or diseases. These markers serve as landmarks in genetic studies, helping researchers identify regions of interest in the genome.

» **Evolutionary and Population Studies:**

- The technique has been used to analyze genetic diversity within and between populations, offering insights into evolutionary relationships and migration patterns.

1.4 Advantages

The RFLP technique offers several significant advantages:

1. **High Accuracy and Reproducibility**:
   - The method provides reliable and precise results, especially for genetic studies requiring detailed information about sequence variation.
2. **Detailed Genetic Variability Analysis**:
   - RFLP enables the identification of even small-scale variations in DNA, making it ideal for studying polymorphisms and constructing genetic linkage maps.
3. **Long-Lasting Utility:**
   - Despite being one of the earliest molecular techniques, RFLP remains relevant in fields requiring high-resolution genetic data.

1.5 Limitations

Despite its advantages, RFLP is not without challenges:

1. **Requirement for High-Quality DNA:**
   - The method demands large quantities of intact DNA, which may not always be available, particularly in forensic or ancient DNA samples.
2. **Labor-Intensive and Time-Consuming:**
   - The multiple steps involved, including restriction digestion, electrophoresis, and Southern blotting, make RFLP a lengthy and labor-intensive process compared to modern techniques.
3. **Limited Use with Degraded Samples:**
   - RFLP is less effective when working with fragmented or degraded DNA, as it relies on intact restriction sites for accurate analysis.
4. **Cost and Complexity:**
   - The need for specialized equipment, radioactive or labeled probes, and technical expertise adds to the complexity and cost of the procedure.

2. Amplified Fragment Length Polymorphism (AFLP)

2.1 Principle

Amplified Fragment Length Polymorphism (AFLP) is a molecular technique that combines restriction digestion and selective polymerase chain reaction (PCR) amplification to generate a unique set of DNA fragments that reflect genetic variation among individuals. The method leverages the ability of restriction enzymes to cut DNA at specific sequences and the power of PCR to selectively amplify a subset of these fragments.

AFLP relies on polymorphisms—genetic differences in DNA sequences—that affect the recognition sites of restriction enzymes or the amplified regions. These polymorphisms result in differences in the number and size of fragments generated, creating a "fingerprint" unique to each individual. The method does not require prior sequence knowledge, making it highly versatile for studying a wide range of organisms.

For example:

- » A mutation that disrupts a restriction enzyme recognition site will prevent digestion at that location, resulting in longer fragments.
- » Similarly, variations in DNA sequences flanking the restriction sites may alter the efficiency of amplification, further contributing to polymorphic patterns.

2.2 Methodology

The AFLP process is a multi-step procedure that integrates molecular biology techniques to analyze genetic polymorphisms:

1. **DNA Digestion:**
    - Genomic DNA is extracted and digested with two types of restriction enzymes:
        - ◊ A rare-cutter enzyme (e.g., *EcoRI*), which generates larger fragments.
        - ◊ A frequent-cutter enzyme (e.g., *MseI*), which produces smaller fragments.
    - This combination ensures the creation of a manageable number of fragments with sufficient diversity.
2. **Adapter Ligation:**
    - Short double-stranded oligonucleotide adapters are ligated to the ends of the restriction fragments. These adapters provide primer-binding sites for subsequent PCR amplification.
3. **Pre-Selective Amplification:**
    - A preliminary PCR amplification is performed using primers complementary to the adapter sequences. These primers may include one or more selective nucleotides, which ensure that only a subset of the fragments is amplified.
4. **Selective Amplification:**
    - A second PCR amplification is carried out using primers with additional selective bases. These bases extend into the restriction fragments, further reducing the number of amplified products and enhancing the resolution of polymorphic patterns.

5. **Fragment Analysis:**
    - The amplified DNA fragments are separated by size using high-resolution techniques such as polyacrylamide gel electrophoresis or capillary electrophoresis.
    - The resulting banding patterns, visualized as a series of peaks or bands, represent the AFLP fingerprint of the sample.

2.3 Applications

AFLP is a versatile tool with broad applications across various scientific disciplines:

» **Genetic Diversity Studies:**
- AFLP is widely used to assess genetic variation within and between populations. It has been instrumental in studying plant and animal biodiversity, helping researchers understand evolutionary processes and population dynamics.

» **Phylogenetic Analysis:**
- By comparing AFLP profiles, researchers can infer evolutionary relationships among species or populations. The technique has been used to resolve taxonomic ambiguities and construct phylogenetic trees.

» **Marker-Assisted Breeding:**
- AFLP markers are valuable in agricultural research, where they assist in identifying and selecting traits of interest, such as disease resistance, drought tolerance, and high yield, in crops and livestock.

» **Conservation Biology:**
- AFLP helps monitor genetic diversity in endangered species, guiding conservation strategies to maintain viable populations.

» **Microbial Ecology:**
- In microbiology, AFLP is used to study genetic variation in microbial communities, aiding in the identification of strains and understanding their ecological roles.

2.4 Advantages

The AFLP technique offers several significant benefits that make it a preferred method in genetic analysis:

1. **High Sensitivity:**
    - AFLP can detect a large number of polymorphisms across the genome without requiring prior sequence information. This makes it suitable for studying organisms with limited genomic resources.
2. **Applicability to Small or Degraded Samples:**
    - The method works effectively with minimal or partially degraded DNA, making it ideal for forensic applications and ancient DNA studies.

3. **Genome-Wide Analysis:**
    - AFLP generates a comprehensive profile of genetic variation, covering both coding and non-coding regions of the genome.
4. **Versatility:**
    - The technique is applicable to a wide range of organisms, including plants, animals, and microorganisms.
5. **Robustness:**
    - AFLP is less sensitive to environmental factors and DNA quality compared to other molecular techniques, ensuring consistent results across diverse sample types.

2.5 Limitations

Despite its advantages, AFLP has certain limitations that must be considered:

1. **Complex Protocol:**
    - The multi-step procedure, involving DNA digestion, adapter ligation, and selective PCR, is labor-intensive and requires technical expertise.
2. **Advanced Equipment Requirements:**
    - AFLP analysis relies on high-resolution separation techniques, such as capillary electrophoresis, which necessitate specialized equipment and reagents.
3. **Reproducibility Issues:**
    - Variability in experimental conditions, such as enzyme activity and PCR settings, can affect the reproducibility of AFLP profiles, making cross-laboratory comparisons challenging.
4. **Limited Quantitative Power:**
    - Unlike sequencing-based methods, AFLP does not provide detailed information about the actual sequences of the amplified fragments.
5. **High Cost:**
    - The cost of restriction enzymes, adapters, and advanced analytical equipment can be prohibitive for large-scale studies or resource-limited laboratories.

Comparative Analysis of AFLP with Other Techniques

Feature	AFLP	RFLP	Microsatellites	RAPD
Sequence Knowledge	Not required	Required for probe design	Required for primer design	Not required
Sensitivity	High	High	Very high	Moderate
Reproducibility	Moderate	High	High	Low
Applicability to Samples	Works with small and degraded DNA	Requires high-quality DNA	Works with small and degraded DNA	Works with low-quality DNA
Ease of Use	Complex	Labor-intensive	Moderate	Simple
Marker Resolution	Genome-wide	Region-specific	Highly polymorphic regions	Random

3. Microsatellites (Simple Sequence Repeats or SSRs)

3.1 Principle

Microsatellites, also known as Simple Sequence Repeats (SSRs) or Short Tandem Repeats (STRs), are DNA sequences consisting of short motifs of 2–6 base pairs (e.g., "AC," "AT," or "GATA") that are repeated in tandem. These sequences are distributed throughout the genome, often in non-coding regions, and exhibit high levels of polymorphism due to variations in the number of repeat units.

The polymorphic nature of microsatellites arises from DNA replication slippage or unequal crossing-over during meiosis, which leads to gains or losses of repeat units. This makes them ideal genetic markers for distinguishing individuals, populations, or species. Their high mutation rate ensures that microsatellites are both abundant and variable, providing a rich source of genetic information. The analysis of microsatellites involves detecting these variations by amplifying the repeat regions using Polymerase Chain Reaction (PCR). Because the size of the amplified products corresponds to the number of repeat units, differences in fragment length can be used to determine genetic diversity or identify individuals.

3.2 Methodology

Microsatellite analysis involves a step-by-step approach to extract, amplify, and analyze genetic variations in SSR loci. Below are the critical steps in the process:

1. **DNA Extraction:**
   - High-quality genomic DNA is isolated from biological samples such as blood, hair follicles, or tissue. The quality of DNA is crucial, as degraded samples may impact amplification efficiency.
2. **Primer Design:**
   - Specific primers are designed to flank the microsatellite loci. These primers are complementary to the sequences adjacent to the repeats and are essential for targeted amplification.
   - Fluorescent labels are often added to one of the primers to facilitate fragment analysis during electrophoresis.
3. **PCR Amplification:**
   - The extracted DNA is amplified using PCR, targeting the microsatellite loci. PCR conditions, such as annealing temperature and cycle number, are optimized to ensure specificity and efficiency.
   - Each allele at a microsatellite locus is amplified, resulting in PCR products of different lengths based on the number of repeats.
4. **Fragment Analysis:**
   - The amplified products are separated by size using high-resolution gel electrophoresis (e.g., polyacrylamide gel) or capillary electrophoresis.
   - The sizes of the fragments are determined by comparing them to a DNA size standard, and the number of repeats is inferred based on fragment length.

- Automated systems using fluorescently labeled primers and capillary electrophoresis produce electropherograms for precise allele sizing.

5. **Data Interpretation:**
   - The resulting profiles display unique patterns for each individual or population. These patterns are analyzed to assess genetic diversity, establish relationships, or identify individuals.

3.3 Applications

Microsatellites have a wide range of applications due to their high polymorphism and ease of analysis. They are particularly valuable in fields that require detailed genetic resolution:

» **Forensic Science**:

- STR profiling is a standard method for human identification in criminal investigations, missing person cases, and disaster victim identification.
- Microsatellites provide robust and reproducible results, even from degraded samples such as bone or hair.

» **Paternity Testing:**

- Microsatellites are highly effective in determining biological parentage due to their Mendelian inheritance patterns.
- By comparing the STR profiles of a child and potential parents, parentage can be established with near-perfect accuracy.

» **Population Genetics:**

- Microsatellites are used to study genetic structure, gene flow, and mating patterns within and between populations.
- They help estimate levels of inbreeding, genetic drift, and population bottlenecks, providing insights into evolutionary processes.

» **Conservation Biology:**

- Microsatellites monitor genetic diversity and inbreeding in endangered species, aiding in the development of conservation strategies.
- They are also used to trace the origins of illegally trafficked wildlife products.

» **Agriculture and Breeding:**

- In crop and livestock breeding programs, microsatellites are employed to identify desirable traits, assess genetic diversity, and track lineage.

» **Medical Research:**

- Microsatellites are used as markers in genetic linkage studies to identify genes associated with hereditary diseases or conditions.

3.4 Advantages

The widespread use of microsatellites is attributed to their numerous advantages:

1. **High Polymorphism:**
    - Microsatellites exhibit significant variability, making them highly effective for distinguishing individuals or closely related groups.
2. **Small DNA Requirement:**
    - The technique requires minimal amounts of DNA, making it suitable for non-invasive sampling or working with degraded DNA.
3. **PCR-Based Analysis:**
    - The reliance on PCR ensures sensitivity and allows for the amplification of specific loci from even small or degraded samples.
4. **Multi-Locus Capability:**
    - Multiple microsatellite loci can be analyzed simultaneously, providing comprehensive genetic profiles.
5. **Versatility:**
    - Microsatellites can be used across a wide range of organisms, from humans and animals to plants and microorganisms.

3.5 Limitations

Despite their strengths, microsatellites have certain limitations that must be addressed in their application:

1. **High Initial Cost:**
    - The development of locus-specific primers and fluorescent labels can be expensive, particularly for non-model organisms.
2. **PCR Sensitivity:**
    - Microsatellite analysis is sensitive to PCR conditions, which can lead to issues such as allele dropouts, stutter bands, or amplification biases.
3. **Complex Data Interpretation:**
    - In polyploid species or populations with high genetic diversity, interpreting microsatellite data can be challenging.
4. **Mutational Hotspots:**
    - The high mutation rate of microsatellites can complicate evolutionary studies by introducing homoplasy, where identical alleles arise independently.
5. **Limited Genome Coverage:**
    - While microsatellites provide high resolution at specific loci, they do not offer the genome-wide coverage of newer techniques like single nucleotide polymorphisms (SNPs) or whole-genome sequencing.

4. Random Amplified Polymorphic DNA (RAPD)

4.1 Principle

Random Amplified Polymorphic DNA (RAPD) is a molecular technique that uses short, arbitrary primers to amplify random DNA sequences in the genome through Polymerase Chain Reaction (PCR). The method does not require prior knowledge of the target DNA sequence, making it a versatile and cost-effective tool for genetic analysis. Polymorphisms in RAPD arise from differences in the presence or absence of primer binding sites due to mutations, insertions, deletions, or rearrangements in the DNA. These variations result in different DNA fragment patterns among individuals or populations.

The core idea of RAPD lies in the random nature of the amplification. Primers of 8–10 nucleotides in length anneal to multiple sites in the genome. If the binding sites are close enough and oriented correctly, amplification occurs, producing fragments of specific sizes. Changes in the DNA sequence, such as mutations at primer binding sites, prevent amplification and lead to polymorphic patterns. These unique banding patterns serve as a "fingerprint" for each individual, strain, or species.

4.2 Methodology

The RAPD process is straightforward and involves the following steps:

1. **DNA Extraction:**
    - Genomic DNA is extracted from the biological sample using standard protocols. The quality and integrity of the DNA are critical for reliable amplification.
2. **PCR Amplification:**
    - Short, random primers are used in PCR to amplify DNA fragments from various regions of the genome. Typically, a single primer is used per reaction to ensure random amplification.
    - PCR conditions, such as annealing temperature and magnesium ion concentration, are carefully optimized to promote specific binding and amplification.
3. **Gel Electrophoresis:**
    - The amplified DNA fragments are separated based on size using agarose gel electrophoresis.
    - The gel is stained with ethidium bromide or another DNA-binding dye, and the banding patterns are visualized under UV light.
4. **Data Analysis:**
    - The banding patterns are compared across samples to assess genetic polymorphisms. The presence or absence of specific bands is scored, and the data are used to calculate genetic distances, construct phylogenetic trees, or identify unique genetic profiles.

4.3 Applications

RAPD is a versatile tool with applications in various fields of biological sciences:

» **Genetic Mapping**:

- RAPD markers are used to identify genes linked to specific traits in plants and animals. They serve as tools for constructing genetic linkage maps, which are essential for studying inheritance patterns.

» **Biodiversity Studies:**

- RAPD helps assess genetic variation within and between populations or species. It is widely used in conservation biology to monitor genetic diversity and understand population structure.

» **Species Identification**:

- RAPD can distinguish closely related species or strains, making it useful in taxonomy, microbial identification, and breeding programs.

» **Pathogen Characterization:**

- RAPD is employed to differentiate pathogenic strains of bacteria, fungi, and viruses, aiding in epidemiological studies and disease management.

» **Agricultural Research:**

- In crop improvement programs, RAPD markers help identify genetic resources, evaluate genetic diversity, and select superior genotypes for breeding.

4.4 Advantages

RAPD offers several benefits, making it a popular choice for genetic studies:

1. **Simplicity:**
   - The technique is easy to perform and requires minimal preparation compared to other molecular methods.
2. **Cost-Effectiveness:**
   - RAPD does not require expensive reagents, specialized equipment, or prior sequence knowledge, making it accessible to resource-limited laboratories.
3. **Wide Applicability:**
   - The method can be applied to a broad range of organisms, including plants, animals, and microorganisms.
4. **No Sequence Information Required:**
   - RAPD is independent of genomic sequence data, allowing its use in non-model organisms.
5. **Rapid Turnaround:**
   - The technique can generate results quickly, making it suitable for preliminary studies or large-scale screenings.

4.5 Limitations

Despite its advantages, RAPD has notable limitations that affect its reliability and applicability:

1. **Reproducibility Issues:**
   - The arbitrary nature of primers and sensitivity to PCR conditions often result in inconsistent amplification patterns across laboratories or experiments.
2. **Limited Resolution:**
   - RAPD provides less detailed information compared to more advanced techniques like microsatellites, AFLP, or next-generation sequencing.
3. **Difficulty in Data Interpretation:**
   - The presence or absence of bands can be influenced by factors unrelated to genetic variation, such as DNA quality, PCR optimization, or primer specificity.
4. **Limited Use in Closely Related Individuals**:
   - RAPD has lower discriminatory power for distinguishing closely related individuals or populations due to its random nature.
5. **Qualitative Nature:**
   - Unlike quantitative methods, RAPD primarily provides binary data (presence or absence of bands), which limits its ability to detect su*Bt*le genetic differences.

2.2 Comparative Analysis of Fingerprinting Techniques and Their Application in Taxonomy

DNA fingerprinting techniques are invaluable tools for studying genetic diversity, evolutionary relationships, and species classification. In taxonomy, these methods allow researchers to uncover genetic similarities and differences across a wide range of organisms, providing a deeper understanding of biodiversity and evolutionary history. Below is a detailed comparative analysis of the most commonly used DNA fingerprinting techniques—RFLP, AFLP, microsatellites (SSRs), and RAPD—highlighting their principles, strengths, limitations, and applications in taxonomy.

Comparative Overview of DNA Fingerprinting Techniques

Technique	Principle	Applications in Taxonomy	Strengths	Limitations
RFLP (Restriction Fragment Length Polymorphism)	Differences in DNA fragment lengths due to variation in restriction enzyme sites.	Species differentiation, genetic diversity studies, evolutionary relationships.	High resolution, reproducibility, detailed genetic variation analysis.	Requires high-quality DNA, labor-intensive, unsuitable for degraded samples.
AFLP (Amplified Fragment Length Polymorphism)	Selective PCR amplification of restriction-digested DNA fragments.	Phylogenetic studies, genetic mapping, population structure analysis.	Detects polymorphisms without prior sequence knowledge, works with degraded DNA, generates genome-wide markers.	Complex and labor-intensive protocol, requires advanced equipment, variability in reproducibility across labs.

Microsatellites (SSRs)	Variation in the number of short tandem repeats (2–6 bp) at specific loci in the genome.	Species identification, population genetics, conservation studies, evolutionary studies.	Highly polymorphic, requires small DNA quantities, high accuracy, and suitability for degraded samples.	High initial cost for primer development, interpretation complexity in polyploid species, sensitive to PCR errors.
RAPD (Random Amplified Polymorphic DNA)	PCR amplification of random DNA sequences using short arbitrary primers.	Preliminary genetic diversity studies, species identification, phylogenetic analyses.	Simple, cost-effective, no prior sequence knowledge required, works with low-quality DNA.	Low reproducibility, limited resolution, qualitative results, difficult data interpretation for closely related taxa.

Detailed Comparative Analysis

1. Resolution and Precision

» **High-Resolution Techniques:**

- **Microsatellites** and **RFLP** provide detailed and precise data about genetic differences, making them ideal for fine-scale taxonomic studies such as species delineation or understanding intraspecific genetic variation.
- **AFLP** offers genome-wide resolution, allowing comprehensive assessment of genetic diversity and evolutionary relationships.

» **Moderate-Resolution Techniques:**

- **RAPD** provides less precise results and is more suited for preliminary studies or broad assessments of genetic variation.

2. Reproducibility

» **Highly Reproducible:**

- **RFLP** and **Microsatellites** are highly reproducible due to their dependence on well-defined genetic markers and optimized protocols.

» **Moderately Reproducible:**

- **AFLP** has moderate reproducibility, as the results can vary depending on experimental conditions such as restriction digestion and selective amplification.

» **Low Reproducibility:**

- **RAPD** is less reproducible due to its reliance on arbitrary primers and sensitivity to PCR conditions, making it unsuitable for studies requiring high precision.

3. Ease of Use and Cost

» **Easy and Cost-Effective:**

- **RAPD** is the simplest and least expensive technique, requiring minimal preparation and no prior sequence information. It is accessible for researchers with limited resources.

» **Moderately Complex and Expensive:**

- **Microsatellites** and **AFLP** require more technical expertise and higher initial costs for primer development and equipment, but they offer higher accuracy and resolution.

» **Complex and Expensive:**

- **RFLP** is labor-intensive and costly due to the need for restriction enzymes, Southern blotting, and radioactive or chemiluminescent probes.

4. Applicability to Degraded Samples

» **Effective for Degraded DNA:**

- **AFLP** and **Microsatellites** are PCR-based techniques that work well with degraded or small amounts of DNA, making them ideal for ancient DNA studies and forensic applications.

» **Limited Use for Degraded DNA**:

- **RFLP** requires high-quality DNA, limiting its application in cases where DNA is degraded or fragmented.

» **Moderate Effectiveness:**

- **RAPD** can work with low-quality DNA but often produces inconsistent results.

5. Prior Sequence Knowledge

» **No Prior Knowledge Require**d:

- **RAPD** and **AFLP** do not require prior information about the genome, making them versatile tools for studying non-model organisms.

» **Requires Prior Knowledge:**

- **Microsatellites** and **RFLP** depend on prior sequence information for designing primers or probes, which can limit their application to well-studied species.

Applications of Fingerprinting Techniques in Taxonomy

1. **Species Identification:**
   - **Microsatellites** and **RFLP** excel in differentiating species due to their high resolution and ability to detect minute genetic variations.
   - **RAPD** is used for rapid and cost-effective species identification in preliminary taxonomic studies.
2. **Phylogenetic Studies:**
   - **AFLP** and **Microsatellites** are widely used in constructing phylogenetic trees and resolving evolutionary relationships, even among closely related taxa.
   - **RFLP** is useful for studying higher-order taxonomic relationships.
3. **Population Genetics:**

- **Microsatellites** and **AFLP** are ideal for analyzing genetic diversity, population structure, and gene flow, making them invaluable in population genetics and conservation biology.

4. **Cryptic Species Detection:**
   - **Microsatellites** and **AFLP** are effective in uncovering cryptic species that cannot be differentiated morphologically.
5. **Hybridization and Introgression**:
   - **RFLP** and **Microsatellites** are used to detect hybridization events and track introgression between species or populations.

Summary Table: Suitability for Taxonomic Applications

Application	**RFLP**	**AFLP**	**Microsatellites**	**RAPD**
Species Identification	High	High	High	Moderate
Phylogenetic Analysis	High	High	Moderate	Moderate
Population Genetics	Moderate	High	High	Low
Conservation Biology	Moderate	High	High	Low
Preliminary Studies	Low	Moderate	Moderate	High
Cost-Effectiveness	Low	Moderate	Moderate	High

2.3 High-Throughput Sequencing Technologies and Phylogenomic Approaches

High-Throughput Sequencing (HTS) Technologies

High-throughput sequencing (HTS), also referred to as next-generation sequencing (NGS), represents a paradigm shift in genomics, enabling researchers to generate vast quantities of genetic data rapidly and cost-effectively. Unlike traditional Sanger sequencing, which processes one DNA fragment at a time, HTS platforms can sequence millions to billions of fragments simultaneously, providing unmatched scalability and precision. These technologies underpin modern phylogenomics by allowing detailed analysis of entire genomes, transcriptomes, and epigenomes, which are essential for reconstructing evolutionary relationships among organisms.

1.1 Key HTS Technologies

1. **Illumina Sequencing (Short-Read Sequencing):** Illumina sequencing is one of the most widely used HTS platforms, leveraging sequencing-by-synthesis (SBS) technology. During the sequencing process, fragmented DNA is attached to flow cell surfaces, amplified into clusters, and sequenced through cyclic nucleotide addition. Each nucleotide emits a unique fluorescence signal, enabling precise base calling.
   - **Advantages:**
     - ◊ High accuracy, short-read lengths (50–300 bp), and scalability.
     - ◊ Ideal for genome resequencing, SNP discovery, and transcriptome studies.
     - ◊ Supports paired-end reads, providing greater context for sequence alignment.

- **Limitations:**
  - ◊ Short read lengths can complicate the assembly of repetitive or complex regions.
  - ◊ Requires high computational resources for data processing and analysis.

2. **PacBio Sequencing (Long-Read Sequencing):** Pacific Biosciences' Single-Molecule Real-Time (SMRT) sequencing technology offers long-read capabilities, producing sequences up to 30 kb in length. This is particularly advantageous for resolving structural variants, repetitive regions, and telomere-to-telomere assemblies.
   - **Advantages:**
     - ◊ Provides full-length sequencing of transcripts, aiding functional annotation.
     - ◊ Facilitates assembly of high-quality reference genomes.
   - **Limitations:**
     - ◊ Higher error rates than Illumina sequencing, though error-correction techniques have improved accuracy.
     - ◊ More expensive per base compared to short-read platforms.
3. **Oxford Nanopore Sequencing (Ultra-Long-Read Sequencing):** Oxford Nanopore Technologies (ONT) employs nanopore-based sequencing, wherein DNA or RNA molecules pass through a nanopore, and changes in electrical current are recorded to determine the sequence. This platform supports portable devices like the MinION, enabling field-based sequencing.
   - **Advantages**:
     - ◊ Generates ultra-long reads (>100 kb), suitable for structural variant detection and rapid whole-genome analysis.
     - ◊ Real-time sequencing allows immediate data acquisition and analysis.
   - **Limitations:**
     - ◊ Lower accuracy compared to short-read technologies, though software-based error correction has improved results.
     - ◊ Higher capital costs for initial setup.
4. **10x Genomics (Linked-Read Sequencing):** By partitioning and barcoding DNA molecules, 10x Genomics enables long-range sequence information using short-read sequencing platforms. This method bridges the gap between short- and long-read sequencing by reconstructing haplotypes and phasing variants.
   - **Applications:**
     - ◊ Haplotyping, genome scaffolding, and structural variant analysis.
     - ◊ Resolving heterozygous regions in complex genomes.
5. **Hi-C and Chromosome Conformation Capture:** Hi-C sequencing provides insights into chromosomal architecture by capturing spatial interactions between genomic regions.

This technique is pivotal for assembling chromosome-scale genomes and studying three-dimensional genome organization.

- **Applications:**
    - ◊ Chromosome-level assembly, genome scaffolding, and epigenetic studies.

6. **RNA-Seq (Transcriptomic Sequencing):** RNA-Seq is used to sequence the transcriptome, providing insights into gene expression, alternative splicing, and regulatory mechanisms. It enables phylogenomic studies by focusing on conserved or lineage-specific transcripts.

1.2 Advantages of HTS Technologies

HTS technologies provide numerous advantages that make them indispensable for phylogenomics:

1. **Massive Data Output:**
    - Capable of sequencing entire genomes in a single run, enabling compre**hensive analyses.**
2. **Cost Efficiency:**
    - The per-base cost of sequencing has decreased significantly, democratizing access to genomics.
3. **Flexibility:**
    - Applicable to a wide range of samples, including DNA, RNA, and even epigenetic marks.
4. **Precision and Sensitivity:**
    - Detects su*Btl*e genetic variations, such as single-nucleotide polymorphisms (SNPs), indels, and structural variants.
5. **Diverse Applications:**
    - HTS supports genomics, transcriptomics, metagenomics, and epigenomics, broadening its utility across disciplines.

1.3 Challenges of HTS Technologies

Despite their transformative potential, HTS technologies face several challenges:

1. **Data Management:**
    - HTS generates vast datasets requiring advanced bioinformatics tools and computational infrastructure for storage, processing, and analysis.
2. **Error Rates:**
    - Long-read platforms like PacBio and Nanopore have higher error rates, necessitating additional error-correction steps.
3. **High Initial Costs:**
    - While sequencing costs have decreased, setting up HTS platforms involves significant capital investment.

4. **Technical Expertise:**
   - Effective use of HTS requires expertise in molecular biology, bioinformatics, and data analysis.

Phylogenomic Approaches

Phylogenomics integrates genomic data with phylogenetic methods to reconstruct evolutionary relationships among organisms. By analyzing whole genomes or large-scale gene datasets, phylogenomics offers unprecedented resolution in deciphering complex evolutionary questions.

2.1 Steps in Phylogenomic Analysis

1. **Data Acquisition:**
   - Genomic data are o*Bt*ained through whole-genome sequencing, transcriptome sequencing, or targeted enrichment methods like exome capture.
   - High-quality DNA or RNA samples are critical for generating reliable datasets.
2. **Data Assembly and Annotation:**
   - Raw sequencing reads are assembled into contigs or scaffolds using bioinformatics tools such as SPAdes or Trinity.
   - Annotation tools like BLAST and InterProScan are used to identify functional elements and conserved regions.
3. **Orthology Identification:**
   - Orthologous genes (genes derived from a common ancestor) are identified using tools like OrthoFinder or OMA.
   - Ensuring accurate orthology detection is essential for robust phylogenetic inference.
4. **Alignment and Phylogenetic Reconstruction:**
   - Orthologous genes are aligned using tools like MAFFT or Clustal Omega.
   - Phylogenetic trees are reconstructed using maximum likelihood (RAxML), Bayesian inference (MrBayes), or coalescent-based methods (ASTRAL).
5. **Analysis and Interpretation:**
   - The resulting phylogenies are analyzed to study speciation events, evolutionary adaptations, and gene family evolution.

2.2 Applications of Phylogenomics

1. **Resolving Complex Phylogenies:**
   - Phylogenomics provides insights into evolutionary relationships among species, including those with conflicting morphological or molecular data.
2. **Cryptic Species Identification:**
   - Genomic data can uncover cryptic species that are morphologically indistinguishable.

3. **Gene Family Evolution:**
   - Analysis of gene duplication and loss events sheds light on functional diversification and adaptation.
4. **Comparative Genomics:**
   - Comparing genomes across taxa reveals conserved elements and lineage-specific innovations.
5. **Microbial Phylogenomics:**
   - HTS-based phylogenomics is instrumental in studying microbial evolution, horizontal gene transfer, and antibiotic resistance mechanisms.

3: Applications of DNA Fingerprinting in Taxonomy and Phylogeny

3.1 Establishing Evolutionary Relationships Among Species

Establishing evolutionary relationships among species is a cornerstone of biological research. It involves reconstructing the "tree of life" to trace how species have diverged and adapted from common ancestors. This process is fundamental for understanding biodiversity, evolutionary processes, and ecological interactions. Evolutionary relationships are determined using morphological, molecular, and computational methods, with molecular data playing a central role in modern studies. Techniques such as phylogenetics, phylogenomics, and comparative genomics allow scientists to unravel the evolutionary history of life with unprecedented precision.

Approaches to Establishing Evolutionary Relationships

The study of evolutionary relationships has evolved significantly, transitioning from morphology-based approaches to molecular and genomic analyses. These methodologies can be grouped into several key categories:

1. Morphological and Anatomical Approaches

Historically, evolutionary relationships were inferred from observable traits, such as body structure, organ systems, and developmental patterns. Comparative anatomy provided insights into homologous structures (traits inherited from a common ancestor) and analogous structures (traits with similar functions but different evolutionary origins).

» **Examples:**
   - The forelimbs of vertebrates (e.g., bats, whales, and humans) share a homologous bone structure despite differing functions.
   - Wings of bats and butterflies are analogous, reflecting convergent evolution.

» **Limitations:**
   - Morphological traits can be influenced by environmental factors, leading to phenotypic plasticity.
   - Convergent evolution and cryptic species complicate the interpretation of morphological data.

2. Molecular and Genetic Approaches

With the advent of molecular biology, DNA, RNA, and protein sequences became primary tools for studying evolutionary relationships. These approaches rely on the premise that genetic similarity reflects evolutionary relatedness.

1. **DNA Sequencing:**
   - Whole-genome sequencing, mitochondrial DNA analysis, and ribosomal RNA (rRNA) sequencing are widely used for phylogenetic studies.
   - Highly conserved regions, such as 16S rRNA genes, provide insights into deep evolutionary relationships, while rapidly evolving regions reveal recent divergences.
2. **Molecular Markers:**
   - Microsatellites, single nucleotide polymorphisms (SNPs), and amplified fragment length polymorphisms (AFLPs) are used to compare genetic diversity and relatedness among species.
3. **Comparative Genomics:**
   - Aligning genomes of different species identifies conserved genes, regulatory elements, and structural variations.
   - For example, comparing the human genome with primates reveals shared evolutionary traits and lineage-specific innovations.

3. Phylogenetic Analysis

Phylogenetics involves constructing evolutionary trees (phylogenies) to visualize relationships among species. Modern phylogenetic methods integrate molecular data with computational tools to infer evolutionary histories.

1. **Tree Construction Methods:**
   - **Maximum Parsimony:** Identifies the simplest tree requiring the fewest evolutionary changes.
   - **Maximum Likelihood:** Estimates the tree most likely to explain the observed data based on a specific model of evolution.
   - **Bayesian Inference:** Uses probabilistic models to infer phylogenies with measures of uncertainty.
2. **Phylogenetic Trees:**
   - Trees represent ancestral relationships, with branches denoting divergence events and nodes indicating common ancestors.
   - Clades (monophyletic groups) consist of all descendants from a single ancestor, while paraphyletic or polyphyletic groups indicate incomplete or artificial groupings.

4. Phylogenomics

Phylogenomics combines genome-wide data with phylogenetic analysis to achieve higher resolution and accuracy. By examining hundreds or thousands of genes, phylogenomics overcomes limitations of single-gene studies, such as incomplete lineage sorting or horizontal gene transfer.

» **Applications:**

- Resolving deep divergences, such as the relationships among major animal phyla.
- Identifying cryptic species with su*Bt*le genetic differences.
- Studying gene family evolution, such as the duplication of opsin genes in vertebrates.

5. Fossil Records and Paleogenomics

Fossil evidence provides a direct record of extinct species and evolutionary transitions. Advances in ancient DNA extraction and sequencing have expanded our ability to study extinct organisms.

» **Examples:**

- Sequencing Neanderthal and Denisovan genomes has clarified their relationship with modern humans.
- Fossilized teeth and bones reveal morphological and genetic traits of early mammals.

» **Limitations:**

- Fossil preservation biases favor species with hard tissues.
- Ancient DNA is often degraded, requiring specialized techniques for analysis.

Molecular Basis of Evolutionary Relationships

1. **Homologous Sequences:**
   - Homology in DNA, RNA, or protein sequences indicates shared ancestry. Orthologs (genes separated by speciation events) and paralogs (genes duplicated within a genome) are key to understanding evolutionary divergence.
2. **Molecular Clocks:**
   - By estimating mutation rates, molecular clocks provide timelines for divergence events. For example, the divergence of humans and chimpanzees is estimated at 5–7 million years ago based on molecular clock analyses.
3. **Horizontal Gene Transfer (HGT):**
   - In microbes, HGT complicates phylogenetic studies by transferring genes between unrelated lineages. Comparative genomics helps identify HGT events and their evolutionary implications.

Applications in Evolutionary Biology

Establishing evolutionary relationships has far-reaching implications across various scientific disciplines:

1. **Biodiversity Studies:**
   - Phylogenetics identifies new species, maps global biodiversity, and informs conservation strategies.
2. **Conservation Biology:**
   - Understanding evolutionary relationships helps prioritize conservation efforts, focusing on evolutionary distinct species or keystone taxa.
3. **Medicine:**
   - Comparative studies of pathogens reveal evolutionary mechanisms underlying drug resistance and virulence.
4. **Agriculture:**
   - Evolutionary relationships among crop species and their wild relatives guide breeding programs for disease resistance and yield improvement.

Limitations and Future Directions

1. **Data Integration:**
   - Combining data from fossils, morphology, and genomics remains challenging, requiring interdisciplinary collaboration.
2. **Incomplete Lineage Sorting:**
   - Conflicting signals in genetic data due to rapid speciation events complicate tree reconstruction.
3. **Computational Demands:**
   - Analyzing genome-scale data requires sophisticated bioinformatics tools and high-performance computing.

3.2 Resolving Taxonomic Ambiguities Using Genetic Data

Taxonomy, the science of classifying and naming organisms, is fundamental to understanding biological diversity and evolutionary relationships. However, the field has long grappled with challenges that obscure clear classification, referred to as taxonomic ambiguities. These ambiguities arise from various factors, such as morphological similarities among unrelated species, the presence of cryptic species, hybridization, and environmental influences on phenotypic traits. Traditional taxonomy, which heavily relied on observable characteristics, often struggled to resolve these issues, leading to incomplete or conflicting classifications.

The advent of molecular biology and genetic data has revolutionized taxonomy, providing objective and quantifiable tools to address these ambiguities. By examining the genetic blueprints of organisms, scientists can uncover evolutionary relationships, delineate species boundaries, and identify cryptic diversity with unparalleled precision. Genetic approaches such as DNA barcoding, comparative genomics, and phylogenetic analyses have become indispensable in resolving complex taxonomic puzzles, transforming taxonomy into a more integrative and robust discipline.

Causes of Taxonomic Ambiguities

Taxonomic ambiguities stem from a variety of biological and methodological factors. Understanding these challenges is essential for effectively applying genetic tools to overcome them.

1. **Cryptic Species:** Cryptic species are genetically distinct groups of organisms that are morphologically indistinguishable. For example, in many amphibians, such as frogs, visually identical individuals from different populations are revealed to be separate species upon genetic analysis.
2. **Convergent Evolution:** Convergent evolution leads to unrelated species developing similar traits due to analogous environmental pressures. This phenomenon complicates classification, as similarities in traits may not reflect shared ancestry. For instance, the wings of bats and birds are structurally different yet functionally similar due to convergent evolution.
3. **Phenotypic Plasticity:** Environmental factors can influence the physical traits of organisms, leading to significant variation within a species. This plasticity often makes it challenging to distinguish between species or populations based on morphology alone.
4. **Hybridization:** Hybridization occurs when closely related species interbreed, producing hybrids that combine traits from both parent species. In plants, hybrids are particularly common and can blur the lines between species.
5. **Incomplete Fossil Records:** In paleontology, the lack of comprehensive fossil evidence often results in speculative or incorrect classification of extinct species.
6. **Intraspecific Variation:** High genetic or morphological diversity within a single species may lead to misclassification as multiple species, especially in taxa with wide geographic distributions.

Role of Genetic Data in Resolving Ambiguities

Genetic data provides a powerful and objective foundation for addressing taxonomic ambiguities. By analyzing DNA, RNA, or protein sequences, researchers can uncover genetic differences that are invisible to traditional morphological methods. Key genetic approaches have transformed taxonomy and offer solutions to specific challenges:

DNA Barcoding:

DNA barcoding is a widely adopted method for species identification and classification. It involves sequencing a standardized region of DNA that serves as a "barcode" for each species. In animals, the mitochondrial *cytochrome c oxidase I* (COI) gene is the most commonly used marker, while in plants, the *matK* and *rbcL* genes are preferred.

Advantages:

» Barcoding provides a universal framework for identifying species across diverse taxa.

» It is particularly useful for distinguishing cryptic species, such as those found in marine ecosystems or tropical forests.

Applications:

- DNA barcoding has been instrumental in identifying over 500 cryptic species in reef fish, significantly enriching our understanding of marine biodiversity.

Phylogenetics:

Phylogenetics uses genetic data to construct evolutionary trees, or phylogenies, which visually represent relationships among species. By comparing conserved genetic regions across taxa, phylogenetic analyses reveal common ancestors and divergence events.

Applications:

- Phylogenetic methods resolved the relationships within the genus *Homo*, clarifying the evolutionary lineage of humans, Neanderthals, and Denisovans.
- In fungi, phylogenetics has been pivotal in distinguishing species complexes previously classified as single species.

Methods:

- Tools like Maximum Likelihood and Bayesian Inference are used to construct robust phylogenies from genomic data.

Comparative Genomics:

Comparative genomics examines entire genomes to identify genetic similarities and differences among species. This approach is particularly effective for resolving ambiguities in groups with high morphological overlap or complex evolutionary histories.

Applications:

- In flowering plants, comparative genomics has revealed that some species thought to be hybrids are ancient polyploids with duplicated genomes.
- Among insects, genome-wide comparisons have uncovered significant cryptic diversity in economically important pests, aiding pest control strategies.

Metagenomics and Environmental DNA (eDNA)

Metagenomics and eDNA allow scientists to analyze genetic material from environmental samples, such as soil, water, or air. This approach is especially valuable for detecting species in challenging environments or in cases where direct sampling is impractical.

Applications:

- eDNA has been used to monitor endangered amphibians in freshwater habitats, revealing the presence of previously undetected populations.
- In microbial taxonomy, metagenomics has identified thousands of new bacterial species from environmental samples.

Molecular Markers: Microsatellites and SNPs

Microsatellites and single nucleotide polymorphisms (SNPs) are high-resolution genetic markers used to study population structure and species boundaries.

Applications:

» Microsatellite analysis resolved the taxonomy of the critically endangered giant panda, confirming its classification as a single species despite significant genetic variation among populations.

Case Studies

1. **Cryptic Species Complex in Frogs:** Genetic analysis of *Rana pipiens* (leopard frogs) revealed multiple cryptic species across North America, reshaping conservation strategies for these amphibians.
2. **Hybrid Sunflowers:** Hybridization among species in the genus *Helianthus* was confirmed through genomic data, revealing that hybrids play a role in adaptive evolution and speciation.
3. **Microbial Diversity in Soils:** Metagenomic studies of soil microbiomes have identified hundreds of new microbial taxa, highlighting the vast, unexplored diversity of microorganisms.

Advantages of Using Genetic Data in Taxonomy

1. **Objectivity and Reproducibility:**
    - Genetic data provides quantifiable evidence for species identification, reducing subjectivity in classification.
2. **High Resolution:**
    - Molecular techniques detect genetic differences at the finest scales, enabling the identification of cryptic and hybrid species.
3. **Universal Applicability:**
    - Genetic methods are applicable across all taxa, including microorganisms, plants, and animals.
4. **Integration with Traditional Methods**:
    - Combining genetic data with morphology and ecology provides a holistic approach to taxonomy.

Limitations of Genetic Approaches

1. **Database Dependency:**
    - Accurate identification relies on comprehensive reference databases, which are still incomplete for many taxa.
2. **Resource-Intensive:**
    - High-throughput sequencing and bioinformatics require substantial financial and computational resources.

3. **Complexity in Hybridization:**
   - Genetic admixture in hybrid species can obscure clear boundaries, complicating classification.

3.3 Case Studies of DNA Fingerprinting in Plants, Insects, and Other Organisms

DNA fingerprinting is a transformative technique that has significantly advanced our understanding of genetics, evolution, and biodiversity. It enables precise identification of species, assessment of genetic diversity, and resolution of complex evolutionary relationships. By using genetic markers such as microsatellites, RAPD (Random Amplified Polymorphic DNA), AFLP (Amplified Fragment Length Polymorphism), and SNPs (Single Nucleotide Polymorphisms), researchers can uncover detailed insights into the genetic makeup of organisms. These tools have been applied to plants, insects, and a variety of other taxa to address issues ranging from conservation to agricultural improvement.

Case Studies in Plants

Plants have been a focal point for DNA fingerprinting due to their ecological, economic, and evolutionary importance. The following case studies illustrate the use of genetic tools in plant research and conservation.

1. Genetic Diversity in Rice (*Oryza sativa*)

Rice (*Oryza sativa*) is one of the world's most important staple crops, feeding over half the global population. Understanding its genetic diversity is crucial for breeding programs aimed at enhancing yield, stress tolerance, and pest resistance.

Objective: To assess genetic variation among rice cultivars to identify traits beneficial for agricultural improvement.

Methodology:

- » Researchers collected over 200 rice accessions, including wild relatives and cultivated varieties.
- » DNA fingerprinting was conducted using microsatellite markers, which are highly polymorphic and distributed throughout the rice genome.
- » Genetic variation was analyzed by amplifying and comparing allelic patterns at specific loci.

Findings:

- » Clear genetic differentiation was observed between the two major subspecies, *indica* and *japonica*, as well as within regional landraces.
- » Microsatellite data revealed specific genomic regions associated with drought tolerance, salinity resistance, and pest resilience.
- » Wild relatives displayed a broader genetic base, offering untapped potential for introgressing beneficial traits into cultivated varieties.

Impact:

- » The study provided a roadmap for crop improvement programs, allowing breeders to develop high-yielding and stress-resistant varieties.
- » The findings highlighted the importance of conserving wild rice species as reservoirs of genetic diversity.

2. Hybrid Origins in Sunflowers (*Helianthus spp.*)

Sunflowers (*Helianthus spp.*) are economically valuable for their oil, seeds, and ornamental uses. The genus also serves as a model for studying hybridization and speciation.

Objective: To investigate the hybrid origins of specific sunflower species and their adaptive traits.

Methodology:

- » AFLP markers were employed to analyze genetic relationships among hybrid and parental sunflower species.
- » Over 1,000 polymorphic markers were generated and compared across populations.

Findings:

- » DNA fingerprinting confirmed that species such as *Helianthus anomalus* arose from hybridization between *H. annuus* and *H. petiolaris*.
- » Adaptive traits, including tolerance to arid conditions and saline soils, were traced to genomic regions inherited from both parental species.

Impact:

- » The research demonstrated how hybridization contributes to ecological diversification and speciation.
- » The findings informed breeding strategies to introduce adaptive traits into cultivated sunflowers.

3. Conservation Genetics of the Endangered Tree *Shorea leprosula*

Objective: To study genetic diversity and connectivity in fragmented populations of *Shorea leprosula*, a commercially valuable but endangered tree in Southeast Asia.

Methodology:

- » Microsatellite markers were used to analyze genetic variation across populations in Malaysia and Indonesia.
- » Gene flow and population structure were assessed to identify genetic bottlenecks.

Findings:

- » Smaller, isolated populations exhibited reduced genetic diversity and signs of inbreeding, while larger populations retained greater genetic variation.
- » Certain forest corridors were identified as critical for maintaining gene flow.

Impact:

- The results guided conservation strategies, including habitat restoration and the establishment of protected corridors to enhance genetic connectivity.
- The study highlighted the importance of genetic monitoring for sustainable forest management.

Case Studies in Insects

Insects play pivotal roles in ecosystems as pollinators, pests, and vectors of disease. DNA fingerprinting has been instrumental in studying their genetic diversity, population structure, and evolutionary relationships.

1. Cryptic Species in Mosquitoes (*Anopheles gambiae* Complex)

The *Anopheles gambiae* complex includes several species that are major vectors of malaria in Africa. Morphological methods alone are insufficient to differentiate these cryptic species.

Objective: To distinguish cryptic species within the *Anopheles gambiae* complex and understand their ecological and behavioral differences.

Methodology:

- RAPD and microsatellite markers were employed to analyze genetic variations among populations.
- Samples were collected from diverse ecological settings, including urban and rural habitats.

Findings:

- The study revealed distinct genetic profiles for species such as *An. gambiae*, *An. arabiensis*, and *An. quadriannulatus*.
- Behavioral differences, such as breeding site preferences and insecticide susceptibility, were correlated with genetic distinctions.

Impact:

- The findings provided a foundation for targeted vector control strategies, including habitat-specific interventions and tailored insecticide programs.

2. Genetic Variation in Honeybees (*Apis mellifera*)

Honeybees (*Apis mellifera*) are essential pollinators in agricultural and natural ecosystems. Their genetic health is vital for maintaining ecological balance and crop productivity.

Objective: To assess genetic diversity and population structure among honeybee subspecies across different geographic regions.

Methodology:

- Microsatellite markers were used to analyze genetic variation in honeybee populations from Europe, Africa, and the Americas.
- DNA fingerprinting revealed gene flow patterns and hybridization events.

Findings:

- Africanized honeybees, hybrids between European and African subspecies, exhibited unique genetic traits that enhanced their resilience to environmental stressors.
- European populations showed reduced genetic diversity due to selective breeding and habitat loss.

Impact:

- The study informed breeding programs aimed at improving disease resistance and adaptability in honeybee populations.
- It also highlighted the need for conservation efforts to preserve genetic diversity in wild honeybee populations.

3. Tracking the Spread of the Fall Armyworm (*Spodoptera frugiperda*)

The fall armyworm is a notorious pest affecting maize and other crops worldwide. Understanding its genetic variation is critical for managing its spread and impact.

Objective: To trace the geographic origin and genetic differentiation of fall armyworm populations.

Methodology:

- SNP analysis was conducted on populations from North and South America, Africa, and Asia.
- DNA fingerprinting identified genetic haplotypes associated with specific host plants.

Findings:

- Two distinct haplotypes, one associated with corn and the other with rice, were identified.
- The pest's rapid spread to Africa and Asia was linked to genetic adaptability and high reproductive rates.

Impact:

- The findings supported the development of region-specific pest management strategies, including biocontrol measures and resistant crop varieties.

Case Studies in Other Organisms

DNA fingerprinting has applications beyond plants and insects, extending to mammals, microorganisms, and even forensic investigations. The following case studies demonstrate its versatility.

1. Conservation Genetics of Tigers (*Panthera tigris*)

Tigers are apex predators whose populations are under severe threat due to habitat loss, poaching, and genetic isolation.

Objective: To study genetic diversity and population structure in wild tiger populations across Asia.

Methodology:

- Microsatellite markers were used to analyze DNA from scat, hair, and tissue samples.
- Genetic profiles were compared across different subspecies, including Bengal, Siberian, and Sumatran tigers.

Findings:

- Significant genetic differentiation was observed between subspecies, emphasizing the need for subspecies-specific conservation strategies.
- Isolated populations exhibited reduced genetic diversity, increasing their vulnerability to extinction.

Impact:

- The study informed conservation policies, including the creation of wildlife corridors and anti-poaching initiatives.

2. Microbial Diversity in Extreme Environments

Microorganisms in extreme environments offer insights into the limits of life and potential biotechnological applications.

Objective: To uncover genetic diversity in extremophilic microbes from hydrothermal vents.

Methodology:

- Metagenomic sequencing was used to analyze DNA from microbial communities.
- Unique genetic markers were identified for thermophilic and chemolithotrophic species.

Findings:

- The study discovered several novel species with adaptations to high temperature, pressure, and chemical extremes.
- Enzymes encoded by these microbes showed potential for industrial applications, such as biofuel production and wastewater treatment.

Impact:

- The findings expanded our understanding of microbial diversity and adaptation.
- They also highlighted the biotechnological potential of extremophilic organisms.

3. Combating Wildlife Trafficking

Illegal wildlife trade poses a significant threat to biodiversity. DNA fingerprinting has become a critical tool in forensic investigations to trace the origin of seized animal products.

Objective: To identify the geographic origin of confiscated ivory and rhino horns.

Methodology:

- DNA extracted from ivory and rhino horn samples was compared with reference databases of wild populations.
- Mitochondrial DNA and microsatellite markers were used for species and population identification.

Findings:

- The analysis pinpointed poaching hotspots in Africa, such as areas in Central and East Africa for elephants and South Africa for rhinos.
- It revealed organized trafficking networks linking poachers to international markets.

Impact:

- DNA fingerprinting has been instrumental in strengthening law enforcement and conservation efforts.
- It also supports international collaborations aimed at curbing wildlife trafficking.

4: Genetic Improvement of Inebriate Tolerance in Natural Enemies

4.1 Definition and Significance of Natural Enemies in Pest Control

Definition of Natural Enemies

Natural enemies, often referred to as biological control agents, are organisms that prey upon, parasitize, or otherwise suppress pest populations in agricultural or natural ecosystems. These include predators, parasitoids, and pathogens, which play a pivotal role in regulating pest populations naturally without the need for chemical interventions.

- **Predators:** Organisms that consume multiple prey items during their lifetime. Examples include lady beetles (*Coccinellidae*), lacewings (*Chrysopidae*), and spiders, which prey on a wide varicty of agricultural pests like aphids, mites, and caterpillars.
- **Parasitoids:** Insects, primarily from the orders Hymenoptera (wasps) and Diptera (flies), whose larvae develop inside or on a host, ultimately killing it. Parasitoids like *Trichogramma* wasps parasitize insect eggs, preventing them from developing into adult pests.
- **Pathogens:** Microorganisms, such as bacteria (*Bacillus thuringiensis*), fungi (*Beauveria bassiana*), viruses, and nematodes, that infect and kill pests, often causing outbreaks that naturally control pest populations.

Characteristics of Natural Enemies

Natural enemies exhibit several key characteristics that make them effective in pest control:

1. **Host Specificity:** Many natural enemies are highly selective, targeting specific pests without harming beneficial organisms or crops. For example, *Trichogramma* wasps parasitize eggs of moths and butterflies, leaving non-target species unaffected.

2. **Reproductive Potential:** Natural enemies, such as parasitoids, can reproduce rapidly under favorable conditions, increasing their populations in response to rising pest densities.
3. **Adaptability:** Predators and parasitoids are often highly adaptable, thriving in diverse environments and coevolving with their hosts.
4. **Persistence:** Certain natural enemies, such as predatory beetles or spiders, establish themselves in ecosystems and provide long-term pest control.

Significance of Natural Enemies in Pest Control

Natural enemies play an essential role in integrated pest management (IPM) by reducing pest populations and minimizing reliance on chemical pesticides. Their significance can be understood through the following perspectives:

1. Environmental Benefits

» **Reduction in Chemical Pesticides:** The use of natural enemies reduces the need for synthetic pesticides, mitigating their harmful effects on non-target organisms, soil health, and water quality.

» **Biodiversity Preservation:** Encouraging natural enemies promotes ecological balance by maintaining diverse and stable ecosystems. This is especially critical in areas prone to pest outbreaks due to monoculture practices.

2. Economic Advantages

» **Cost-Effectiveness:** Biological control is often less expensive in the long term compared to chemical control, particularly when natural enemies establish self-sustaining populations.

» **Sustainable Agriculture:** Natural enemies reduce crop losses due to pests while minimizing the costs associated with repeated pesticide applications.

3. Reduced Resistance in Pests

Overreliance on chemical pesticides has led to the evolution of pesticide-resistant pest populations. Natural enemies provide an alternative, non-chemical method of control, reducing selection pressure for resistance and ensuring long-term effectiveness.

4. Specificity and Safety

Unlike broad-spectrum pesticides, natural enemies often target specific pests, reducing collateral damage to beneficial insects, pollinators, and other wildlife. For example:

» *Bacillus thuringiensis* (*Bt*) targets caterpillar pests without affecting beneficial insects like bees.

» Predatory mites selectively feed on pest mites while sparing other organisms.

5. Adaptation to Local Environments

Natural enemies coevolve with their hosts and local ecosystems, making them well-suited for biological control in their native or similar environments. For example, introducing predatory beetles to combat aphid populations has proven effective in both temperate and tropical climates.

Examples of Natural Enemies in Pest Control

1. **Lady Beetles (*Coccinellidae*):**
   - Prey on aphids, whiteflies, and other soft-bodied pests.
   - Widely used in greenhouses and open fields for aphid control.
2. ***Trichogramma* Wasps:**
   - Parasitize eggs of moths and butterflies.
   - Effective against lepidopteran pests in crops such as cotton, corn, and rice.
3. **Green Lacewings (*Chrysopidae*):**
   - Larvae, commonly known as "aphid lions," are voracious predators of aphids, mealybugs, and thrips.
4. **Entomopathogenic Fungi:**
   - *Beauveria bassiana* infects and kills insect pests, such as beetles, whiteflies, and termites, through spore germination.
5. **Predatory Mites:**
   - Control pest mites in orchards, vineyards, and greenhouses by feeding on their eggs and immature stages.

Challenges in Using Natural Enemies

Despite their benefits, natural enemies face challenges that can limit their effectiveness:

1. **Habitat Loss and Fragmentation:**
   - Agricultural intensification reduces habitats for natural enemies, affecting their populations and effectiveness.
2. **Pesticide Use:**
   - Residual pesticides can inadvertently harm natural enemies, disrupting biological control efforts.
3. **Climatic Factors:**
   - Extreme temperatures, humidity, and other environmental conditions can limit the survival and efficacy of certain natural enemies.
4. **Compatibility Issues:**
   - Introducing non-native natural enemies can sometimes result in unintended ecological consequences, such as competition with local species or harm to non-target organisms.

Integration into Pest Management

To maximize the effectiveness of natural enemies, they are often integrated into pest management strategies, such as:

1. **Habitat Augmentation:**
   - Providing habitats, such as hedgerows or flowering plants, supports the populations of natural enemies by offering shelter and alternative food sources.
2. **Conservation Biological Control:**
   - Minimizing pesticide use and adopting sustainable farming practices help preserve natural enemy populations.
3. **Augmentative Releases:**
   - Releasing natural enemies at critical times, such as during pest outbreaks, boosts their impact on pest populations.
4. **Classical Biological Control:**
   - Introducing natural enemies from a pest's native region to control invasive species, as demonstrated by the introduction of *Vedalia beetles* to control cottony cushion scale in citrus orchards.

4.2 Challenges Posed by Pesticides and Insecticides to the Survival of Natural Enemies

The use of chemical pesticides and insecticides has been a cornerstone of modern agriculture for decades, providing an efficient means to control pest populations and enhance crop yields. However, this reliance on chemical controls has had unintended consequences, particularly for natural enemies such as predators, parasitoids, and pathogens that play critical roles in regulating pest populations. These beneficial organisms often fall victim to the non-specific action of pesticides, disrupting ecosystems and undermining sustainable pest management strategies. The survival and efficacy of natural enemies are compromised by the direct toxic effects of pesticides, as well as by indirect consequences such as habitat disruption, food scarcity, and physiological stress.

Direct Effects of Pesticides on Natural Enemies

1. **Toxicity**:
   - Many pesticides are broad-spectrum, targeting a wide range of insects without distinguishing between pests and beneficial organisms. For instance:
     - ◊ Organophosphates and pyrethroids, commonly used in agriculture, are highly toxic to natural predators like lady beetles (*Coccinellidae*) and parasitoid wasps (*Trichogramma* spp.).
     - ◊ Sublethal doses can also impair natural enemies by reducing their longevity, reproduction, and mobility.

2. **Mortality Rates:**

    - Contact pesticides applied directly to crops can kill natural enemies present on the foliage, soil, or within pest-host habitats. For example:

        ◊ Predatory mites that regulate spider mite populations are highly susceptible to pesticide sprays.

3. **Developmental Delays:**

    - Exposure to pesticides during critical developmental stages can result in delayed maturation or morphological abnormalities in natural enemies, reducing their effectiveness as biological control agents.

Indirect Effects of Pesticides on Natural Enemies

1. **Reduction in Prey or Host Availability:**

    - Pesticides can decimate pest populations that serve as food or hosts for natural enemies, leading to starvation or reduced reproduction rates. For example:

        ◊ Parasitoid wasps rely on specific insect hosts for reproduction. The elimination of these hosts disrupts their life cycle.

2. **Habitat Disruption:**

    - Repeated pesticide applications can degrade habitats by reducing plant diversity, altering microclimates, and eliminating alternative food sources like pollen and nectar.
    - This is particularly detrimental to generalist predators and parasitoids that rely on diverse ecosystems for survival.

3. **Behavioral Alterations:**

    - Sublethal pesticide exposure can affect the foraging, mating, and dispersal behaviors of natural enemies. For instance:

        ◊ Predatory beetles exposed to neonicotinoids exhibit impaired prey detection and reduced hunting efficiency.

4. **Physiological Stress:**

    - Even when not lethal, pesticides can induce oxidative stress or disrupt metabolic pathways in natural enemies, weakening their resilience to environmental challenges.

Case Studies: Impact of Pesticides on Natural Enemies

1. **Impact on Lady Beetles (*Coccinellidae*):**

    - Studies have shown that pyrethroid insecticides significantly reduce the survival rates of lady beetles, which are key predators of aphids.
    - Sublethal exposure has been linked to reduced egg-laying and impaired larval development.

2. **Decline in Parasitoid Populations:**
   - *Trichogramma* wasps, widely used in biological control, suffer high mortality rates when exposed to common pesticides like carbamates.
   - The disruption of their parasitic behavior results in unchecked pest population growth.
3. **Harm to Pollinators and Predators:**
   - Neonicotinoids, applied as seed treatments or foliar sprays, have been implicated in declines of pollinators like bees and predatory insects such as lacewings (*Chrysopidae*).

Long-Term Ecological Consequences

1. **Pest Resurgence:**
   - ◊ The elimination of natural enemies creates ecological vacuums that allow pest populations to rebound rapidly, often requiring higher doses of pesticides for control.
2. **Secondary Pest Outbreaks:**
   - Natural enemies regulate secondary pests that are not the primary target of pesticides. Their decline can lead to outbreaks of previously minor pests, exacerbating crop losses.
3. **Biodiversity Loss:**
   - Widespread pesticide use reduces overall biodiversity, affecting not only natural enemies but also non-target organisms that contribute to ecosystem health.
4. **Development of Pesticide Resistance:**
   - Overuse of pesticides accelerates resistance in pest populations, reducing the long-term efficacy of chemical control and increasing reliance on unsustainable practices.

Mitigating the Impact of Pesticides on Natural Enemies

1. **Selective Pesticides:**
   - Developing and using pesticides with high specificity for target pests minimizes collateral damage to natural enemies. For example:
     - ◊ Insect growth regulators (IGRs) primarily affect pest species while sparing beneficial insects.
2. **Reduced Application Rates:**
   - Applying pesticides at the lowest effective dose reduces the exposure risk for natural enemies.
3. **Timing of Applications:**
   - Targeting pesticide applications to periods when natural enemies are least active or abundant (e.g., early morning or late evening) can reduce their impact.
4. **Habitat Enhancement:**
   - Creating refuges, such as hedgerows, flower strips, or cover crops, provides safe habitats for natural enemies and alternative food sources.

5. **Adoption of Integrated Pest Management (IPM):**
   - Combining biological control with cultural, mechanical, and chemical control methods ensures that pesticides are used judiciously and only when necessary.
6. **Research and Monitoring:**
   - Continuous monitoring of pesticide effects on natural enemies and further research into their ecological roles can guide sustainable pest management practices.

4.3 Genetic Engineering Approaches to Enhance Insecticide Tolerance

Insecticides, while essential for controlling pest populations, often pose significant risks to beneficial organisms such as natural predators, parasitoids, and pollinators. Enhancing insecticide tolerance in these beneficial organisms through genetic engineering represents a novel strategy to sustain their populations in agricultural ecosystems. By modifying their genetic makeup, scientists aim to develop organisms that can withstand exposure to common insecticides, ensuring their survival and continued efficacy in pest management programs.

Genetic Engineering Approaches

Genetic engineering employs advanced molecular tools to alter the genetic structure of organisms, enabling the introduction or modification of traits such as insecticide tolerance. The following approaches have been explored to enhance tolerance in beneficial organisms:

1. Overexpression of Detoxification Genes

Mechanism:

- » Insecticides are often detoxified by enzymes such as cytochrome P450 monooxygenases, glutathione-S-transferases (GSTs), and carboxylesterases.
- » Genetic engineering can upregulate the expression of these detoxification enzymes, allowing organisms to metabolize and neutralize insecticides more effectively.

Example:

- » In transgenic wasps, the overexpression of cytochrome P450 genes has been shown to confer resistance to organophosphate and pyrethroid insecticides, enabling them to parasitize pests in treated fields.

2. Introduction of Exogenous Resistance Genes

Mechanism:

- » Genes from insecticide-resistant pest populations or other organisms are introduced into the genome of beneficial species.
- » These genes encode proteins that neutralize or prevent the binding of insecticides to their target sites.

Example:

- » In lady beetles, the introduction of genes encoding modified acetylcholinesterase (AChE) has provided resistance to carbamate and organophosphate insecticides.

3. Gene Editing with CRISPR-Cas9

Mechanism:

- » CRISPR-Cas9 allows precise editing of the genome to introduce mutations that confer resistance.
- » Target sites for insecticides, such as voltage-gated sodium channels or acetylcholinesterase, are altered to prevent binding by the insecticide.

Example:

- » In parasitoid wasps, CRISPR-Cas9 has been used to modify the sodium channel gene, conferring resistance to pyrethroid insecticides while maintaining their natural pest control efficacy.

4. RNA Interference (RNAi) for Targeted Protection

Mechanism:

- » RNAi silences genes critical for insecticide susceptibility, such as those encoding insecticide-binding proteins.
- » Beneficial organisms are genetically engineered to produce small interfering RNAs (siRNAs) that block the expression of these susceptibility genes.

Example:

- » RNAi has been used to enhance tolerance in pollinators like bees by silencing genes involved in susceptibility to neonicotinoids.

5. Symbiont-Mediated Protection

Mechanism:

- » Beneficial organisms often harbor microbial symbionts that can be genetically engineered to metabolize insecticides.
- » These engineered symbionts produce enzymes that break down toxic compounds before they affect the host.

Example:

- » In predatory mites, symbiotic bacteria have been engineered to degrade organophosphate insecticides, ensuring the mites' survival in treated environments.

Advantages of Genetic Engineering for Insecticide Tolerance

1. **Sustained Biological Control:** Enhanced tolerance ensures the survival of beneficial organisms in pesticide-treated environments, maintaining their role in pest management.
2. **Reduced Pesticide Resistance in Pests:** With robust populations of natural enemies, pest populations are suppressed without relying solely on chemical control, slowing the development of resistance.
3. **Target-Specific Protection:** Genetic modifications can be tailored to specific insecticides, minimizing risks to non-target organisms.
4. **Environmental Safety:** By reducing the need for repeated pesticide applications, genetic engineering promotes sustainable agriculture and ecological balance.

Challenges and Ethical Considerations

1. **Off-Target Effects:** Genetic modifications may have unintended consequences on other physiological processes or behaviors of the organism.
2. **Ecological Impact:** The introduction of genetically engineered organisms into ecosystems must be carefully monitored to avoid disrupting existing ecological interactions.
3. **Public Perception:** The use of genetically modified organisms (GMOs) in biological control raises ethical and regulatory concerns that require transparent communication and oversight.

4.4 Case Studies on Improved Tolerance in Beneficial Organisms

Case Study 1: Enhanced Insecticide Tolerance in Lady Beetles (*Coccinellidae*)

Objective: To develop lady beetles with increased tolerance to organophosphate and pyrethroid insecticides used in aphid control.

Methodology:

- Genes encoding detoxification enzymes, such as cytochrome P450 and carboxylesterases, were introduced into the genome of *Hippodamia convergens*, a common predatory lady beetle.
- CRISPR-Cas9 was used to modify the sodium channel gene, reducing susceptibility to pyrethroids.

Findings:

- Genetically engineered lady beetles exhibited a significant increase in survival rates when exposed to field doses of insecticides.
- Behavioral studies confirmed that enhanced tolerance did not compromise their predatory efficiency or reproduction.

Impact:

- The engineered lady beetles successfully controlled aphid populations in pesticide-treated fields, reducing crop losses and promoting sustainable pest management.

Case Study 2: Genetic Modification of Parasitoid Wasps (*Trichogramma* spp.)

Objective: To enhance the survival of *Trichogramma* wasps in pesticide-treated agricultural landscapes.

Methodology:

- Cytochrome P450 genes from insecticide-resistant pest species were introduced into *Trichogramma chilonis*.
- RNAi technology was employed to suppress genes associated with insecticide susceptibility.

Findings:

- Modified wasps displayed resistance to carbamates and pyrethroids, with survival rates exceeding 90% in treated fields.
- Parasitism rates on lepidopteran pest eggs remained consistent, indicating no loss of biological control efficacy.

Impact:

- Farmers reported reduced dependency on chemical insecticides due to the effectiveness of these engineered parasitoids in pest suppression.

Case Study 3: Symbiont-Enhanced Tolerance in Predatory Mites

Objective: To develop predatory mites resistant to organophosphate insecticides used in orchards and vineyards.

Methodology:

- Symbiotic bacteria within the mites were genetically engineered to produce enzymes capable of breaking down organophosphates.
- The engineered mites were then tested in pesticide-treated environments for survival and pest control efficacy.

Findings:

- Predatory mites harboring the engineered symbionts maintained high survival rates and effectively controlled pest mite populations.
- The engineered symbionts showed no adverse effects on the mites' physiology or behavior.

Impact:

- This approach provided a non-invasive alternative to direct genetic modification of the host, addressing ethical concerns while ensuring ecological compatibility.

Case Study 4: Pollinator Protection Against Neonicotinoids

Objective: To protect honeybees (*Apis mellifera*) from the harmful effects of neonicotinoid insecticides.

Methodology:

- RNAi was used to silence genes involved in the susceptibility of bees to neonicotinoids.
- Genetically engineered bees were tested for foraging behavior, reproduction, and colony health in treated fields.

Findings:

- Engineered bees exhibited reduced mortality rates and maintained normal foraging and reproductive behaviors.
- Colonies thrived in pesticide-treated environments, with no observable decline in population size.

Impact:

- This breakthrough demonstrated the feasibility of protecting pollinators without reducing pesticide efficacy against target pests.

5: Integration of DNA Fingerprinting and Genetic Improvement in Pest Management

Pest management is a critical aspect of agricultural productivity and environmental sustainability. The integration of advanced molecular techniques like DNA fingerprinting with genetic improvement strategies has revolutionized how we manage pest populations and protect crops.

5.1 The Role of DNA Fingerprinting in Identifying and Enhancing Traits in Natural Enemies

DNA fingerprinting, a molecular technique that identifies unique genetic patterns, has become an indispensable tool in pest management. It provides detailed insights into the genetic makeup of natural enemies, such as predators, parasitoids, and pathogens, enabling researchers to identify beneficial traits and optimize their effectiveness in biological control programs.

Identifying Natural Enemies with DNA Fingerprinting

One of the foremost applications of DNA fingerprinting in pest management is the identification and characterization of natural enemies. This process involves analyzing genetic markers to differentiate between species, strains, or populations.

1. **Species Identification:**
   - Many natural enemies, such as parasitoid wasps and predatory mites, belong to cryptic species complexes that are morphologically similar but genetically distinct. DNA fingerprinting enables precise identification, ensuring the correct species is employed in pest control.
   - Example: Studies on *Trichogramma* wasps, widely used to control lepidopteran pests, have identified cryptic species with varying levels of host specificity and effectiveness through microsatellite and RAPD markers.
2. **Population Structure Analysis:**
   - Understanding the genetic diversity and population structure of natural enemies is

crucial for assessing their adaptability and performance in different environments.

- Example: Microsatellite analysis of lady beetle (*Coccinella septempunctata*) populations revealed genetic differences linked to environmental conditions, guiding their use in specific agricultural regions.

3. **Tracking Biological Control Agents:**
   - DNA fingerprinting allows researchers to track introduced natural enemies in the field, ensuring their establishment and effectiveness.
   - Example: Genetic markers were used to monitor the spread of *Harmonia axyridis* (Asian lady beetle) introduced for aphid control.

Enhancing Traits in Natural Enemies

DNA fingerprinting facilitates the identification of genetic traits associated with pest suppression, enabling targeted breeding or genetic modification to enhance these traits.

1. **Host Specificity:**
   - Genetic markers help identify natural enemies with a high degree of host specificity, reducing risks to non-target species and enhancing pest control efficiency.
   - Example: Molecular studies on *Aphidius colemani*, a parasitoid of aphids, revealed genetic variations influencing host range.
2. **Reproductive Efficiency:**
   - Traits like fecundity and parasitism rates can be linked to specific genetic markers, aiding in the selection of high-performing strains.
   - Example: DNA fingerprinting identified genetic variations in *Phytoseiulus persimilis*, a predatory mite, linked to reproductive success.
3. **Pesticide Tolerance:**
   - DNA fingerprinting identifies genetic factors contributing to insecticide resistance, allowing for the development of natural enemies that can thrive in pesticide-treated environments.
   - Example: Genetic analysis of *Trichogramma* species revealed resistance-associated markers, guiding the selection of tolerant strains.

Applications in Conservation and Augmentation

The use of DNA fingerprinting extends to the conservation and augmentation of natural enemies in biological control programs.

1. **Conservation of Genetic Diversity:**
   - Maintaining genetic diversity is essential for the adaptability and resilience of natural enemy populations. DNA fingerprinting helps monitor diversity levels and prevent inbreeding.

- Example: Studies on *Chrysoperla* lacewings used genetic markers to assess diversity in laboratory-reared populations.

2. **Hybridization for Trait Enhancement:**
   - Controlled hybridization between strains or species with desirable traits is facilitated by genetic profiling.
   - Example: Hybrid lines of *Encarsia formosa*, a parasitoid wasp, were developed to improve pest suppression and environmental tolerance.

5.2 Combined Use of Molecular Tools for Effective Biological Control Strategies

The integration of DNA fingerprinting with other molecular tools has opened new avenues for designing and implementing effective biological control strategies. By combining various techniques, researchers can achieve a comprehensive understanding of pest-natural enemy interactions and optimize control measures.

Synergistic Use of Molecular Techniques

1. **DNA Barcoding and Phylogenetics:**
   - DNA barcoding and phylogenetic analyses provide a taxonomic framework for identifying natural enemies and understanding their evolutionary relationships.
   - Example: DNA barcoding revealed evolutionary divergence among parasitoid wasps, guiding their classification and deployment.
2. **Genomic and Transcriptomic Studies:**
   - Whole-genome and transcriptome sequencing enable the identification of genes involved in critical traits like host specificity, reproduction, and environmental tolerance.
   - Example: Transcriptomic analysis of *Nasonia vitripennis*, a parasitoid wasp, identified genes linked to venom production and parasitism efficiency.
3. **RNA Interference (RNAi):**
   - RNAi is used to silence specific genes in pests, making them more vulnerable to natural enemies.
   - Example: RNAi-targeted suppression of pest defense genes enhanced the effectiveness of entomopathogenic fungi.
4. **Microbiome Studies:**
   - The role of symbiotic microbes in the physiology and behavior of natural enemies is increasingly recognized. Metagenomic studies identify beneficial microbes that can be manipulated for improved pest control.
   - Example: Symbiotic bacteria in predatory mites were shown to enhance tolerance to environmental stressors.

Field-Level Applications

1. **Predicting Pest-Natural Enemy Dynamics:**
   - Molecular tools are used to model interactions between pests and their natural enemies under varying environmental conditions.
   - Example: DNA-based monitoring of pest and predator populations provided insights into ecosystem dynamics in apple orchards.
2. **Enhancing Compatibility with IPM:**
   - Integrating molecularly optimized natural enemies into IPM systems reduces reliance on chemical pesticides.
   - Example: Pesticide-tolerant strains of predatory mites were developed for use in IPM programs for greenhouse crops.

Challenges and Solutions

1. **Scalability:** While molecular techniques provide detailed insights, scaling these approaches for large agricultural systems remains a challenge. Advances in high-throughput sequencing and automated data analysis are addressing these limitations.
2. **Cost Considerations:** The high cost of molecular tools can limit their adoption. Collaborative efforts and government support for research and implementation are essential for wider adoption.

5.3 Future Trends and Innovations in DNA-Based Taxonomy and Pest Control

As technology evolves, DNA-based approaches in taxonomy and pest control are poised to become more sophisticated and impactful. Emerging trends and innovations will further enhance our ability to manage pests sustainably and effectively.

Trends in DNA-Based Taxonomy

1. **Environmental DNA (eDNA) Monitoring:**
   - eDNA analysis enables the detection of pests and natural enemies from environmental samples like soil, water, and air, providing real-time monitoring capabilities.
   - Example: eDNA techniques were used to detect invasive pests in agricultural ecosystems, enabling early intervention.
2. **High-Resolution Taxonomy:**
   - Advances in single-cell genomics and metagenomics are providing unprecedented resolution in classifying cryptic species and microbial natural enemies.
   - Example: Single-cell sequencing identified novel fungal pathogens of agricultural pests.
3. **Automated Taxonomic Tools:**
   - Machine learning and AI are being integrated with DNA-based taxonomy to automate species identification and classification.

- Example: AI-driven platforms analyzed DNA barcode data to classify parasitoid wasps with high accuracy.

Innovations in Pest Control

1. **Gene Editing for Pest Management:**
   - CRISPR-Cas9 and related technologies are being used to develop genetically modified natural enemies with enhanced traits, such as pesticide tolerance and increased reproduction rates.
   - Example: Gene-edited lady beetles were developed to withstand neonicotinoid exposure.
2. **Synthetic Biology:**
   - Synthetic biology enables the design of custom biological control agents, such as engineered pathogens or symbionts with specific pest-targeting capabilities.
   - Example: Synthetic microbes were created to produce insecticidal compounds only in the presence of specific pest pheromones.
3. **Integrated Pest Monitoring Systems:**
   - Combining DNA-based pest identification with IoT-enabled sensors provides real-time data on pest and natural enemy populations.
   - Example: Smart traps with DNA analysis capabilities were deployed to monitor pest outbreaks in orchards.
4. **Biocontrol Optimization through AI:**
   - AI algorithms analyze genetic and ecological data to optimize the deployment of natural enemies in complex agricultural systems.
   - Example: AI models predicted the best timing and location for releasing predatory mites based on field conditions and pest dynamics.

Ethical and Regulatory Considerations

1. **Biosafety:**
   - Ensuring that genetically modified natural enemies do not harm non-target organisms or disrupt ecosystems is a priority.
   - Regulatory frameworks must adapt to address the unique challenges posed by emerging technologies.
2. **Public Acceptance:**
   - Transparent communication and stakeholder engagement are essential for gaining public trust in DNA-based pest control innovations.

Chapter 8

DNA Diagnostics and Sf Transgenic Technology

1. Introduction

The integration of molecular biology and biotechnology into insect and pest management has revolutionized how we address challenges in agriculture, health, and ecological balance. Central to these advances are innovations in DNA diagnostics, Sf transgenic technology, and the understanding of insect biology at a molecular level. These tools and insights have provided unprecedented precision in identifying, studying, and combating pests and pathogens, improving crop yields, and reducing dependency on chemical pesticides.

DNA diagnostics, one of the foundational tools of modern biotechnology, involves the use of molecular methods to detect specific genetic markers. These diagnostics are widely used to identify pathogens, insect species, and genetic variations that influence resistance or susceptibility to various agents. In agriculture, DNA-based diagnostics have become indispensable for early detection of crop diseases, pest infestations, and the presence of resistance traits in both pests and crops. Techniques such as polymerase chain reaction (PCR), quantitative PCR (qPCR), and CRISPR-based detection systems have advanced diagnostics, making them faster, more accurate, and more accessible. These tools allow for targeted interventions, reducing the need for broad-spectrum insecticides and enhancing the sustainability of pest management strategies.

An equally transformative area of study is the immune system of insects, which differs significantly from that of vertebrates. While vertebrates possess both innate and adaptive immunity, insects rely exclusively on innate immune mechanisms, including humoral and cellular responses. These mechanisms, though simpler than those of vertebrates, are highly efficient and well-suited to the life cycles and ecological roles of insects. Understanding the immune systems of insects provides insights into their survival strategies, their vulnerabilities to pathogens, and the potential for exploiting these weaknesses in pest control. For instance, advances in insect immunology have paved the way for designing targeted biopesticides and enhancing the effectiveness of biological control agents.

The molecular basis of metamorphosis is another area where advances in molecular biology have provided deep insights. Metamorphosis, a process unique to insects and a few other animal groups, involves dramatic changes in physiology, morphology, and behavior, driven by complex hormonal

DOI: 10.1201/9781003757405-8

and genetic regulation. Hormones such as ecdysteroids and juvenile hormones play critical roles in coordinating the transitions from larval to pupal and adult stages. The genes and pathways involved in these processes not only illuminate fundamental aspects of insect development but also offer targets for biotechnological applications. For example, disrupting key genes involved in metamorphosis can be used as a strategy to control pest populations by preventing them from reaching reproductive maturity.

Among the most impactful technologies in pest management is the development of transgenic insects and crops, particularly those targeting the fall armyworm (*Spodoptera frugiperda*), a notorious agricultural pest. Sf transgenic technology involves the genetic modification of *Spodoptera* or the plants it infests to reduce its impact on crop yields. The fall armyworm, with its remarkable adaptability and rapid spread, poses a significant threat to global agriculture. Transgenic crops expressing specific resistance traits or targeting the pest's biology through RNA interference (RNAi) and other gene-silencing techniques have shown promise in controlling its populations. These innovations highlight the potential of genetic engineering to address even the most resilient agricultural pests.

The role of baculoviruses in insect molecular biology and pest management is another cornerstone of this field. Baculoviruses, naturally occurring pathogens of insects, are used both as biological control agents and as tools for gene expression systems in research and industry. Their host specificity and safety to non-target organisms make them ideal candidates for eco-friendly pest management. Advances in molecular biology have allowed for the genetic engineering of baculoviruses to enhance their virulence and effectiveness, further cementing their role in integrated pest management programs.

Insecticide resistance, however, remains a persistent challenge in the fight against pests. The overuse of chemical pesticides has led to the evolution of resistance mechanisms in pest populations, reducing the efficacy of traditional control methods. Resistance management strategies in transgenic crops, such as refuge planting, gene stacking, and monitoring, have been developed to counteract this issue. These strategies aim to delay the onset of resistance and ensure the long-term sustainability of transgenic technologies. By combining genetic tools with ecological approaches, researchers are working to create more resilient pest management systems.

The intersection of these disciplines—DNA diagnostics, insect immune systems, metamorphosis, transgenic technology, and resistance management—represents a holistic approach to tackling the challenges posed by insect pests. Each innovation contributes to a deeper understanding of insect biology and ecology, paving the way for more sustainable, precise, and effective pest control methods. As these technologies continue to evolve, their integration promises to transform agriculture and environmental management, ensuring food security and ecological balance in the face of growing global challenges.

A. Principles and Methodologies in DNA-Based Diagnostics

DNA-based diagnostics rely on molecular techniques to identify and analyze genetic material, enabling the detection of diseases, pathogens, and genetic traits with precision. This approach combines the fundamental principles of molecular genetics with advanced technologies to offer rapid, sensitive, and accurate diagnostic solutions. These methods are now integral to healthcare, agriculture, and environmental sciences, shaping the future of diagnostics.

1. Principles of DNA-Based Diagnostics

The principles underlying DNA-based diagnostics are rooted in the basic properties of nucleic acids, enzymatic processes, and molecular interactions. These principles ensure that DNA or RNA can be selectively targeted, amplified, and detected with high specificity.

1.1 Complementary Base Pairing

Complementary base pairing, discovered through the work of Watson and Crick, is the cornerstone of DNA-based diagnostics. The specific hydrogen bonding between adenine (A) and thymine (T), and between guanine (G) and cytosine (C), enables the selective binding of nucleic acid sequences.

1. **Applications in Probes:**
   - Single-stranded DNA or RNA probes are designed to hybridize specifically with complementary sequences in a sample.
   - Example: Fluorescence in situ hybridization (FISH) uses labeled probes to detect specific DNA sequences in chromosomes.
2. **Role in PCR:**
   - Primers, short sequences of single-stranded DNA, bind to complementary regions in target DNA to initiate polymerase chain reactions (PCR).
3. **Specificity:**
   - The high specificity of base pairing ensures that only the target sequence is amplified or detected, minimizing false positives.

1.2 Nucleic Acid Amplification

Amplification is a vital principle in DNA diagnostics, as it increases the concentration of target nucleic acids to detectable levels. This is especially important for diagnosing conditions where the target DNA or RNA is present in low abundance.

1. **Polymerase Chain Reaction (PCR):**
   - Amplifies specific DNA sequences exponentially through cycles of denaturation, annealing, and extension.
   - Applications:
     - ◊ Detecting genetic mutations in hereditary diseases.
     - ◊ Identifying viral DNA, such as in HIV or HPV diagnostics.
2. **Isothermal Amplification:**
   - Amplifies DNA without thermal cycling, using enzymes that operate at a constant temperature.
   - Example: Loop-mediated isothermal amplification (LAMP) is used in field diagnostics for malaria and COVID-19.

3. **Signal Amplification:**
    - Some techniques amplify detection signals rather than the nucleic acid itself, using chemiluminescence or fluorescence.

1.3 Sequence-Specific Recognition

The unique sequence of DNA or RNA in each organism or pathogen allows for targeted diagnostic approaches. This specificity is achieved through probes, primers, and enzymes tailored to recognize particular genetic sequences.

1. **Molecular Beacons:**
    - Probes that fluoresce upon binding to a complementary sequence, enabling real-time monitoring.
    - Used in real-time PCR (qPCR) for quantifying viral loads in infections.
2. **CRISPR-Cas Systems:**
    - Engineered CRISPR systems, such as SHERLOCK and DETECTR, use Cas enzymes to recognize and cleave target DNA or RNA sequences.
    - Example: Rapid detection of SARS-CoV-2 in clinical samples.

1.4 Enzymatic Reactions

Enzymes are integral to DNA diagnostics, catalyzing key processes such as amplification, synthesis, and cleavage.

1. **DNA Polymerase:**
    - Synthesizes new DNA strands in PCR and sequencing reactions.
    - High-fidelity polymerases ensure accurate replication of target sequences.
2. **Reverse Transcriptase:**
    - Converts RNA into complementary DNA (cDNA), enabling the analysis of RNA viruses and gene expression.
3. **Endonucleases:**
    - Restriction enzymes cut DNA at specific sites, aiding in genetic mapping and cloning.

1.5 Detection Principles

Detection of DNA or RNA relies on visual, electrical, or molecular signals generated by probes or enzymes.

1. **Fluorescence:**
    - Probes emit light upon hybridization with the target sequence, as seen in real-time PCR.
2. **Colorimetric Detection:**
    - Enzymatic reactions produce color changes, often used in ELISA-based DNA detection systems.

3. **Electrochemical Detection:**
   - Measures changes in electrical signals caused by DNA hybridization, used in biosensors.

B. Applications in Identifying Pathogens, Genetic Disorders, and Pest Species

DNA-based diagnostics have become a cornerstone in various scientific fields, providing unparalleled precision and sensitivity for identifying pathogens, diagnosing genetic disorders, and managing pest species. With advancements in molecular biology, these tools have revolutionized healthcare, agriculture, and environmental sciences.

1. Identifying Pathogens

Pathogens are responsible for a wide range of diseases affecting humans, animals, and plants. Accurate identification is essential for effective diagnosis, treatment, and prevention. DNA-based diagnostics offer unmatched specificity, speed, and sensitivity, enabling the detection of pathogens at early stages.

1.1 Bacterial Pathogens

DNA diagnostics have significantly enhanced our ability to identify bacterial infections, often outperforming traditional culture-based methods.

1. **Polymerase Chain Reaction (PCR):**
   - **Mechanism:** PCR amplifies bacterial DNA, making even minute quantities detectable.
   - Applications:
     - ◊ **Clinical Diagnostics:** Detects *Mycobacterium tuberculosis* for early tuberculosis diagnosis, often from sputum samples.
     - ◊ **Food Safety:** Identifies contamination by *Salmonella*, *Listeria monocytogenes*, and *Escherichia coli* in food products.
   - **Advantages:** Rapid, sensitive, and specific, requiring only a few hours to deliver results.
2. **Quantitative PCR (qPCR):**
   - **Use in Disease Management:** Measures bacterial load in chronic infections such as *Helicobacter pylori* in gastritis.
   - **Example:** Monitoring antibiotic-resistant strains, such as methicillin-resistant *Staphylococcus aureus* (MRSA), to guide treatment.
3. **Whole-Genome Sequencing (WGS):**
   - **Scope:** Offers a comprehensive view of bacterial genomes, identifying species, virulence factors, and antibiotic resistance genes.
   - **Impact:** In outbreaks, WGS has been used to trace the source of bacterial contamination, such as during the 2011 *Escherichia coli* outbreak in Europe.

1.2 Viral Pathogens

Viruses are challenging to diagnose due to their rapid mutation rates and minimal structural components. DNA-based diagnostics address these challenges effectively.

1. **Real-Time PCR (qPCR):**
   - Detects and quantifies viral DNA or RNA, providing real-time insights into viral load.
   - **Applications**:
     - ◊ **HIV Diagnosis:** Monitors viral load for assessing antiretroviral therapy efficacy.
     - ◊ **COVID-19 Testing:** Became the global standard for detecting SARS-CoV-2 in respiratory samples.
2. **CRISPR-Based Detection Systems:**
   - **SHERLOCK (Specific High Sensitivity Enzymatic Reporter UnLOCKing):**
     - ◊ Uses CRISPR technology to detect viral RNA with extreme specificity.
     - ◊ **Example:** Rapid detection of Zika virus during outbreaks.
   - **DETECTR (DNA Endonuclease-Targeted CRISPR Trans Reporter):**
     - ◊ Effective in identifying SARS-CoV-2 variants with minimal equipment.
3. **Next-Generation Sequencing (NGS):**
   - Enables comprehensive analysis of viral genomes, including mutations and recombination events.
   - **Example:** Tracking the evolution of SARS-CoV-2 variants such as Delta and Omicron.

1.3 Fungal Pathogens

Fungal infections, often life-threatening in immunocompromised individuals, require precise identification for effective treatment.

1. **DNA Barcoding:**
   - Targets conserved genetic regions like the internal transcribed spacer (ITS) for species identification.
   - **Example:** Differentiating between *Candida albicans* and non-*albicans* species in bloodstream infections.
2. **LAMP (Loop-Mediated Isothermal Amplification):**
   - Provides a rapid, field-friendly diagnostic tool for fungal pathogens in agriculture and clinical settings.
   - **Example:** Detecting *Aspergillus fumigatus*, a common cause of invasive aspergillosis.
3. **Metagenomic Approaches**:
   - Identifies mixed fungal communities in complex environments.
   - **Impact:** Improved diagnostics for polymicrobial infections in clinical and agricultural samples.

1.4 Parasitic Pathogens

Parasites pose unique diagnostic challenges due to their complex life cycles and morphological similarities. DNA-based methods provide precise and rapid solutions.

1. **PCR-Based Detection:**
   - Amplifies parasite-specific DNA, enabling early diagnosis.
   - **Example:** Detecting *Plasmodium falciparum* DNA for malaria diagnosis, even in asymptomatic cases.
2. **Environmental DNA (eDNA):**
   - Tracks parasitic infections in water and soil samples.
   - **Example:** Monitoring schistosomiasis transmission through eDNA from *Schistosoma* larvae.
3. **Digital PCR:**
   - Allows precise quantification of parasite DNA in clinical samples.
   - **Impact:** Enables monitoring of low-abundance infections, such as *Toxoplasma gondii* in immunocompromised patients.

2. Identifying Genetic Disorders

DNA-based diagnostics have revolutionized genetic medicine by providing tools to identify, assess, and manage hereditary conditions. These methods offer insights into the genetic basis of diseases, enabling personalized treatment strategies.

2.1 Single-Gene Disorders

Many genetic disorders are caused by mutations in single genes, which DNA diagnostics can identify with precision.

1. **Sanger Sequencing:**
   - **Application:** Detects point mutations in genes like BRCA1 and BRCA2, associated with breast and ovarian cancer.
   - **Example:** Identifying the F508del mutation in the *CFTR* gene for cystic fibrosis diagnosis.
2. **TaqMan Assays:**
   - Detect single nucleotide polymorphisms (SNPs) in genetic diseases.
   - **Example:** Detecting HBB mutations for sickle cell anemia diagnosis.
3. **Multiplex PCR:**
   - Simultaneously identifies multiple mutations in genes.
   - **Example:** Diagnosing Duchenne muscular dystrophy by detecting deletions in the *DMD* gene.

2.2 Chromosomal Disorders

Chromosomal abnormalities, such as trisomies and microdeletions, are major causes of developmental and genetic disorders. DNA diagnostics have advanced the detection of these abnormalities.

1. **Non-Invasive Prenatal Testing (NIPT):**
   - Analyzes cell-free fetal DNA from maternal blood to detect chromosomal aneuploidies.
   - **Example:** Screening for Down syndrome (trisomy 21) during early pregnancy.
2. **Fluorescence In Situ Hybridization (FISH):**
   - Detects structural chromosomal abnormalities.
   - **Example: Identifying the Philadelphia chromosome in chronic myeloid leuke**mia (CML).
3. **Array Comparative Genomic Hybridization (aCGH):**
   - Identifies genome-wide copy number variations.
   - **Example:** Diagnosing microdeletion syndromes like Prader-Willi syndrome.

2.3 Polygenic and Multifactorial Disorders

Polygenic disorders, resulting from the interplay of multiple genes and environmental factors, require advanced DNA diagnostics for risk assessment.

1. **Genome-Wide Association Studies (GWAS):**
   - Identifies SNPs associated with complex diseases.
   - **Example:** Discovering genetic markers linked to type 2 diabetes.
2. **Epigenetic Profiling:**
   - Analyzes DNA methylation patterns linked to cancer and neurodevelopmental disorders.
3. **NGS Panels:**
   - Screen for mutations across multiple genes associated with polygenic traits.
   - **Example:** Comprehensive cancer panels for hereditary cancer risk assessment.

3. Identifying Pest Species

DNA diagnostics play a critical role in managing pest populations, ensuring agricultural productivity, and protecting ecosystems. These tools enable the precise identification of pest species, monitor their resistance to pesticides, and support the development of sustainable management strategies.

3.1 Morphologically Similar Species

Pest species often exhibit morphological similarities, making traditional identification methods unreliable. DNA-based diagnostics overcome these limitations.

1. **DNA Barcoding:**
   - Uses mitochondrial markers like COI (cytochrome oxidase I) to identify species.

- **Example:** Differentiating between *Spodoptera frugiperda* (fall armyworm) and *Spodoptera exigua* (beet armyworm).

2. **Species-Specific PCR:**
   - Targets DNA unique to specific pest species.
   - **Example:** Rapid identification of *Helicoverpa armigera* in cotton fields.

3.2 Invasive Pests

Early detection of invasive pests is crucial to prevent their establishment and spread. DNA diagnostics provide reliable and rapid identification tools.

1. **Environmental DNA (eDNA):**
   - Detects pest DNA in environmental samples.
   - **Example:** Monitoring the presence of *Drosophila suzukii* (spotted wing drosophila) in fruit crops.
2. **NGS for Community Analysis:**
   - Simultaneously identifies multiple pest species in agricultural ecosystems.
   - **Example:** Detecting invasive nematodes in soil using high-throughput sequencing.

3.3 Monitoring Pest Resistance

Pest resistance to pesticides poses a significant challenge in agriculture. DNA diagnostics help monitor resistance genes in pest populations.

1. **PCR-Based Resistance Testing:**
   - Detects mutations associated with pesticide resistance.
   - **Example:** Identifying kdr mutations in *Anopheles* mosquitoes resistant to pyrethroids.
2. **CRISPR-Based Detection:**
   - Screens for resistance alleles in pest populations.
   - **Example:** Detecting herbicide resistance in *Amaranthus* species.

C. Advances in Technologies: PCR, qPCR, and CRISPR-Based Diagnostics in Molecular Entomology

The integration of cutting-edge molecular technologies into entomology has redefined the scope and precision of insect studies. Molecular diagnostics, such as Polymerase Chain Reaction (PCR), Quantitative PCR (qPCR), and CRISPR-based diagnostics, have become indispensable tools for entomologists. These techniques have advanced our understanding of insect genetics, disease vector dynamics, pest control strategies, and evolutionary biology.

1. Polymerase Chain Reaction (PCR) in Entomology

Polymerase Chain Reaction (PCR) is a cornerstone technique in molecular biology, enabling the amplification of specific DNA sequences from even minuscule amounts of genetic material. Since its introduction in 1983 by Kary Mullis, PCR has revolutionized entomology by providing a rapid and reliable method for the genetic analysis of insects and their associated pathogens.

Mechanism of PCR

The PCR process involves repeated cycles of:

1. **Denaturation**: Heating the DNA sample to ~94–98°C to separate the double strands.
2. **Annealing**: Cooling the sample to ~50–65°C, allowing primers to bind to their complementary sequences on the target DNA.
3. **Extension**: Raising the temperature to ~72°C, enabling DNA polymerase to synthesize a new DNA strand by adding nucleotides.

This cycle is repeated multiple times, leading to the exponential amplification of the target DNA sequence. In entomology, this method allows researchers to analyze specific gene regions in insect genomes with unparalleled precision.

Applications in Molecular Entomology

PCR has become a fundamental tool in molecular entomology, serving a wide array of applications:

1. **Taxonomic Identification:**
   - Differentiating cryptic species that are morphologically similar but genetically distinct.
   - Amplifying mitochondrial DNA (e.g., COI gene) for DNA barcoding and phylogenetic studies.
2. **Insect-Pathogen Interactions:**
   - Detecting pathogens carried by insect vectors, such as *Plasmodium* spp. in mosquitoes or *Borrelia* in ticks.
   - Investigating microbial symbionts in insects like aphids and their role in host fitness.
3. **Invasive Species Monitoring:**
   - Identifying invasive pest species through rapid genetic screening.
   - Tracking genetic variations to determine the origin and spread of invasive populations.
4. **Pest Resistance Studies:**
   - Detecting mutations associated with insecticide resistance (e.g., kdr mutations in mosquitoes).
   - Analyzing genes related to detoxification enzymes in pests.

Limitations of PCR

Despite its versatility, PCR has certain limitations:

» **Contamination Sensitivity**: Even trace amounts of foreign DNA can lead to false results.

» **Dependency on Prior Knowledge**: Requires specific primers tailored to the target sequence.

» **Semi-Quantitative Nature:** Traditional PCR does not provide quantitative data, necessitating supplementary techniques.

2. Quantitative PCR (qPCR) in Entomology

Quantitative PCR (qPCR), also known as real-time PCR, has enhanced the utility of traditional PCR by enabling real-time quantification of DNA. This advancement has become critical in studies requiring precise measurements of gene expression, pathogen load, or DNA concentration in insect samples.

Mechanism of qPCR

qPCR employs fluorescence-based detection to monitor DNA amplification as it occurs. The two primary detection methods are:

1. **Dye-Based Detection:**
    - Dyes such as SYBR Green intercalate with double-stranded DNA, and the fluorescence increases as the DNA amplifies.
2. **Probe-Based Detection:**
    - Fluorescent probes (e.g., TaqMan) are sequence-specific and emit fluorescence upon cleavage during DNA amplification.

Key metrics in qPCR include the **Threshold Cycle (Ct)**, representing the cycle number at which fluorescence surpasses the background. The Ct value correlates with the initial amount of target DNA or RNA in the sample.

Applications in Molecular Entomology

qPCR's precision and quantitative capabilities have expanded its use in entomological research:

1. **Pathogen Load Quantification:**
    - Measuring the intensity of infection in insect vectors, such as the malaria parasite load in *Anopheles* mosquitoes.
    - Quantifying viral loads in agricultural pests like whiteflies infected with plant viruses.
2. **Gene Expression Analysis:**
    - Examining the expression of stress-related genes under environmental challenges (e.g., temperature or pesticide exposure).
    - Investigating genes involved in insect immune responses to pathogens.
3. **Monitoring Transgene Expression:**
    - Assessing the expression levels of transgenes in genetically modified insects used in pest control strategies.
    - Verifying the stability of transgene expression across generations.
4. **Population Genetics:**
    - Quantifying allele frequencies related to pesticide resistance in field populations.

- Tracking genetic diversity within and between insect populations.

Advantages of qPCR

» **Quantitative Accuracy**: Allows precise measurement of DNA/RNA.

» **Multiplexing Capability**: Simultaneous detection of multiple targets in a single reaction.

» **High Sensitivity**: Effective in detecting low-abundance genetic material.

Challenges in qPCR for Entomology

» **High Costs**: Requires specialized equipment and reagents.

» **Complexity in Primer/Probe Design**: Demands rigorous validation to ensure specificity.

» **Dependence on Quality Samples**: Poorly preserved samples can lead to inaccurate quantification.

3. CRISPR-Based Diagnostics in Entomology

CRISPR technology, initially celebrated for its gene-editing capabilities, has found a novel application in molecular diagnostics. CRISPR-based diagnostics leverage the specificity of CRISPR-associated (Cas) proteins to detect target DNA or RNA sequences with exceptional accuracy and speed.

Mechanism of CRISPR Diagnostics

CRISPR diagnostics utilize guide RNAs (gRNAs) to direct Cas proteins (e.g., Cas12, Cas13) to the target genetic sequence. Upon recognition of the target, the Cas enzyme activates collateral cleavage of reporter molecules, generating a detectable signal.

Key diagnostic platforms include:

1. **SHERLOCK**:
   - Employs Cas13 to detect RNA targets.
   - Highly sensitive and capable of single-molecule detection.
2. **DETECTR:**
   - Uses Cas12 for DNA detection.
   - Facilitates rapid detection of pathogens and genetic markers.

Applications in Molecular Entomology

CRISPR diagnostics are opening new avenues for precision and speed in insect research:

1. **Pathogen Detection in Vectors:**
   - Rapid identification of arboviruses like dengue and Zika in mosquito vectors.
   - Detection of bacterial symbionts that influence vector competence (e.g., *Wolbachia*).
2. **Pest Control Innovations:**
   - Identifying genetic markers for susceptibility to gene drive systems.

- Screening for resistance genes in pest populations.

3. **Biodiversity and Conservation:**
    - Monitoring genetic diversity in endangered insect species.
    - Detecting invasive species in real time to mitigate ecological impacts.
4. **Environmental Surveillance:**
    - Screening for DNA traces of insects in environmental samples using CRISPR-based portable devices.

Advantages of CRISPR Diagnostics

» **Rapid Turnaround**: Results available within 30–60 minutes.

» **High Specificity**: Target recognition relies on precise gRNA matching.

» **Portability**: Suitable for field-based applications in remote locations.

Challenges and Limitations

» **Standardization Issues**: Protocols require optimization for consistency.

» **Regulatory Approval**: Validation for widespread use in diagnostics is ongoing.

» **Scalability**: Adapting CRISPR diagnostics for high-throughput applications in entomology is a work in progress.

Synergistic Potential and Future Perspectives

The integration of PCR, qPCR, and CRISPR diagnostics has the potential to revolutionize molecular entomology. For example:

» PCR and qPCR can provide foundational insights into genetic sequences and expression levels.

» CRISPR diagnostics can add rapid, field-deployable solutions for on-site detection of pathogens, invasive species, or resistance markers.

Future innovations will likely involve the fusion of these technologies with microfluidics, biosensors, and artificial intelligence to create comprehensive diagnostic platforms tailored to the needs of entomologists. These tools promise to enhance pest management, vector control, and conservation efforts, ensuring sustainable solutions for ecological and agricultural challenges.

2: Insect Immune Systems in Comparison to Vertebrates

Insects, as some of the most successful and diverse organisms on Earth, rely on an immune system that is vastly different yet functionally comparable in many ways to that of vertebrates. While vertebrates possess both innate and adaptive immunity, insects depend solely on innate immunity. Despite this limitation, their immune systems are remarkably effective and multifaceted, offering a rapid and robust defense against a wide range of pathogens, including bacteria, fungi, viruses, and parasites.

2.1. Overview of the Insect Immune System

The insect immune system is a prime example of evolutionary ingenuity. Unlike vertebrates, insects lack adaptive immunity, meaning they cannot produce antibodies, T cells, or B cells. However, their innate immune system compensates by employing a diverse array of cellular and humoral mechanisms that are both highly efficient and specialized. The immune responses in insects can be broadly divided into two main components: **humoral immunity** and **cellular immunity**, both of which operate in a complementary manner to combat pathogens.

Key Features of the Insect Immune System

1. **Exclusively Innate**: The insect immune system relies entirely on innate immunity, with no capacity for adaptive immune responses or long-term memory.
2. **Rapid and Non-Specific**: Innate responses are immediate and broadly effective against a variety of pathogens.
3. **Systemic and Localized**: Immune responses can occur both locally at the infection site and systemically throughout the body.
4. **Highly Conserved Pathways**: The core signaling pathways and effector mechanisms are conserved across different insect species.

Components of Insect Innate Immunity

Insect immunity is organized into two primary arms, each comprising several interconnected processes:

1. **Humoral Immunity**: Mediated by soluble factors in the hemolymph.
2. **Cellular Immunity**: Mediated by specialized immune cells, or hemocytes.

1. Humoral Immunity in Insects

Humoral immunity involves soluble proteins and enzymes that circulate in the hemolymph. These molecules are responsible for detecting pathogens, amplifying immune signals, and neutralizing or killing invaders.

a. Antimicrobial Peptides (AMPs)

AMPs are short peptides produced primarily by the fat body (the insect equivalent of the vertebrate liver) in response to infection. These peptides serve as the frontline defense against microbial invaders.

» **Diversity and Function:**

- AMPs are highly diverse and often pathogen-specific.
- They function by disrupting microbial membranes, interfering with vital processes, or inhibiting growth.

» **Examples:**

- **Cecropins**: Target Gram-negative bacteria.
- **Defensins**: Effective against Gram-positive bacteria by disrupting their membranes.

- **Drosomycin**: Specific for fungal pathogens.

» **Regulation:**

- AMP production is tightly regulated by immune signaling pathways, primarily the Toll and Imd pathways.

b. Enzymatic Cascades

The insect immune system employs several enzymatic cascades that amplify the immune response and directly target pathogens:

» **Phenoloxidase Cascade:**

- Key to the process of melanization, which encapsulates and neutralizes pathogens.
- Phenoloxidase catalyzes the production of melanin, creating a physical and chemical barrier around invaders.

» **Lysozymes**:

- Hydrolytic enzymes that degrade the peptidoglycan layer of bacterial cell walls.
- Found in hemolymph and digestive tracts, lysozymes protect insects from ingested pathogens.

c. Immune Signaling Pathways

The recognition of pathogens triggers specific signaling pathways that regulate the expression of immune effectors like AMPs:

1. **Toll Pathway**:
   - Activated primarily by fungal and Gram-positive bacterial infections.
   - Shares homology with the vertebrate Toll-like receptor pathway.
2. **Imd (Immune Deficiency) Pathway:**
   - Responds to Gram-negative bacteria and regulates the production of AMPs like cecropins.
3. **JAK-STAT Pathway**:
   - Involved in antiviral immunity and responses to some parasites.
4. **RNA Interference (RNAi)**:
   - Acts as a specific antiviral defense by degrading viral RNA.
5. **Melanization Pathway**:
   - Produces melanin at the site of infection, effectively encapsulating and immobilizing pathogens.

d. Hemolymph Coagulation

Hemolymph coagulation serves as a physical barrier to prevent the spread of pathogens. Proteins in the hemolymph form clots at the site of injury or infection, sealing wounds and trapping invaders.

2. Cellular Immunity in Insects

Cellular immunity involves the actions of hemocytes, which are the primary immune cells in insects. Hemocytes are responsible for direct interactions with pathogens through phagocytosis, encapsulation, and nodulation.

a. Types of Hemocytes

Insects possess several types of hemocytes, each with specialized functions:

1. **Plasmatocytes:**
   - Analogous to macrophages in vertebrates.
   - Engulf small pathogens through phagocytosis.
2. **Granulocytes:**
   - Secrete antimicrobial molecules and mediate the encapsulation of large invaders.
3. **Lamellocytes:**
   - Specialized hemocytes in certain insects (e.g., *Drosophila*) for encapsulating large pathogens, such as parasitic wasps.
4. **Prohemocytes:**
   - Undifferentiated hemocytes that can differentiate into other types as needed.

b. Mechanisms of Cellular Immunity

1. **Phagocytosis:**
   - Hemocytes engulf and digest small pathogens such as bacteria and yeast.
   - Involves the recognition of PAMPs (Pathogen-Associated Molecular Patterns) via PRRs (Pattern Recognition Receptors).
2. **Encapsulation:**
   - Large invaders, such as parasitic eggs, are encapsulated by layers of hemocytes.
   - The encapsulated invaders are often melanized to neutralize them.
3. **Nodulation:**
   - Hemocytes aggregate to form nodules around microbial infections, containing and eliminating pathogens.
4. **Cytotoxicity:**
   - Hemocytes release reactive oxygen species (ROS) and antimicrobial factors to kill pathogens.

Comparison to Vertebrate Immune Systems

Insects and vertebrates have evolved distinct immune strategies, but their systems share striking parallels despite differences in complexity and mechanisms.

Similarities

1. **Innate Immunity**:
   - Both systems have conserved innate immune mechanisms, such as the Toll and Imd pathways in insects and the Toll-like receptor pathways in vertebrates.
2. **Recognition of PAMPs:**
   - Both systems use PRRs to detect PAMPs like lipopolysaccharides and peptidoglycan.
3. **Phagocytosis:**
   - Both insects and vertebrates employ immune cells for engulfing and digesting pathogens.
4. **Effector Molecules:**
   - AMPs in insects function similarly to defensins in vertebrates, targeting microbial membranes.

Differences

1. **Adaptive Immunity**:
   - Vertebrates possess adaptive immunity, characterized by T cells, B cells, and antibodies, whereas insects lack these components.
2. **Memory**:
   - Vertebrates have immunological memory, enabling faster and stronger responses upon re-infection, while insects rely on immediate innate responses.
3. **Complexity:**
   - Vertebrate immune systems are more complex, with specialized organs (e.g., thymus, spleen) and a lymphatic system, which are absent in insects.
4. **Effector Mechanisms:**
   - Insects rely heavily on AMPs and melanization, whereas vertebrates utilize a broader range of effector mechanisms, including complement activation and antibody-mediated neutralization.

Evolutionary Insights and Applications

The study of insect immunity provides valuable evolutionary insights and practical applications:

1. **Evolutionary Conserved Mechanisms**:
   - Comparative studies reveal the evolutionary origins of innate immunity and its conservation across taxa.
2. **Pest Control:**
   - Targeting insect immune pathways, such as RNAi, offers novel pest management strategies.

- **Vector Control**:
- Enhancing the immune competence of disease vectors (e.g., mosquitoes) through microbial symbionts like *Wolbachia* can reduce pathogen transmission.

3. **Biomedical Applications:**
   - Insect-derived AMPs are being explored for their potential as antibiotics against multidrug-resistant bacteria.

2.2. Key Differences Between Insect and Vertebrate Immune Systems

The immune systems of insects and vertebrates represent two distinct evolutionary strategies for defense against pathogens. Insects rely exclusively on innate immunity, while vertebrates possess both innate and adaptive immunity. These fundamental differences, alongside unique features within each system, shape how each group responds to infections.

1. Presence of Adaptive Immunity

The most prominent distinction between insects and vertebrates lies in the presence or absence of adaptive immunity.

a. Adaptive Immunity in Vertebrates

» Vertebrates possess an adaptive immune system characterized by specificity, diversity, and memory.

» **Key Components**:

- **B Cells**: Produce antibodies that bind to specific antigens, neutralizing pathogens or marking them for destruction.
- **T Cells**: Mediate cellular immunity, including the destruction of infected cells and coordination of immune responses.

» **Immunological Memory**:

- The vertebrate adaptive immune system generates long-term memory after exposure to pathogens or vaccines, allowing faster and stronger responses upon re-infection.

b. Absence of Adaptive Immunity in Insects

» Insects lack adaptive immunity, relying entirely on innate mechanisms for pathogen defense.

» **No Antibodies or Memory**:

- Insects do not produce antibodies or retain memory of prior infections. Their immune responses are non-specific and do not improve with repeated exposure to the same pathogen.

Implications:

» Vertebrates benefit from enhanced protection over time due to immunological memory, whereas insects must rely on rapid, non-specific responses to eliminate infections immediately.

2. Nature of Innate Immunity

Both insects and vertebrates possess innate immunity, but its structure and function differ significantly.

a. Similarities in Innate Immunity

» **Pattern Recognition**:

- Both insects and vertebrates use pattern recognition receptors (PRRs) to detect pathogen-associated molecular patterns (PAMPs), such as lipopolysaccharides, peptidoglycan, and viral RNA.

» **Signaling Pathways**:

- Immune pathways such as the Toll pathway in insects and Toll-like receptor pathways in vertebrates share evolutionary homology.

» **Effector Molecules**:

- Antimicrobial peptides (AMPs) in insects and defensins in vertebrates disrupt microbial membranes and inhibit pathogen growth.

b. Differences in Innate Immunity

» **Effector Mechanisms**:

- Insects rely heavily on processes like melanization and phenoloxidase activity, which are absent in vertebrates.
- Vertebrates use a complement system that lyses pathogens and enhances phagocytosis, which insects lack.

» **Cellular Components**:

- Insects utilize hemocytes for phagocytosis, encapsulation, and nodulation, while vertebrates rely on specialized immune cells such as neutrophils, macrophages, and dendritic cells.

» **Systemic vs. Localized Responses**:

- Insects primarily deploy systemic antimicrobial peptides and local melanization, whereas vertebrates activate both systemic and highly localized responses through cytokines and chemokines.

3. Immune System Complexity

The structural and organizational complexity of the immune system varies significantly between insects and vertebrates.

a. Immune Organs

» **Vertebrates**:

- Vertebrates possess specialized immune organs such as the thymus, spleen, lymph nodes, and bone marrow.

- The lymphatic system facilitates the transport of immune cells and molecules throughout the body.

» **Insects**:

- Insects lack dedicated immune organs. Their immune responses are mediated by circulating hemocytes in the hemolymph and immune proteins produced by the fat body.
- The fat body serves as a multifunctional organ, analogous to the vertebrate liver, producing immune effectors like antimicrobial peptides.

b. Diversity of Immune Cells

» **Vertebrates**:

- The immune cell repertoire includes lymphocytes (T cells, B cells, natural killer cells) and myeloid cells (macrophages, neutrophils, dendritic cells).

» **Insects**:

- Hemocyte diversity is limited, with primary cell types including plasmatocytes, granulocytes, and lamellocytes.
- Hemocytes lack the specialization seen in vertebrate immune cells.

4. Immune System Memory

The ability to "remember" previous infections is a hallmark of vertebrate adaptive immunity, a feature absent in insects.

a. Vertebrate Memory

» Adaptive immunity enables vertebrates to recognize and mount enhanced responses against previously encountered pathogens.

» Memory B cells and T cells persist for years, providing long-lasting immunity.

» Vaccination leverages this memory to protect against diseases.

b. Insect Immunity Without Memory

» Insects rely solely on innate immunity, which does not retain memory of prior exposures.

» While insects lack true immunological memory, some studies suggest "priming" effects where prior exposure enhances subsequent responses to the same pathogen, though this is not comparable to vertebrate adaptive immunity.

Implications:

» Vertebrates can develop targeted and long-term immunity through vaccination, a strategy not applicable to insects.

5. Speed and Specificity of Responses

The speed and specificity of immune responses differ between insects and vertebrates due to their underlying mechanisms.

a. Rapid Responses in Insects

- Innate immunity in insects is fast-acting, with immediate activation upon pathogen detection.
- Antimicrobial peptides and melanization mechanisms are deployed within hours to neutralize infections.
- Responses are non-specific, targeting broad categories of pathogens rather than specific strains.

b. Balancing Speed and Specificity in Vertebrates

- Vertebrate innate immunity provides an initial rapid response, while adaptive immunity takes days to fully develop but offers high specificity.
- Antibodies and T cells target specific antigens, ensuring efficient pathogen elimination with minimal collateral damage.

Comparative Advantage:

- Insects are highly effective at combating acute infections due to their rapid response, while vertebrates excel in long-term immunity and pathogen specificity.

6. Pathogen Recognition

Both systems recognize pathogens through specialized molecules, but the mechanisms differ.

a. Pattern Recognition Molecules

- **Insects**:
  - PRRs like peptidoglycan recognition proteins (PGRPs) and β-glucan recognition proteins detect PAMPs.
  - These PRRs activate downstream pathways, such as Toll and Imd, leading to immune responses.
- **Vertebrates**:
  - Toll-like receptors (TLRs) and other PRRs detect a wide array of PAMPs and initiate innate immune responses.

b. Diversity of Recognition Mechanisms

- Vertebrates possess a more diverse array of recognition molecules due to the presence of adaptive immunity.
- Antibodies in vertebrates can bind to virtually any antigen, offering unparalleled flexibility compared to the limited PRRs in insects.

7. Effector Molecules and Processes

Insects and vertebrates deploy different effector molecules and processes to eliminate pathogens.

a. Effector Molecules

» **Insects**:

- Antimicrobial peptides, phenoloxidase enzymes, and melanin are the primary effectors.

» **Vertebrates**:

- Antibodies, complement proteins, cytokines, and defensins play key roles.

b. Processes

» **Insects**:

- Melanization: Encapsulation of pathogens with melanin, which is unique to insects.
- Nodulation: Hemocyte aggregation around microbial infections.

» **Vertebrates:**

- Phagocytosis by macrophages and neutrophils.
- Antibody-mediated neutralization and opsonization.

8. Evolutionary Perspective

The immune systems of insects and vertebrates reflect their distinct evolutionary trajectories:

» Insects evolved rapid, energy-efficient immune mechanisms suitable for short lifespans and high reproductive rates.

» Vertebrates developed complex, adaptive systems to provide long-term protection in environments with diverse and persistent pathogens.

2.3. Evolutionary Adaptations and Limitations of Insect Immunity

Insects, as one of the most ancient and diverse groups of organisms, have evolved immune systems that ensure their survival across a wide range of environments. Their immune systems, though limited to innate mechanisms, exhibit remarkable efficiency and flexibility, allowing them to combat a variety of pathogens. However, this reliance on innate immunity also imposes certain limitations, which influence their evolutionary trajectory and ecological success.

1. Evolutionary Adaptations of Insect Immunity

The innate immune system of insects is a product of millions of years of evolution, honed to provide immediate and effective protection against pathogens. Several adaptations have contributed to their success:

1.1. Streamlined and Energy-Efficient Defense Mechanisms

» **Immediate Response**:

- Insects rely on a pre-programmed, rapid immune response to detect and neutralize pathogens. This immediacy is crucial for short-lived organisms like insects, where delays in pathogen defense could be fatal.

» **Energy Efficiency**:

- Unlike adaptive immunity, which is energetically costly due to the production of antibodies and memory cells, innate immunity involves relatively low metabolic costs. This is advantageous for insects, which often face resource limitations.

1.2. Robust Pathogen Recognition Systems

» **Pattern Recognition Receptors (PRRs)**:

- PRRs, such as Peptidoglycan Recognition Proteins (PGRPs) and β-glucan recognition proteins, allow insects to detect conserved Pathogen-Associated Molecular Patterns (PAMPs) on pathogens.
- These receptors are highly conserved across insect species, ensuring broad-spectrum pathogen detection.

» **Toll and Imd Pathways**:

- The evolutionary refinement of Toll and Imd pathways enables effective discrimination between different classes of pathogens, such as Gram-positive and Gram-negative bacteria.

1.3. Adaptation to Diverse Pathogens

» **Antimicrobial Peptides (AMPs)**:

- The evolution of a diverse repertoire of AMPs (e.g., cecropins, defensins, drosomycin) allows insects to target bacteria, fungi, and viruses effectively.
- AMPs are modular and rapidly evolvable, enabling insects to adapt to emerging pathogens.

» **RNA Interference (RNAi)**:

- RNAi provides a specific mechanism for combating viral infections by degrading viral RNA, a unique adaptation among invertebrates.

1.4. Phenoloxidase and Melanization

» **Encapsulation and Melanin Production**:

- Phenoloxidase-driven melanization encapsulates and neutralizes large pathogens like parasitic eggs.
- This process not only immobilizes the pathogen but also generates reactive oxygen species (ROS) that contribute to pathogen killing.

1.5. Behavioral Immunity

» **Grooming and Hygienic Behavior**:

- Many insects exhibit behaviors that reduce pathogen exposure, such as grooming and selective nesting.

» **Social Immunity**:

- In social insects like ants and bees, group behaviors such as allogrooming and collective hygiene help prevent the spread of infections within colonies.

2. Limitations of Insect Immunity

While the insect immune system is highly effective, it is not without constraints. These limitations arise from their evolutionary reliance on innate immunity and the absence of adaptive immune components.

2.1. Lack of Adaptive Immunity

» **Absence of Immunological Memory**:

- Insects cannot generate immunological memory, meaning their responses to recurring infections are not enhanced over time.
- This limits their ability to develop long-term protection against persistent or evolving pathogens.

» **No Antibodies**:

- The lack of antibodies restricts their ability to target specific antigens with precision, unlike vertebrates.

2.2. Limited Immune Diversity

» **Fixed PRR Repertoire**:

- Insects rely on a fixed set of PRRs to detect pathogens. While effective for broad recognition, this system is less adaptable to rapidly evolving pathogens.

» **Restricted Effector Molecules**:

- The relatively small number of AMPs and immune effectors limits their ability to combat highly diverse or novel pathogens.

2.3. Vulnerability to Immune Evasion

» **Pathogen Evasion Strategies**:

- Some pathogens have evolved mechanisms to evade or suppress insect immunity, such as mimicking host molecules or inhibiting immune signaling pathways.

» **Parasitic Manipulation**:

- Parasitic wasps, for example, inject venom that suppresses the host insect's immune system, allowing their eggs to develop unchallenged.

2.4. Energetic Trade-Offs

» **Costs of Immune Activation**:

- While energy-efficient compared to adaptive immunity, activating innate immune responses still incurs significant metabolic costs, diverting resources from reproduction and growth.

» **Trade-Offs with Longevity**:

- Immune responses can lead to tissue damage or oxidative stress, potentially reducing the insect's lifespan.

2.5. Inability to Address Chronic Infections

» **Focus on Acute Infections**:

- Insects are well-equipped to deal with acute infections but struggle with chronic or long-term infections due to the absence of sustained immune responses.

2.4. Implications for Pest Control and Transgenic Technologies

The unique characteristics of insect immunity have profound implications for pest control strategies and the development of transgenic technologies. By understanding and leveraging the strengths and weaknesses of insect immune systems, researchers and practitioners can devise innovative methods for managing pest populations and reducing the spread of insect-borne diseases.

1. Pest Control Strategies Targeting Immune Pathways

The limitations of insect immunity make it an attractive target for pest control strategies aimed at exploiting their vulnerabilities.

1.1. Immune Suppression

» **Pathogen-Enhanced Control**:

- Introducing pathogens or parasites that evade or suppress insect immunity can effectively reduce pest populations.
- For example, fungal biopesticides exploit the insect immune system's inability to combat certain fungal infections effectively.

» **RNAi-Based Immune Disruption**:

- RNA interference can be used to silence genes involved in immune pathways, such as those encoding key components of the Toll or Imd pathways, rendering insects more susceptible to pathogens.

1.2. Overloading the Immune System

» **Pathogen Overload**:

- Introducing multiple pathogens simultaneously can overwhelm the insect immune system, leading to higher mortality rates.

» **Immune Exhaustion**:

- Chronic exposure to pathogens or immune-stimulating agents can deplete an insect's resources, reducing their reproductive capacity and lifespan.

1.3. Targeting Symbionts

» **Symbiont Disruption**:

- Many insects rely on symbiotic microbes for immunity and other physiological functions. Disrupting these relationships can impair immune responses and overall fitness.
- For example, eliminating *Wolbachia* symbionts in mosquitoes can reduce their resistance to viral infections.

2. Transgenic Technologies for Pest and Vector Control

Advances in genetic engineering have opened new avenues for manipulating insect immune systems to control pests and vectors of disease.

2.1. Engineering Immune-Competent Vectors

» **Enhanced Immunity**:

- Transgenic mosquitoes have been engineered to overexpress immune effectors, such as antimicrobial peptides, to reduce their ability to transmit pathogens like *Plasmodium* (malaria) or dengue virus.

» **Symbiont Engineering**:

- Genetic modification of symbiotic bacteria, such as *Wolbachia*, can enhance their ability to suppress pathogen transmission.

2.2. Gene Drives

» **Immune-Based Gene Drives**:

- Gene drives can be used to spread immune-modulating genes through pest populations, rendering them less capable of transmitting diseases or surviving in certain environments.

» **Self-Limiting Populations**:

- Gene drives can also be designed to limit pest reproduction by disrupting immune genes critical for reproductive success.

2.3. Synthetic Immune Pathways

» **Artificial Pathogen Resistance**:

- Synthetic biology techniques can be used to introduce entirely novel immune pathways into insect vectors, providing them with enhanced resistance to pathogens.

3. Integrated Pest Management (IPM) Applications

Understanding insect immunity contributes to the development of more effective Integrated Pest Management (IPM) strategies:

» **Biocontrol Agents**:

- Using natural enemies, such as parasitoid wasps or entomopathogenic fungi, to exploit the weaknesses in insect immune systems.

» **Behavioral Manipulation**:

- Exploiting social immunity behaviors, such as grooming, to introduce pathogens into pest populations.

4. Challenges and Ethical Considerations

The use of immune-targeting and transgenic technologies raises several challenges and ethical questions:

» **Resistance Development**:

- Overuse of immune-targeting strategies may lead to the evolution of resistant insect populations.

» **Ecological Impact**:

- Altering insect immunity could have unintended effects on ecosystems, particularly for beneficial insects like pollinators.

» **Regulatory and Public Acceptance**:

- The deployment of genetically modified organisms (GMOs) in pest control requires rigorous regulatory approval and public trust.

3: Molecular Basis of Metamorphosis

Metamorphosis, the process by which insects undergo significant morphological and physiological changes during their life cycle, is a hallmark of insect development. It is governed by intricate molecular and hormonal mechanisms, including the interplay of ecdysteroids, juvenile hormones (JH), and genetic pathways. This chapter delves into the molecular basis of metamorphosis, exploring the hormonal regulation and genetic pathways that drive insect development.

3.1. Hormonal Regulation of Metamorphosis: Roles of Ecdysteroids and Juvenile Hormones

Hormones play a pivotal role in orchestrating the metamorphic changes in insects. The transition from one developmental stage to another—larva, pupa, and adult—is tightly regulated by the balance and timing of two primary hormones: **ecdysteroids** and **juvenile hormones (JH)**.

1. Ecdysteroids: The Molting Hormones

Ecdysteroids, particularly 20-hydroxyecdysone (20E), are steroid hormones synthesized by the prothoracic gland. These hormones are the primary drivers of molting and metamorphosis.

a. Synthesis and Regulation

» **Source**:

- Ecdysteroids are produced in the prothoracic glands under the stimulation of prothoracicotropic hormone (PTTH), secreted by the brain.

» **Activation**:

- PTTH triggers a signaling cascade via receptor tyrosine kinases, ultimately activating enzymes involved in ecdysteroid biosynthesis.

» **Conversion**:

- The precursor, ecdysone, is converted into the active form, 20-hydroxyecdysone, in peripheral tissues.

b. Mechanism of Action

» Ecdysteroids exert their effects through the ecdysone receptor (EcR) complex, which is a heterodimer of EcR and ultraspiracle protein (USP), the insect homolog of the vertebrate retinoid X receptor (RXR).

» **Transcriptional Regulation:**

- Binding of 20E to EcR/USP activates a cascade of transcription factors, including early genes such as *E74*, *E75*, and *Broad-Complex (BR-C)*.
- These early genes regulate the expression of late genes responsible for cellular changes during molting and metamorphosis.

c. Role in Molting and Metamorphosis

» Ecdysteroids initiate and coordinate:

- **Apoptosis** of larval tissues.
- **Proliferation and differentiation** of imaginal discs, which form adult structures.
- Transition from larva to pupa and pupa to adult.

2. Juvenile Hormones: The Stage-Determining Hormones

Juvenile hormones (JHs), secreted by the corpora allata, are sesquiterpenoids that prevent precocious metamorphosis and maintain larval characteristics.

a. Synthesis and Regulation

» JH synthesis is regulated by environmental and internal cues, including nutritional status and neural signals.

» The corpora allata produce JH, which is transported via the hemolymph to target tissues.

b. Mechanism of Action

» JH functions through the Methoprene-tolerant (Met) receptor, a member of the bHLH-PAS family of transcription factors.

» Binding of JH to Met forms a complex with Taiman (Tai), which modulates the transcription of JH-responsive genes.

c. Role in Development

» JH levels are high during larval stages, preventing the premature development of pupal and adult structures.

» A decrease in JH levels during the final larval instar allows the action of ecdysteroids to trigger metamorphosis.

» A transient surge in JH during the pupal stage is required for proper adult differentiation.

3. Interplay Between Ecdysteroids and Juvenile Hormones

The dynamic interplay between ecdysteroids and JH ensures precise regulation of metamorphosis:

1. **Larval Stages:**
    - High JH levels suppress the metamorphic effects of ecdysteroids, ensuring successive molts between larval stages.
2. **Metamorphic Transition:**
    - A decline in JH levels during the last larval instar permits ecdysteroids to initiate pupation.
3. **Adult Differentiation:**
    - Coordinated ecdysteroid surges and JH regulation during the pupal stage ensure proper formation of adult structures.

4. Environmental and Genetic Influences

» **Environmental Signals**:

- Photoperiod, temperature, and nutrition influence hormone levels, modulating the timing and progression of metamorphosis.

» **Genetic Regulation**:

- Genes such as *Krüppel*, *Ultrabithorax*, and *Broad-Complex* integrate hormonal signals and regulate developmental transitions.

3.2. Genetic and Molecular Pathways Driving Insect Development and Metamorphosis

Insect metamorphosis is governed by intricate genetic networks and molecular pathways that translate hormonal signals into developmental outcomes. These pathways coordinate cellular proliferation, differentiation, and apoptosis to ensure proper stage transitions.

1. Key Genetic Players in Metamorphosis

Several genes play critical roles in regulating metamorphosis:

» **Early Genes**:

- Induced by ecdysteroids, these include *E74*, *E75*, and *Broad-Complex* (BR-C), which regulate downstream genes involved in tissue remodeling.

» **Homeotic Genes**:

- Homeobox (Hox) genes, such as *Ultrabithorax* and *Antennapedia*, determine the identity and differentiation of body segments.

» **Imaginal Disc Genes**:

- Genes such as *wingless* and *decapentaplegic* (dpp) regulate the growth and patterning of imaginal discs, which form adult structures like wings and legs.

2. Hormone-Sensitive Pathways

The interaction between hormones and genetic pathways ensures coordinated development:

» **Ecdysteroid Pathway**:
   - The ecdysone receptor (EcR) complex regulates transcriptional cascades that control molting and tissue differentiation.

» **Juvenile Hormone Pathway**:
   - The Met-Tai complex integrates JH signals, influencing gene expression involved in stage determination and adult differentiation.

3. Programmed Cell Death and Tissue Remodeling

Metamorphosis involves extensive tissue remodeling, driven by:

» **Apoptosis**:
   - Triggered by ecdysteroids, apoptosis eliminates larval tissues, such as the midgut and salivary glands.

» **Autophagy**:
   - Cellular recycling mechanisms repurpose larval materials for adult tissue formation.

4. Imaginal Discs and Adult Structure Formation

Imaginal discs are clusters of undifferentiated cells in larvae that give rise to adult structures:

» **Growth and Patterning**:
   - Controlled by genes like *wingless*, *hedgehog*, and *engrailed.*

» **Differentiation**:
   - Ecdysteroid surges activate genes that drive imaginal disc differentiation into wings, legs, and eyes.

5. Epigenetic Regulation

Epigenetic mechanisms, including histone modifications and non-coding RNAs, modulate gene expression during metamorphosis:

» **Histone Acetylation**:
   - Promotes transcription of ecdysteroid-responsive genes.

» **MicroRNAs**:
   - Fine-tune gene expression involved in tissue remodeling and hormone signaling.

6. Cross-Species Conservation and Divergence

» **Conserved Pathways**:
   - Many genetic pathways, such as Hox and Wnt signaling, are conserved across insect species.

» **Species-Specific Adaptations**:

- Variations in hormone sensitivity and gene expression patterns underlie species-specific metamorphic processes.

3.4. Evolutionary Perspectives on Complete Versus Incomplete Metamorphosis

Metamorphosis represents one of the most fascinating evolutionary innovations in the insect world, enabling diversification into a wide range of ecological niches. Insects are categorized into two major groups based on their type of metamorphosis: **complete metamorphosis (holometaboly)** and **incomplete metamorphosis (hemimetaboly)**. The evolutionary significance of these developmental strategies sheds light on the ecological advantages and genetic adaptations associated with each.

1. Complete Metamorphosis (Holometaboly)

Holometaboly is characterized by four distinct life stages: egg, larva, pupa, and adult. This type of development occurs in insects such as butterflies, beetles, flies, and bees.

a. Key Features of Holometaboly

» **Distinct Stages**:

- Each stage has a unique morphology and ecological role. Larvae are specialized for feeding, while adults are specialized for reproduction and dispersal.

» **Pupal Stage**:

- The pupa serves as a transitional stage where extensive remodeling occurs, allowing for the development of adult structures.

b. Evolutionary Advantages

» **Niche Partitioning**:

- The larval and adult stages occupy different ecological niches, reducing intraspecific competition for resources.

» **Specialization**:

- Larvae can evolve specialized feeding structures, while adults can develop wings and sensory systems optimized for reproduction and dispersal.

» **Adaptability**:

- The separation of growth and reproduction allows for greater flexibility in adapting to environmental changes.

c. Genetic and Molecular Basis

» **Hormonal Regulation**:

- The transition between stages is controlled by ecdysteroids and juvenile hormones, with distinct hormonal profiles for larval growth, pupation, and adult emergence.

» **Imaginal Discs:**

- These clusters of undifferentiated cells in larvae form the basis for adult structures, driven by transcriptional cascades initiated by ecdysteroids.

2. Incomplete Metamorphosis (Hemimetaboly)

Hemimetaboly involves three stages: egg, nymph, and adult. This type of development is seen in insects like grasshoppers, crickets, and dragonflies.

a. Key Features of Hemimetaboly

» **Gradual Development:**

- Nymphs resemble miniature adults and gradually acquire adult characteristics, such as wings and reproductive organs, through successive molts.

» **No Pupal Stage:**

- Development occurs without a quiescent pupal stage, with growth and differentiation occurring incrementally.

b. Evolutionary Advantages

» **Energy Efficiency:**

- The absence of a pupal stage reduces the metabolic costs associated with complete metamorphosis.

» **Rapid Maturity:**

- Hemimetabolous insects can mature more quickly, providing an advantage in environments where rapid reproduction is critical.

c. Genetic and Molecular Basis

» **Direct Development:**

- Growth and differentiation are regulated by ecdysteroids and juvenile hormones but lack the dramatic remodeling seen in holometaboly.

» **Conserved Pathways:**

- Many of the same hormonal and genetic pathways are present but are deployed differently to allow gradual morphological changes.

3. Evolutionary Transition from Hemimetaboly to Holometaboly

Holometaboly is thought to have evolved from hemimetaboly approximately 300 million years ago. This transition is associated with key evolutionary innovations:

» **Imaginal Discs:**

- The evolution of imaginal discs allowed for the separation of larval and adult morphologies, facilitating niche specialization.

» **Hormonal Shifts:**

- Changes in the timing and regulation of juvenile hormones and ecdysteroids enabled the development of a distinct pupal stage.

» **Genetic Modifications:**

- Evolutionary changes in Hox genes and other developmental regulators likely played a role in the divergence of these developmental modes.

4. Comparative Ecological and Evolutionary Success

» **Holometaboly:**

- Holometabolous insects account for over 85% of all insect species, highlighting their ecological success.
- Their ability to occupy diverse niches and adapt to changing environments is a key factor in their evolutionary dominance.

» **Hemimetaboly:**

- While less diverse, hemimetabolous insects thrive in stable environments where rapid development and simplicity offer advantages.

3.5. Applications of Molecular Knowledge in Pest Control and Biotechnology

Advances in understanding the molecular basis of metamorphosis have opened new frontiers in pest control and biotechnology. By targeting hormonal pathways, genetic regulators, and developmental processes, researchers can develop innovative strategies to manage pest populations, enhance beneficial insects, and create bio-inspired technologies.

1. Pest Control Applications

Molecular insights into metamorphosis provide opportunities to disrupt the life cycles of pest insects, reducing their impact on agriculture, public health, and ecosystems.

a. Hormonal Disruption

» **Juvenile Hormone Analogs:**

- Synthetic compounds that mimic or inhibit juvenile hormone activity can prevent proper development, leading to sterility or death.
- Example: Methoprene, a juvenile hormone analog, is used to control mosquito larvae.

» **Ecdysteroid Inhibitors:**

- Compounds that interfere with ecdysteroid synthesis or signaling can block molting and metamorphosis.
- Example: Azadirachtin, derived from neem, disrupts ecdysteroid pathways in pests.

b. RNA Interference (RNAi)

» **Gene Silencing**:

- RNAi can be used to target genes critical for hormonal signaling or metamorphic pathways, effectively disrupting development.
- Example: Silencing genes involved in chitin synthesis can impair exoskeleton formation, leading to mortality.

c. Targeting Imaginal Disc Development

» **Inhibiting Adult Formation**:

- Compounds or genetic interventions that disrupt imaginal disc development can prevent the formation of functional adults, reducing reproductive success.

2. Enhancing Beneficial Insects

Molecular knowledge of metamorphosis can also be applied to promote the growth and productivity of beneficial insects, such as pollinators and biocontrol agents.

a. Pollinator Health

» **Hormonal Modulation**: Optimizing juvenile hormone levels can enhance the health and longevity of pollinators like honeybees.

» **Disease Resistance**: Genetic engineering can improve the immune responses of beneficial insects, protecting them from pathogens and parasites.

b. Mass Rearing of Biocontrol Agents

» **Synchronizing Development**: Controlling hormonal pathways can standardize the development of predatory insects or parasitoids used in biological control programs.

» **Enhancing Performance**: Genetic modifications can improve traits such as fecundity, predation efficiency, and environmental tolerance.

3. Biotechnology Applications

The molecular basis of insect metamorphosis has inspired innovations in biotechnology, ranging from materials science to synthetic biology.

a. Bio-Inspired Materials

» **Silk Production**:

- Understanding the genetic regulation of silk production in metamorphosing insects like silkworms has led to advances in textile and biomedical applications.

» **Chitin-Based Products**:

- Insights into chitin synthesis and remodeling during molting have enabled the development of biodegradable materials, such as films and coatings.

b. Synthetic Biology

» **Genetic Circuits:** The transcriptional cascades driving metamorphosis serve as models for designing genetic circuits in synthetic biology.

» **Programmable Development:** Synthetic biology tools can mimic metamorphic processes to create programmable organisms or bio-factories.

c. Drug Development

» **Hormone Analogs:** Juvenile hormone and ecdysteroid analogs are being explored for therapeutic applications, such as wound healing and cancer treatment.

4. Integrated Pest Management (IPM)

Molecular knowledge enhances IPM strategies by combining traditional methods with precision-targeted interventions:

» **Hormonal Baits:** Hormone-based traps can attract and sterilize pests, reducing populations without widespread chemical use.

» **Genetic Sterilization:** CRISPR-based gene drives can be used to introduce sterility or disrupt developmental genes in pest populations.

4: Sf Transgenic Technology and Its Implications

The **Spodoptera frugiperda**, commonly known as the fall armyworm (FAW), has emerged as one of the most devastating agricultural pests globally. Advances in transgenic technologies have opened new avenues for its management, providing tools for pest control, crop protection, and deeper insights into insect biology.

4.1. *Spodoptera frugiperda* (Fall Armyworm) and Its Economic Impact

Spodoptera frugiperda is a highly polyphagous and migratory pest that poses significant threats to global food security. Native to the Americas, this pest has now spread to Africa, Asia, and parts of Europe, exacerbating its economic impact.

1. Biology and Life Cycle of Spodoptera frugiperda

» **Morphology:**

- Adult fall armyworms are nocturnal moths with a wingspan of 32–40 mm. They exhibit sexual dimorphism, with males displaying distinctive white and brown patterns on their forewings.
- Larvae are greenish-brown with stripes and a characteristic inverted "Y" marking on the head.

» **Life Cycle:**

- The lifecycle consists of egg, larval, pupal, and adult stages, with the larval stage being the most destructive.

- Under optimal conditions, FAW completes its lifecycle in 30–40 days, enabling multiple generations per year.

2. Feeding Habits and Damage

» **Host Range**:

- FAW is a polyphagous feeder, attacking over 350 plant species, including maize, rice, sorghum, cotton, and sugarcane.

» **Feeding Behavior**:

- Larvae feed on leaves, stems, and reproductive structures, causing extensive defoliation and damage to crops.
- They exhibit a "window-pane" feeding pattern during early instars, progressing to more aggressive feeding in later stages.

3. Economic Impact

» **Crop Losses**:

- FAW infestations cause annual crop losses valued at billions of dollars globally, with maize being the most affected crop.
- In Africa alone, FAW damages are estimated to cause losses of $2.5–$6 billion annually in maize production.

» **Increased Pesticide Use**:

- Farmers often resort to intensive pesticide applications, leading to increased production costs, environmental damage, and pesticide resistance.

4. Global Spread and Invasiveness

» **Migration and Invasiveness**:

- FAW's ability to migrate over long distances contributes to its rapid spread and establishment in new regions.

» **Resistance Development**:

- The pest has developed resistance to multiple chemical classes of insecticides and transgenic *Bt* crops, compounding control challenges.

5. Challenges in Conventional Management

» **Pesticide Resistance**:

- Over-reliance on chemical pesticides has led to the emergence of resistant FAW populations.

» **Ecological Concerns**:

- Broad-spectrum pesticides disrupt ecosystems, affecting non-target organisms, including pollinators and natural enemies.

» **Climate Change**:

- Rising temperatures and changing weather patterns are expected to expand FAW's range, further complicating management strategies.

4.2. Development and Applications of Sf Transgenic Technology

Transgenic technology targeting Spodoptera frugiperda offers a promising solution to mitigate its impact. These technologies employ genetic modifications to disrupt FAW biology, enhance pest resistance in crops, and reduce reliance on chemical pesticides.

1. Development of Sf Transgenic Technology

Advancements in genetic engineering and molecular biology have facilitated the creation of transgenic tools to manage FAW.

a. Genomic Insights

» **Genome Sequencing**:

- The sequencing of the FAW genome has provided critical insights into its biology, including genes associated with pesticide resistance, detoxification, and reproduction.

» **CRISPR-Cas9**:

- CRISPR-based genome editing has enabled precise modifications in FAW genes, offering new avenues for functional studies and pest management.

b. RNA Interference (RNAi)

» **Mechanism**:

- RNAi involves the introduction of double-stranded RNA (dsRNA) targeting essential FAW genes, leading to gene silencing and mortality.

» **Applications**:

- Targeting genes involved in chitin synthesis, digestion, and detoxification has shown promise in disrupting larval development and survival.

c. Transgenic Crops

» ***Bt* Crops**:

- Genetically engineered crops expressing *Bacillus thuringiensis* (*Bt*) toxins have been widely adopted to control FAW.
- Example: *Bt* maize produces Cry1Ab and Cry1F proteins, which target FAW larvae.

» **Stacked Traits**:

- New generations of transgenic crops combine multiple *Bt* toxins with RNAi constructs for enhanced efficacy.

d. Synthetic Biology Approaches

» **Gene Drives:**

- Gene drive systems, powered by CRISPR, can spread genetic modifications through FAW populations, leading to population suppression or replacement.

» **Synthetic Peptides:**

- Engineered peptides that disrupt critical physiological processes in FAW are being explored as biopesticides.

2. Applications of Sf Transgenic Technology

Transgenic technologies offer versatile applications in pest control, crop protection, and research.

a. Pest Control Strategies

1. **Population Suppression:**
    - Transgenic FAW with lethal gene constructs can be released to reduce wild populations.
2. **Sterile Insect Techniques:**
    - Genetic modifications can induce sterility in male FAW, reducing reproduction rates.

b. Enhancing Crop Resistance

» **Improved *Bt* Crops:**

- New *Bt* crops are designed to combat FAW resistance by incorporating novel Cry and Vip proteins.

» **RNAi Crops:**

- Crops expressing dsRNA targeting FAW-specific genes offer an environmentally friendly alternative to chemical pesticides.

c. Resistance Management

» **Stacked Traits:**

- Combining multiple modes of action in transgenic crops reduces the likelihood of resistance development.

» **Gene Monitoring:**

- Transgenic tools can track resistance alleles in FAW populations, informing adaptive management strategies.

d. Functional Genomics

» **Gene Function Studies:**

- CRISPR and RNAi technologies enable the study of gene functions in FAW, providing insights into resistance, detoxification, and reproduction.

» **Target Discovery**:

- Identifying novel genetic targets facilitates the development of next-generation control strategies.

3. Broader Implications of Sf Transgenic Technology

Transgenic approaches targeting FAW have implications that extend beyond pest control, influencing agriculture, ecology, and biotechnology.

a. Environmental Benefits

» **Reduced Pesticide Use**:

- Transgenic crops and RNAi-based solutions minimize the reliance on chemical pesticides, reducing environmental contamination and harm to non-target organisms.

» **Preservation of Biodiversity**:

- Targeted genetic approaches protect beneficial insects, such as pollinators and natural enemies.

b. Economic Impact

» **Yield Protection**:

- Transgenic technologies prevent crop losses, ensuring food security and economic stability for farmers.

» **Cost-Effectiveness**:

- Long-term benefits of transgenic crops outweigh the initial costs, providing sustainable pest management solutions.

c. Ethical and Regulatory Considerations

» **Biosafety**:

- The release of transgenic organisms raises concerns about ecological balance, gene flow, and non-target effects.

» **Public Perception**:

- Acceptance of transgenic technologies depends on effective communication of their safety and benefits.

d. Advancing Biotechnology

» **Research Applications**:

- Transgenic FAW serve as models for studying insect physiology, gene function, and resistance mechanisms.

» **Platform for Innovation**:

- Insights from FAW transgenics inspire novel approaches in synthetic biology and pest management.

4.3. Case Studies on the Use of Sf Transgenics for Pest Resistance

The development and deployment of **Spodoptera frugiperda (fall armyworm, FAW) transgenic technologies** have resulted in significant advancements in pest resistance. Real-world applications have provided valuable insights into their efficacy, limitations, and broader impacts. Below are detailed case studies demonstrating the utility of Sf transgenics in pest resistance.

Case Study 1: *Bt* Crops for Fall Armyworm Control in Maize

Background: *Bt* crops, genetically engineered to express *Bacillus thuringiensis* (*Bt*) toxins, have been one of the most successful transgenic solutions for controlling FAW. These crops produce insecticidal proteins that target larval pests, reducing damage to plants and the need for chemical pesticides.

Implementation

» **Cry Proteins**:

- Maize varieties expressing Cry1Ab and Cry1F proteins were among the first to target FAW populations.
- Cry proteins bind to specific receptors in the midgut epithelium of FAW larvae, causing pore formation, cell lysis, and eventual death.

» **Stacked Traits**:

- To address resistance, new *Bt* crops combined multiple Cry proteins (e.g., Cry1Ac + Cry2Ab) and incorporated Vip proteins, such as Vip3Aa, which exhibit a different mode of action.

Results

- Significant reductions in FAW infestation and associated crop losses.
- Decreased reliance on synthetic insecticides, lowering production costs and environmental harm.

Challenges

» **Resistance Evolution**:

- Prolonged use of Cry1F-expressing maize led to resistance in FAW populations, necessitating the development of alternative toxins and resistance management strategies.

» **Geographical Variation**:

- Effectiveness varied across regions due to differences in FAW genetics and resistance allele frequencies.

Case Study 2: RNA Interference (RNAi) Crops for Targeted Gene Silencing

Background: RNA interference (RNAi) is an emerging technology that enables targeted silencing of essential FAW genes by delivering double-stranded RNA (dsRNA) through transgenic plants.

Implementation

» **Target Genes**:
  - Genes involved in chitin synthesis, digestion, and detoxification were selected for silencing.
  - For example, silencing the *chitin synthase* gene impaired exoskeleton formation, leading to developmental defects and mortality.

» **Delivery Mechanism**:
  - Maize plants were engineered to express dsRNA, which is ingested by FAW larvae during feeding.

Results

» High larval mortality and reduced crop damage were observed in controlled field trials.

» The specificity of RNAi reduced the risk of harming non-target organisms.

Challenges

» **Environmental Stability**:
  - The degradation of dsRNA in the environment and within insect guts limited the duration of protection.

» **Delivery Challenges**:
  - Ensuring sufficient dsRNA uptake by FAW larvae remained a technical hurdle in some field conditions.

Case Study 3: Gene Drive Systems for Population Suppression

Background: Gene drive systems, powered by CRISPR-Cas9, offer a method to propagate genetic modifications through FAW populations, potentially leading to population suppression.

Implementation

» **Lethal Gene Drives**:
  - Gene constructs designed to disrupt reproductive genes were introduced into laboratory populations of FAW.

» **Population Replacement**:
  - Gene drives carrying sterility-inducing genes were developed to replace wild populations with non-reproductive individuals.

Results

» Rapid spread of gene drives in controlled populations demonstrated the potential for population suppression.

» Field-level applications are under investigation for practical deployment.

Challenges

» **Ethical and Regulatory Concerns**:

- Gene drives raise concerns about ecological consequences and the potential for unintended spread to non-target populations.

» **Resistance Evolution**:

- The emergence of resistance alleles that block gene drive propagation poses a significant challenge.

4.4. Ecological and Agricultural Implications of Using Transgenic Spodoptera

While transgenic technologies targeting Spodoptera frugiperda offer significant benefits, their use has complex ecological and agricultural implications. These implications must be carefully considered to balance the advantages of pest control with potential risks to ecosystems, non-target species, and agricultural sustainability.

1. Ecological Implications

a. Impact on Non-Target Organisms

» **Pollinators and Beneficial Insects**:

- While *Bt* crops and RNAi are designed to be species-specific, unintended effects on beneficial insects, such as pollinators and natural enemies of FAW, have been observed in some cases.
- For example, Cry protein ingestion by non-target Lepidopterans has been documented in regions with high *Bt* crop adoption.

» **Predators and Parasitoids**:

- Reduced FAW populations may affect predators and parasitoids that rely on FAW as a food source, potentially disrupting food webs.

b. Biodiversity Effects

» **Loss of Genetic Diversity**:

- Gene drive systems could lead to significant reductions in FAW populations, raising concerns about the unintended loss of genetic diversity and its ecological consequences.

» **Resistant Pest Species**:

- Evolutionary pressure exerted by transgenics may select for resistant pest species, potentially leading to secondary pest outbreaks.

c. Environmental Contamination

» **Transgene Persistence**:

- Transgenic DNA or RNA may persist in the environment, potentially affecting non-target species or soil microbiomes.

» **Gene Flow:**

- The possibility of transgene escape into wild Spodoptera populations or related species raises concerns about ecological balance.

2. Agricultural Implications

a. Enhanced Crop Protection

» **Yield Stability:**

- Transgenic crops expressing *Bt* proteins or RNAi constructs have significantly reduced crop losses, improving yield stability and farmer incomes.

» **Reduced Pesticide Use:**

- The adoption of transgenic technologies has decreased the reliance on chemical pesticides, lowering production costs and minimizing health risks to farmers.

b. Resistance Management

» **Challenges of Resistance Evolution:**

- Resistance to Cry1F in FAW populations has already been documented, highlighting the need for integrated resistance management strategies.

» **Stacked Traits:**

- Combining multiple modes of action in transgenic crops is essential to delay resistance development and ensure long-term efficacy.

c. Economic Considerations

» **Cost of Technology:**

- The high cost of transgenic seeds and the associated regulatory requirements may limit access for small-scale farmers.

» **Market Dynamics:**

- Increased adoption of transgenic crops may influence global agricultural markets, potentially creating dependencies on specific technologies.

3. Sustainability and Ethical Considerations

a. Balancing Risks and Benefits

» The ecological risks associated with transgenic Spodoptera technologies must be weighed against the benefits of reduced crop losses and pesticide use.

» Ongoing monitoring and adaptive management are critical to minimize unintended consequences.

b. Public Perception and Acceptance

» Public concerns about the safety and environmental impact of transgenic technologies influence regulatory frameworks and adoption rates.

» Transparent communication and stakeholder engagement are essential to build trust and ensure responsible use.

5: Molecular Biology of Baculoviruses

Baculoviruses are large, double-stranded DNA viruses that primarily infect insects, especially those in Lepidoptera, Hymenoptera, and Diptera. They are renowned for their specificity to insect hosts, making them valuable tools for pest management and platforms for biotechnological applications.

5.1. Structure, Life Cycle, and Host Interactions of Baculoviruses

Baculoviruses are distinguished by their complex structure, unique replication strategies, and intricate interactions with their insect hosts. Their life cycle includes both extracellular and intracellular phases, enabling efficient propagation and dissemination within host populations.

1. Structure of Baculoviruses

Baculoviruses are large, rod-shaped viruses with a genome size ranging from 80 to 180 kilobase pairs (kbp). Their structural components include:

a. Viral Envelope

» Baculoviruses are enveloped viruses with a lipid bilayer derived from host cell membranes.

» The envelope contains glycoproteins, such as GP64, which facilitate attachment to host cells and viral entry.

b. Nucleocapsid

» The rod-shaped nucleocapsid houses the viral genome and is composed of a protein shell made of capsid proteins like VP39.

» The nucleocapsid protects the DNA and aids in its delivery into host cells.

c. Occlusion Bodies (OBs)

» Baculoviruses are unique in forming proteinaceous occlusion bodies (OBs), which encapsulate viral particles and protect them in the environment.

» Types of OBs:

- **Polyhedra** (formed by nucleopolyhedroviruses, NPVs).
- **Granules** (formed by granuloviruses, GVs).

» OBs are composed of polyhedrin or granulin proteins and play a key role in the environmental stability and oral infectivity of the virus.

d. Genome

- » The circular double-stranded DNA genome encodes 90–180 genes, depending on the species.
- » Genes are organized into early, late, and very late transcriptional cascades, controlling viral replication and host manipulation.

2. Life Cycle of Baculoviruses

The baculovirus life cycle is divided into two distinct phases: **budded virus (BV) phase** and **occlusion-derived virus (ODV) phase**, each serving specific roles in viral propagation.

a. Entry and Initial Infection

1. **Oral Infection:**
    - Insets ingest occlusion bodies (OBs) while feeding on contaminated surfaces.
    - OBs dissolve in the alkaline pH of the insect midgut, releasing occlusion-derived virions (ODVs).
2. **Invasion of Midgut Epithelium**:
    - ODVs attach to and fuse with midgut epithelial cells, delivering the viral genome into the host cell nucleus.
3. **Primary Replication**:
    - Viral replication begins in the nucleus, producing progeny nucleocapsids.

b. Budded Virus (BV) Phase

- » **Dissemination**:
    - Budded viruses (BVs) are released from infected midgut cells into the hemolymph, facilitating systemic infection.
- » **Structural Features**:
    - BVs have a single nucleocapsid enclosed in a membrane containing GP64 or F proteins, crucial for cell-to-cell transmission.
- » **Host Range**:
    - The BV phase determines the host range and tissue tropism of the virus.

c. Occlusion-Derived Virus (ODV) Phase

- » **OB Formation**:
    - In later stages of infection, ODVs are embedded in occlusion bodies (polyhedra or granules) within the nuclei of infected cells.
- » **Environmental Stability**:
    - OBs protect the virus in harsh environmental conditions, enabling long-term persistnce and transmission to new hosts.

» **Horizontal Transmission**:

- Infected insects release OBs into the environment upon death, spreading the virus to other individuals.

d. Host Cell Manipulation

1. **Subversion of Host Transcription:**
   - Early viral genes suppress host immune responses and hijack cellular machinery for viral replication.
2. **Cytoskeletal Remodeling**:
   - Viral proteins manipulate the host cytoskeleton to facilitate viral assembly and egress.
3. **Induction of Host Death**:
   - Baculoviruses often induce apoptosis or liquefaction of the host body to maximize OB release.

3. Host Interactions

Baculoviruses exhibit complex interactions with their insect hosts, influencing infection dynamics and host behavior.

a. Specificity to Insect Hosts

» Baculoviruses are highly specific to their insect hosts, infecting primarily Lepidoptera.

» Host specificity is determined by viral glycoproteins, receptor interactions, and the midgut environment.

b. Host Immune Evasion

1. **Immune Suppression:**
   - Early viral proteins, such as *p35*, inhibit apoptosis to prolong host cell viability.
2. **Subversion of Host Antiviral Responses**:
   - Baculoviruses inhibit RNA interference (RNAi), a major antiviral mechanism in insects, by expressing suppressor proteins.
3. **Melanization Inhibition**:
   - Some baculoviruses produce enzymes that block the melanization response, preventing encapsulation of the virus.

c. Manipulation of Host Behavior

» **Climbing Behavior**:

- Baculoviruses induce "tree-top disease," causing infected larvae to climb to elevated positions before death, enhancing OB dispersal.

» **Feeding Suppression**:

- Late-stage infection suppresses feeding, limiting the host's energy expenditure and focusing resources on viral replication.

d. Co-Evolution with Hosts

» Baculoviruses and their hosts have undergone co-evolution, with selective pressures driving host resistance mechanisms and viral counter-adaptations.

» Examples of host resistance include midgut barrier modifications and immune gene diversification.

4. Molecular Insights into Baculovirus Replication

a. Temporal Gene Expression

» **Early Genes**:

- Involved in DNA replication and immune evasion.
- Example: *IE-1* acts as a master regulator of viral transcription.

» **Late Genes**:

- Encode structural proteins for nucleocapsid assembly.
- Example: *vp39* encodes a major capsid protein.

» **Very Late Genes:**

- Responsible for occlusion body formation.
- Example: *polyhedrin* and *granulin* genes.

b. Viral-Host Interactions at the Molecular Level

» Baculoviruses exploit host RNA polymerases for early gene transcription and encode their own polymerase for late-stage replication.

» Host proteostasis pathways are hijacked to ensure efficient viral protein synthesis.

5.2. Applications of Baculoviruses in Pest Management and Gene Expression Systems

» Ba**culoviruses are widely util**ized for pest management due to their host specificity and environmental safety. They are also prominent tools in molecular biology as expression systems for producing recombinant proteins. This dual functionality underscores their versatility in agricultural and biotecnological applications.

1. Applications in Pest Management

Baculoviruses are effective biopesticides, offering a sustainable alternative to chemical insecticides. Their specificity to insect pests minimizes harm to non-target organisms, including pollinators and natural enemies.

a. Biopesticide Development

- **Mode of Action**:
  - Baculoviruses infect pests when ingested, targeting the midgut cells and spreading systemically to cause host mortality. The infection culminates in host liquefaction, releasing occlusion bodies (OBs) that persist in the environment.
- **Targeted Pests**:
  - Baculoviruses have been used to control major agricultural pests such as:
    - ◊ *Helicoverpa armigera* (cotton bollworm).
    - ◊ *Spodoptera frugiperda* (fall armyworm).
    - ◊ *Plutella xylostella* (diamondback moth).
- **Commercial Biopesticides**:
  - Products like *Helicoverpa armigera Nucleopolyhedrovirus* (HearNPV) and *Spodoptera litura Nucleopolyhedrovirus* (SpltNPV) have been developed and deployed successfully.

b. Environmental Benefits

- **Safety for Non-Target Organisms**:
  - Baculoviruses infect only specific insect hosts, ensuring minimal impact on pollinators, predators, parasitoids, and vertebrates.
- **Biodegradability**:
  - These viruses degrade naturally in the environment, leaving no harmful residues.

c. Advantages Over Chemical Insecticides

- **Resistance Management**:
  - Baculoviruses provide an alternative to chemical pesticides in areas with resistant pest populations.
- **Integrated Pest Management (IPM)**:
  - Baculoviruses are compatible with IPM strategies, combining biological control with cultural and chemical practices for sustainable pest management.

d. Challenges in Field Application

- **Environmental Stability**:
  - Baculoviruses degrade under UV radiation and high temperatures, requiring formulations with UV protectants for field efficacy.
- **Slower Action**:
  - Compared to chemical pesticides, baculoviruses act more slowly, which can limit their use in high-pressure pest situations.

2. Applications in Gene Expression Systems

Baculoviruses are powerful expression vectors widely used in molecular biology and biotechnology to produce high-value proteins for research, therapeutics, and industrial applications.

a. Baculovirus Expression Vector Systems (BEVS)

» **Overview:**

- BEVS is based on the *Autographa californica Multiple Nucleopolyhedrovirus* (AcMNPV), engineered to express recombinant proteins in insect cells.

» **Mechanism:**

- A target gene is inserted into the viral genome under the control of a strong promoter, such as the polyhedrin promoter, enabling high-level expression in host cells like *Spodoptera frugiperda* Sf9 or Sf21 cell lines.

» **Advantages:**

- High protein yield and post-translational modifications similar to those in eukaryotes.
- Rapid production and scalability for commercial applications.

b. Applications in Research and Medicine

1. **Recombinant Protein Production:**
   - Production of complex proteins such as enzymes, antibodies, and vaccines.
   - Example: The human papillomavirus (HPV) vaccine, Gardasil, is produced using BEVS.
2. **Viral Vector Systems:**
   - Baculoviruses are used as vectors for gene delivery in mammalian cells, facilitating research in gene therapy and vaccine development.
3. **Functional Genomics:**
   - BEVS enables the functional characterization of genes and proteins, including enzyme activities, binding interactions, and signaling pathways.

c. Industrial Applications

» **Biomanufacturing:**

- Baculovirus systems are employed for the large-scale production of industrial enzymes and biopharmaceuticals.

» **Agrobiotechnology:**

- BEVS is used to produce bioinsecticides and plant-based vaccines.

5.3. Genetic Engineering of Baculoviruses for Enhanced Efficacy in Pest Control

Advances in genetic engineering have transformed baculoviruses into more potent and versatile tools for pest management. These modifications enhance their infectivity, speed of action, and environmental stability, addressing some of the challenges associated with their field application.

1. Enhancing Baculovirus Efficacy

a. Improving Infection Speed

» **Incorporation of Insect-Specific Toxins**:

- Baculoviruses have been engineered to express insect-selective toxins, such as *Bacillus thuringiensis* (*Bt*) Cry proteins, to accelerate host mortality.
- Example: A baculovirus expressing a scorpion toxin showed faster action against *Helicoverpa armigera*.

» **Disruption of Host Defense**:

- Modifications targeting host immune pathways, such as RNA interference suppressors, enhance viral replication and efficacy.

b. Enhancing Environmental Stability

» **UV Resistance**:

- Genetic modifications or protective formulations improve viral stability under UV exposure, extending their effectiveness in the field.

» **Heat Tolerance**:

- Engineering heat-stable occlusion body proteins ensures viral viability in high-temperature environments.

c. Expanding Host Range

» **Receptor Engineering**:

- Altering viral envelope proteins, such as GP64, allows baculoviruses to infect a broader range of pest species.

» **Hybrid Viruses**:

- Combining genetic elements from multiple baculovirus species expands host specificity.

2. Incorporating RNA Interference (RNAi)

a. Dual-Action Baculoviruses

» Baculoviruses expressing dsRNA targeting essential pest genes provide a two-pronged attack by inducing viral infection and silencing critical host pathways.

b. Advantages of RNAi-Expressing Baculoviruses

» Increased lethality and specificity.

» Potential to overcome resistance by targeting pest-specific genes.

3. Synthetic Biology Approaches

a. Modular Design

- Synthetic biology enables the design of modular baculoviruses with customizable traits, such as toxin expression, host specificity, and environmental stability.

b. Gene Drives

- CRISPR-based gene drives incorporated into baculoviruses could spread desirable genetic modifications through pest populations, enhancing long-term control.

4. Ecological and Agricultural Considerations

a. Reduced Non-Target Impact

» Engineered baculoviruses are designed to target pests with minimal risk to beneficial insects, ensuring ecological balance.

b. Regulatory and Public Acceptance

» Safety and ethical considerations surrounding genetically modified viruses require rigorous testing and transparent communication to gain public trust.

6: Insecticide Resistance and Resistance Management Strategies in Transgenic Crops

The challenge of insecticide resistance in pest populations is critical in modern agriculture. This phenomenon threatens the efficacy of pest control strategies and puts global food security at risk. The advent of transgenic crops, such as *Bt* crops, has provided innovative tools to manage resistance, yet their effectiveness depends on strategic implementation and management.

6.1. Mechanisms of Insecticide Resistance: Behavioral, Physiological, and Genetic Factors

Insecticide resistance emerges when pest populations adapt to survive exposure to control measures, whether chemical, biological, or transgenic. Behavioral changes, physiological modifications, or genetic evolution can drive this adaptation.

1. Behavioral Resistance

Behavioral resistance involves changes in pest activity or preferences that reduce their exposure to insecticides. These adaptations often occur quickly and can be challenging to detect.

a. Avoidance of Treated Areas

Pests may evolve to avoid treated surfaces altogether. For example, some mosquito populations (*Anopheles gambiae*) have shifted their resting habits from indoor walls treated with insecticides to outdoor locations, thereby reducing their exposure.

b. Feeding Time Modifications

Certain pests alter their feeding schedules to minimize exposure. For instance, insects feeding on *Bt* crops may shorten feeding periods to ingest less toxin, allowing them to survive despite the crop's transgenic defenses.

c. Oviposition Shifts

Some pests avoid laying eggs on insecticide-treated surfaces, reducing the vulnerability of their offspring. This behavior is particularly evident in agricultural pests like moths, which seek refuge areas within mixed cropping systems.

d. Temporal Activity Shifts

Pests may shift their active periods to times when insecticides are less effective. For example, nocturnal activity patterns in pests like the diamondback moth (*Plutella xylostella*) can circumvent daytime pesticide applications.

2. Physiological Resistance

Physiological resistance arises from changes within the pest's body that neutralize or avoid the effects of insecticides. This includes enhanced detoxification, target-site modifications, and reduced penetration.

a. Enhanced Detoxification

One of the most common resistance mechanisms is the overproduction of detoxification enzymes, such as:

» **Cytochrome P450 Monooxygenases**:

- These enzymes metabolize insecticides into non-toxic compounds. Elevated levels of P450s are a significant cause of resistance to pyrethroids in pests like *Helicoverpa armigera*.

» **Glutathione-S-Transferases (GSTs)**:

- GSTs break down organophosphates and other insecticides, as observed in resistant mosquito populations.

» **Carboxylesterases:**

- These enzymes hydrolyze ester bonds in pesticides like carbamates, rendering them inactive.

b. Target-Site Insensitivity

Mutations in the molecular targets of insecticides reduce binding efficacy. Examples include:

» **kdr (Knockdown Resistance) Mutations:**

- Changes in sodium channel proteins confer resistance to pyrethroids, as seen in resistant mosquito species.

» **Acetylcholinesterase Mutations:**

- Alterations in the acetylcholinesterase enzyme reduce susceptibility to organophosphates and carbamates.

c. Reduced Penetration

Structural changes, such as thickened cuticles, reduce the absorption of insecticides into the pest's body. This mechanism has been reported in pests like *Spodoptera frugiperda*.

d. Sequestration

Some pests produce proteins that bind and sequester toxins, preventing them from reaching critical sites. For example, binding proteins in the midgut of resistant insects can neutralize *Bt* Cry toxins.

3. Genetic Resistance

Genetic resistance involves heritable changes in pest populations that enhance survival under insecticide pressure. These changes can be driven by singl-gene mutations, polygenic effects, or even horizontal gene transfer.

a. Single-Gene Mutations

Point mutations in critical genes often confer resistance. For example:

» Mutations in the ryanodine receptor gene confer resistance to diamide insecticides in pests like the diamondback moth.

b. Polygenic Resistance

Resistance may arise from the combined effects of multiple genes, each contributing incrementally to overall resistance. This phenomenon is common in pests exposed to various insecticides over time.

c. Gene Amplification

Increased copies of detoxification genes enhance resistance. For instance, the amplification of P450 genes has been observed in resistant populations of *Helicoverpa armigera.*

d. Horizontal Gene Transfer

Rarely, resistance genes can be transferred between species through environmental or symbiotic vectors, creating new resistance challenges.

4. Synergistic Resistance Mechanisms

Pests often exhibit multiple resistance mechanisms simultaneously, compounding their ability to survive. For example, a pest may both avoid treated areas (behavioral) and metabolize the toxin effectively (physiological), making control strategies less effective.

6.2. Development and Spread of Resistance in Pest Populations

Resistance development and spread are influenced by evolutionary dynamics, agricultural practices, and environmental factors. Identifying these drivers is crucial to mitigate resistance.

1. Evolutionary Drivers of Resistance

a. Selection Pressure

Repeated exposure to insecticides imposes strong selection pressure, favoring individuals with resistance traits. For instance, continuous application of pyrethroids in cotton fields has selected for resistant *Helicoverpa armigera* populations.

b. High Reproductive Rates

Pests with short generation times and high fecundity, like aphids and whiteflies, can develop resistance rapidly. Each generation provides new opportunities for resistance mutations to arise and spread.

c. Genetic Variability

High genetic diversity within pest populations increases the likelihood of resistance alleles being present and spreading under selective pressure.

2. Spread of Resistance

a. Gene Flow

Migration and dispersal of resistant pests allow resistance genes to spread across regions and crops. For example, the migration of *Spodoptera frugiperda* from the Americas to Africa introduced resistance alleles to new regions.

b. Founder Effects

Small populations of resistant pests establishing in new areas can rapidly dominate, especially if no competition exists from susceptible individuals.

c. Cross-Resistance

Resistance to one insecticide can confer cross-resistance to related compounds, further complicating control efforts.

3. Monitoring Resistance Spread

a. Genetic Markers

Identifying resistance-associated mutations, such as kdr mutations in mosquitoes, enables proactive resistance management.

b. Geographic Surveillance

Mapping resistance hotspots using GIS tools provides insights into regional resistance dynamics.

c. Laboratory Bioassays

Regular bioassays to test pest susceptibility to insecticides help track resistance trends and inform control strategies.

6.3. Role of Transgenic Crops in Managing Resistance (e.g., *Bt* Crops)

Transgenic crops expressing insecticidal proteins, such as *Bt* toxins, have revolutionized pest control. They offer targeted pest management, reduce reliance on chemical pesticides, and contribute to sustainable agriculture. However, resistance management is crucial to preserve their efficacy.

1. *Bt* Crops and Their Mechanisms

a. Cry Proteins

Bt crops express Cry toxins that target pest-specific gut receptors, disrupting midgut cells and causing mortality.

b. Specificity

Different *Bt* proteins target specific pests:

» **Cry1Ac** targets Lepidopteran pests like *Helicoverpa armigera.*

» **Cry3Bb** targets Coleopteran pests like rootworms.

2. Challenges in *Bt* Resistance

Resistance to *Bt* crops has been reported in several pests, including:

» **Cry1F Resistance:**

- Documented in *Spodoptera frugiperda* populations in the Americas.

» **Cry2Ab Resistance:**

- Observed in *Helicoverpa zea* in cotton-growing regions.

3. Strategies to Combat *Bt* Resistance

a. Gene Pyramiding

Combining multiple *Bt* toxins in a single crop delays resistance development by targeting multiple pathways simultaneously.

b. Refuge Planting

Planting non-*Bt* crops alongside *Bt* fields provides a refuge for susceptible pests, diluting resistance alleles.

c. Monitoring and Adaptive Management

Regular monitoring of resistance trends informs adaptive strategies, such as alternating *Bt* traits or integrating chemical and biological controls.

6.4. Integrated Resistance Management Strategies: Refuges, Gene Stacking, and Monitoring

Integrated Resistance Management (IRM) is essential to sustain the effectiveness of transgenic crops. IRM combines genetic, cultural, and chemical strategies to delay resistance evolution and maintain pest control efficacy.

1. Refuge Strategies

a. High-Dose/Refuge Strategy

Non-*Bt* refuges near *Bt* fields maintain a population of susceptible pests to dilute resistance alleles in mating.

b. Refuge Designs

» **Block Refuges:**

- Separate blocks of *Bt* and non-*Bt* crops.

» **Seed Mixtures:**

- *Bt* and non-*Bt* seeds mixed within the same field.

2. Gene Stacking

a. Combining Toxins

Crops expressing multiple *Bt* proteins with different modes of action reduce resistance risks.

b. Synergistic Effects

Stacking genes for complementary traits, such as resistance to multiple pest species, enhances durability.

3. Resistance Monitoring

a. Genetic Surveillance

Tracking resistance alleles in pest populations allows for early intervention.

b. Field Testing

Field trials and bioassays assess the efficacy of IRM strategies.

References

Abbott WS. 1925. A method of computing the effectiveness of an insecticide. *Journal of Economic Entomology*, 18(2): 265-267.

Abeles, R. H., Frey, P. A., & Jencks, W. P. (1992). Biochemistry. Jones and Bartlett, Boston.

Abraham, C., & Schepartz, B. (1967). Biochemistry (4th ed.). W. B. Saunders Co., Philadelphia and London.

Alberts, B., Bray, D., Lewis, J., Raff, M., Roberts, K., & Watson, J. D. (1994). Molecular Biology of the Cell (3rd ed.). Garland Publishing, Inc., New York/London.

Armstrong, F. B. (1989). Biochemistry (3rd ed.). Oxford University Press, New York.

Baldwin, E. (1963). Dynamic Aspects of Biochemistry (4th ed.). Cambridge University Press, Cambridge, U.S.A.

Baldwin, E. (1967). The Nature of Biochemistry (2nd ed.). Cambridge University Press, London.

Bennet, T. P., & Frieden, E. (1969). Modern Topics in Biochemistry. The McMillan Co., New York.

Berg, J. M., Tymoczko, J. L., & Stryer, L. (2002). Biochemistry (5th ed.). W. H. Freeman and Company, New York.

Bhattacharya TK, Kumar P, and Sharma A. 2007. *Animal Biotechnology*. 1st Ed., Kalyani Publication, New Delhi.

Bonner, J. E. (1966). Plant Biochemistry. Academic Press, Inc., New York.

Bravo A, Gill SS, and Soberón M. 2007. Mode of action of Bacillus thuringiensis Cry and Cyt toxins and their potential for insect control. *Toxicon*, 49(4): 423-435.

Burke, J. D. (1970). Cell Biology. The Williams & Wilkins Company, Baltimore, U.S.A.

CABI. 2013. *Insect Resistance Management in Transgenic Crops*. CAB International, Wallingford.

Casida JE and Quistad GB. 1998. Golden age of insecticide research: Past, present, or future? *Annual Review of Entomology*, 43(1): 1-16.

Conn, E. E., Stumpf, P. K., Bruening, G., & Doi, R. H. (1987). Outlines of Biochemistry (5th ed.). John Wiley and Sons, Singapore.

Cooper D, Eleftherianos I, and Joyce S. 2009. The insect immune system: A comparative perspective. *The Open Biology Journal*, 2(1): 1-14.

Cunningham, E. B. (1978). Biochemistry: Mechanism of Metabolism. McGraw-Hill, New York.

Darnell, J. E., Lodish, H. F., & Baltimore, D. (1990). Molecular Cell Biology (2nd ed.). Scientific American Books, New York.

Datta, S. P., & Ottaway, J. H. (1969). Biochemistry (2nd ed.). Bailliere Tindall and Cassell, London.

Devlin, R. M. (1975). Plant Physiology (3rd ed.). Van Nostrand Reinhold Co., New York.

Devlin, T. M. (Ed.). (2002). Textbook of Biochemistry with Clinical Correlations (5th ed.). Wiley-Liss, New York.

Doby, G. (1965). Plant Biochemistry. Interscience Publishers, London and New York.

Edwards, N. A., & Hassall, K. A. (1980). Biochemistry and Physiology of the Cell. McGraw-Hill Book Company (UK) Ltd., London.

Fairley, J. L., & Kilgour, G. L. (1966). Essentials of Biological Chemistry (2nd ed.). Van Nostrand Reinhold Co., New York.

Fasman, G. D. (Ed.). (1976). Handbook of Biochemistry and Molecular Biology (3rd ed.). CRC Press, Boca Raton, Fla.

Finar, I. L. (1973, 1975). Organic Chemistry (Vol. I, 6th ed.; Vol. II, 5th ed.). Longman, London.

Flickinger, C. J., Brown, J. C., Kutchai, H. C., & Ogilvie, J. W. (1979). Medical Cell Biology. W. B. Saunders Company, Philadelphia.

Florkin, M., & Stotz, E. H. (Eds.). (1962–1967). Comprehensive Biochemistry. American Elsevier Publishing Co., Inc., New York.

Freeland, P. W. (1985). Problems in Theoretical Advanced Level Biology (1st ed.). Hodder & Stoughton, London.

Freifelder, D. M. (1978). Molecular Biology and Biochemistry: Problems and Applications. W. H. Freeman and Company, San Francisco.

Fruton, J. S., & Simmonds, S. (1958). General Biochemistry (2nd ed.). John Wiley and Sons, Inc., New York.

Gahan LJ, Gould F, and Heckel DG. 2001. Identification of a gene associated with *Bt* resistance in *Heliothis virescens*. *Science*, 293(5531): 857-860.

Gill SS, Cowles EA, and Pietrantonio PV. 1992. The mode of action of Bacillus thuringiensis endotoxins. *Annual Review of Entomology*, 37(1): 615-636.

Green, J. H. (1978). Basic Clinical Physiology (3rd ed.). ELBS and Oxford University Press, London.

Gullan PJ and Cranston PS. 2014. *The Insects: An Outline of Entomology*. 5th Ed., Wiley-Blackwell, Hoboken, NJ.

Hagedon HH, Hilderbrand JG, Kidwell MG, and Law JH. 1990. *Molecular Insect Science*. Plenum Press, New York.

Happer, H. A., Rodwell, V. W., & Mayes, P. A. (1979). Review of Physiological Chemistry (17th ed.). Lange Medical Pub., California.

Hartman, P. E., & Suskind, S. R. (1969). Gene Action (2nd ed.). Prentice-Hall, Inc., Englewood Cliffs, N.J., U.S.A.

Heimpel GE and Mills NJ. 2017. *Biological Control: Ecology and Applications*. Cambridge University Press, Cambridge.

Hoy MA. 2003. *Insect Molecular Genetics: An Introduction to Principles and Applications*. 2nd Ed., Academic Press, New York.

Hutchison WD, Burkness EC, Mitchell PD, Moon RD, Leslie TW, Fleischer SJ, Abrahamson M, et al. 2010. Areawide suppression of European corn borer with *Bt* maize reaps savings to non-*Bt* maize growers. *Science*, 330(6001): 222-225.

Jain, J. L. (1990). Elementary Biochemistry (2nd ed.). S. Chand & Co. Ltd., New Delhi.

Karlson, P. (1975). Introduction to Modern Biochemistry (4th ed.). Academic Press, Inc., New York.

Kingsman, S. M., & Kingsman, A. J. (1988). Genetic Engineering (1st ed.). Blackwell Scientific Publications, Oxford/London.

Kumar R and Singh M. 2020. Advances in CRISPR/Cas-based genome editing tools for insect pest management. *Frontiers in Physiology*, 11: 1234.

Lehninger, A. L., Nelson, D. L., & Cox, M. M. (2000). Principles of Biochemistry (3rd ed.). Macmillan Publishers, Worth, New York.

Li X, Schuler MA, and Berenbaum MR. 2007. Molecular mechanisms of metabolic resistance to synthetic and natural xenobiotics. *Annual Review of Entomology*, 52(1): 231-253.

Mahler, H. R., & Cordes, E. H. (1968). Basic Biological Chemistry (2nd ed.). Harper and Row, Publishers Inc., New York.

Mallette, M. F., Clagett, C. O., Phillips, A. T., & McCarl, R. L. (1979). Introductory Biochemistry (1st ed.). Robert E. Krieger Publishing Co., New York.

Marsland, D. (1964). Principles of Modern Biology (4th ed.). Holt, Rinehart, and Winston, New York.

Marvier M, McCreedy C, Regetz J, and Kareiva P. 2007. A meta-analysis of effects of *Bt* crops on honey bees and natural enemies. *Science*, 316(5830): 1475-1477.

Mathews, C. K., van Holde, K. E., & Ahern, K. G. (2000). Biochemistry (3rd ed.). Pearson Education, California.

Matthews RW. 2011. *Insect Behavior*. Springer, Dordrecht.

Mazur, A., & Harrow, B. (1971). Textbook of Biochemistry (10th ed.). W. B. Saunders Co., Philadelphia/London.

McElroy, W. D. (1961). Cellular Physiology and Biochemistry. Englewood Cliffs, N.J., U.S.A.

McGilvery, R. W., & Goldstein, D. (1983). Biochemistry: A Functional Approach (3rd ed.). W. B. Saunders Co., London/Toronto.

Mertz, E. T. (1967). Elementary Biochemistry (3rd ed.). Burgess Pub. Co., Minneapolis.

Metzler, D. E. (1977). Biochemistry. Academic Press, Inc., New York.

Oakeshott J and Whitten MA. 1994. *Molecular Approaches to Fundamental and Applied Entomology*. Springer Verlag.

Olson DM, Dively GP, and Romeis J. 2009. Monitoring non-target effects of insect-resistant GM crops: *Bt* maize and natural enemies in the field. *Environmental Biosafety Research*, 8(2): 73-93.

Orten, J. M., & Neuhaus, O. W. (1970). Biochemistry (8th ed.). The C. V. Mosby Co., Saint Louis.

Palli SR and Kapitskaya M. 2006. Insect hormonal regulation and applications in pest management. *Pesticide Biochemistry and Physiology*, 85(2): 101-114.

Poinar GO. 2011. *Silk, Spiders, and Insect Biotechnology*. Wiley-Blackwell, Oxford.

Rawn, J. D. (1990). Biochemistry. Neil Patterson Publishers, Burlington, N.C.

Rechcigl JE and Rechcigl NA. 1998. *Biological and Biotechnological Control of Insect Pests*. Lewis Publishers, North Carolina.

Ren X and Hoiczyk E. 2014. Bacillus thuringiensis as a platform for genetic engineering and biopesticide production. *Current Opinion in Biotechnology*, 30: 46-51.

Rethel, F. J. (1967). Concepts in Biochemistry. McGraw-Hill Book Co., New York.

Rosenberg, E., & Cohen, I. R. (1983). Microbial Biology (1st ed.). Holt-Saunders, Japan.

Routh, J. I., Eyman, D. P., & Burton, D. J. (1976). General, Organic, and Biochemistry (2nd ed.). W. B. Saunders Co., Philadelphia.

Roy U and Saxena V. 2007. *A Handbook of Genetic Engineering*. 1st Ed., Kalyani Publishers, New Delhi.

Shelton AM, Zhao J-Z, and Roush RT. 2002. Economic, ecological, food safety, and social consequences of the deployment of *Bt* transgenic plants. *Annual Review of Entomology*, 47(1): 845-881.

Singh BD. 2008. *Biotechnology (Expanding Horizons)*. Kalyani Publishers, New Delhi.

Singh P. 2007. *Introductory to Biotechnology*. 2nd Ed., Kalyani Publishers, New Delhi.

Smith, E. L., Hill, R. L., Lehman, I. R., Lefkowitz, R. J., Handler, P., & White, A. (1983). Principles of Biochemistry: General Aspects (7th ed.). McGraw-Hill International Book Co., London/Tokyo.

Smith, E. L., Hill, R. L., Lehman, I. R., Lefkowitz, R. J., Handler, P., & White, A. (1983). Principles of Biochemistry: Mammalian Biochemistry (7th ed.). McGraw-Hill International Book Co., London/Tokyo.

Stephenson, W. K. (1973). Concept in Biochemistry: A Programmed Text. Wiley Eastern Pvt. Ltd., New Delhi.

Street, H. E. (1966). Plant Metabolism. Pergamon Press Ltd., Oxford.

Tabashnik BE, Gassmann AJ, Crowder DW, and Carrière Y. 2008. Insect resistance to *Bt* crops: Evidence, theory, and implications. *Annual Review of Entomology*, 53(1): 47-70.

Van Emden HF and Service MW. 2004. *Pest and Vector Control*. Cambridge University Press, Cambridge.

Voet, D., & Voet, J. G. (1995). Biochemistry (2nd ed.). Wiley, New York.

Watson, J. D., Hopkins, N. H., Roberts, J. W., Steitz, J. A., & Weiner, A. M. (1987). Molecular Biology of the Gene (4th ed.). Benjamin/Cummings Publishing Company Inc., California.

West, E. S., Todd, W. R., Mason, H. S., & van Bruggen, J. T. (1966). Textbook of Biochemistry (4th ed.). Macmillan Co.

Wiel, J. H. (1990). General Biochemistry (6th ed.). Wiley Eastern Limited, New Delhi.

Wilson K and Cotter SC. 2013. Host-parasite interactions and the evolution of immune defense in insects. *Annual Review of Entomology*, 58(1): 225-247.

Wood, W. B., Wilson, J. H., Benbow, R. M., & Hood, L. E. (1981). Biochemistry–A Problems Approach (2nd ed.). Benjamin/Cummings, Menlo Park, California.

Woods, R. A. (1980). Biochemical Genetics (2nd ed.). Chapman and Hall, London.

Yudkin, M., & Offord, R. (1973). Comprehensible Biochemistry. Longman.

Zubay, G. L., Parson, W. W., & Vance, D. E. (1995). Principles of Biochemistry (1st ed.). Wm. C. Brown Publishers, Dubuque.

[illegible] concerns and information please contact [illegible]
[illegible] representative [illegible] Taylor & Francis Verlag
GmbH, Kaufingerstraße 24, 80331 München, Germany

For Product Safety Concerns and Information please contact our
EU representative GPSR@taylorandfrancis.com Taylor & Francis
Verlag GmbH, Kaufingerstraße 24, 80331 München, Germany